International Conference e-Learning and Digital Learning (ELDL2023) and International Conference on Sustainability, Technology and Education (STE2023)

Held at the 17th Multi-Conference on Computer Science and Information Systems (MCCSIS2023)

Porto, Portugal
15-18 July 2023

Editors:

Miguel Baptista Nunes
Pedro Isaías
Tomayess Issa
Theodora Issa

ISBN: 978-1-7138-8466-8

Printed from e-media with permission by:

Curran Associates, Inc.
57 Morehouse Lane
Red Hook, NY 12571

Some format issues inherent in the e-media version may also appear in this print version.

Printed with permission by Curran Associates, Inc. (2024)

For permission requests, please contact International Association for Development of the Information Society (IADIS) at the address below.

International Association for Development of the Information Society (IADIS)
Rua Sao Sebastiao Da Pedreira
No. 100, 30, 1050-209
Lisbon, Portugal

Phone: 351 21 3151373
Fax: 351 21 3151244

secretariat@iadis.org

Additional copies of this publication are available from:

Curran Associates, Inc.
57 Morehouse Lane
Red Hook, NY 12571 USA
Phone: 845-758-0400
Fax: 845-758-2633
Email: curran@proceedings.com
Web: www.proceedings.com

TABLE OF CONTENTS

FULL PAPERS

<u>SHORT PAPERS</u>

REFLECTION

FULL PAPERS

A SURVEY WITH FUTURE PRIMARY SCHOOL TEACHERS ON USING GEOGEBRA FOR 2D AND 3D SHAPES

Vittoria Bonanzinga
Dipartimento di Ingegneria dell'Informazione, delle Infrastrutture e dell'Energia Sostenibile, Università Mediterranea di Reggio Calabria, Italy

ABSTRACT

This article presents a quantitative and qualitative analysis conducted in Italy, regarding the planning of didactic units on the part of 23 university students. The degree course is in Primary Education Sciences, for future elementary school teachers, in the field of geometry and in particular of 2D and 3D shapes, integrating traditional teaching with the use of GeoGebra software. The data collected through a questionnaire administered to prospective primary school teachers present the many positive aspects of using the GeoGebra software for teaching and learning 2D and 3D geometric shapes and some negative aspects perceived by the sample considered.

KEYWORDS

Geogebra, Primary School, 2D And 3D Shapes, University Education

1. INTRODUCTION

The Ministry of Education and Research in Italy has promoted the creation of "Innovative Learning Environments", i.e. learning environments and spaces equipped with innovative technological resources, capable of integrating the use of technologies into teaching. An innovative learning environment is characterized by the centrality of students with their active engagement, by cooperative learning, with teachers who are open to acknowledging students' motivations and sensitive to individual differences, and who emphasize formative feedback. Such an environment is flexible, adaptable, and allows people and information to connect. Moreover, it strengthens the interaction between students and teachers, between contents and resources. Many studies have been conducted on teaching and learning processes in various disciplines (Adair, 2009), (Bruner1, 1957), (Bruner2, 1961), (Bruner3, 1973) (Bruner4, 1996), (Dewey1, 1933), (Dewey2, 1938), (Feiman1, 1990), (Feiman2, 2001), (Munna and Kalam, 2021). Literature on the teaching and learning of geometry in primary school is very broad, and different aspects can be considered, from learning difficulties to assessment, (Jirotková et al. 2019), (Bonanzinga et al. 2022). We focus our attention on Geometry and 2 and 3 dimensional (3D) shapes and on its teaching and learning in primary school. Geometry is often considered a difficult subject to understand and several software programs have been developed to aid teaching and learning processes, such as Cabri-Geometry, Geometer's Sketchpad and GeoGebra. The latter is used worldwide from elementary school to university, (Tatar and Zengin, 2016), (Velichova, 2011), (Zulnaidi et al., 2020), (Putra, et al. 2021) (Kokol-Voljc, 2007), (Bonanzinga, 2023). Numerous studies conducted in the United Kingdom, Luxembourg, Portugal, Lithuania, Hungary, the Czech Republic, Slovakia, Malaysia, Indonesia, Vietnam, Nepal, Africa and many other countries show the effectiveness of its use in learning geometry, (Bulut et al., 2015), (Žilinskienė, and Demirbilek, 2015), (Gunčaga and Žilková 2019), (Tay and Wonkyi, 2018) et al. GeoGebra software originated from a master's thesis project of Markus Hohenwarter at the University of Salzburg in 2002. It was designed to combine features of interactive geometry software (e.g. Cabri Geometry, Geometer's Sketchpad) and computer algebra systems (e.g. Derive, Maple) in a single, integrated and easy-to use system for teaching and learning mathematics (Hohenwarter and Preiner, 2007). Geometry and algebra are central to mathematics and have been called its "two formal pillars" (Atiyah, 2001). Over 20 years, GeoGebra has developed into an open-source project with a group of 15 developers and over 100 translators all over the world. GeoGebra has been translated into 52 languages and used in 190 countries. It has grown from a small student project to an international organization. At the end of 2007, the International GeoGebra Institute (IGI) was established to offer teacher training and support, to develop teaching materials and software, to conduct research, to outreach to less well-off communities. The GeoGebra Mobile application also allows its use on mobile devices, such as smartphones and tablets and therefore greater use by teachers and students who do not necessarily need to have a computer or a classroom equipped to use the dynamic geometry software GeoGebra, (Alkhateeb, et al. 2019), (Bonanzinga, V. et al. 2020). The Italian National Guidelines (MIUR 2012) for primary school suggest a conscious and motivated use of calculators and computers which must be appropriately encouraged from the first years of primary school, for example to verify the correctness of mental and written calculations and to explore the world of numbers and shapes. In this article, we present a survey conducted in Italy on the planning of lessons on flat and solid geometric figures by a sample of 23 students of a Primary Education Sciences degree course, future elementary school teachers, integrating traditional teaching with the use of GeoGebra software. The objective of this research is twofold; to evaluate future teachers in the fruitful use of technology, in particular dynamic geometry software (DGS), for (i) their training and (ii) teaching. Students were asked to assess their knowledge of software used for learning/teaching. Only 3 had any knowledge so we wanted to assess the skill of acquisition and time of acquisition. Numerous studies have demonstrated the effectiveness of using DGS in mathematics teaching and learning. This study investigates the time needed to acquire the basic knowledge of GeoGebra software for calculating the perimeter and area of plane figures and the volume of solid figures and the perception of future elementary school teachers regarding the obstacles and difficulties in their usage. Our aim was to ask if future teachers are ready for the use of specific technology in teaching. In detail, we asked if tutorials, prepared by expert primary school teachers, help train future teachers to better understand geometric concepts and help in teaching using new technology.

2. PARTICIPANTS

Twenty-three students of an Italian university (21 females and 2 males), aged between 21 and 46 years, participated in the study: 3 students had worked with the GeoGebra application before the proposed activity, while the other 20 had never used it before. Only two students knew about software other than GeoGebra, namely MATLAB and WeSchool. Twelve students were enrolled in the second year of the course, 6 in the third and 5 in the fourth year. They were asked on a scale of 1 (not at all) to 5 (very much) about their interest in mathematics.

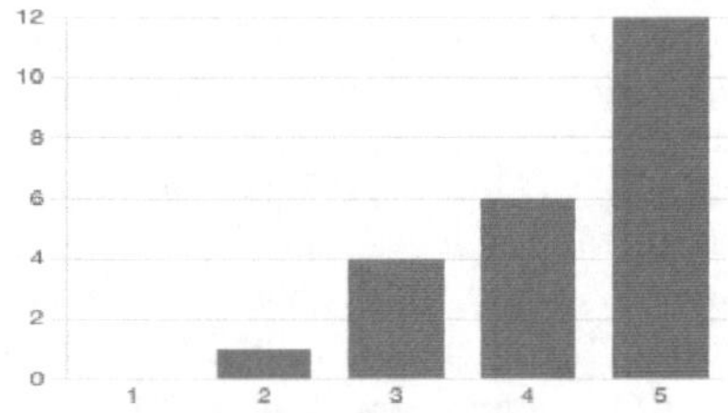

Figure 1. Interest in mathematics

Regarding the use of the computer, it was noted that most students had access to this resource at home. On assessing the students' entrance exam to the course, we ascertained that level of math skill was quite low and therefore this sample provided an opportunity to evaluate the use of GeoGebra with a relatively naïve population.

3. METHODOLOGY

This study can be considered in line with the above mentioned larger GeoGebra projects on developing a model to support teachers' mathematical, didactic, and technological knowledge by integrating DGS GeoGebra software. In this present paper, the researchers present a case study of prospective elementary teachers' experiences with GeoGebra for 2D and 3D geometric shapes. First, the students were given 5 links relevant to GeoGebra: 1 link was to download the program; 3 links were for tutorials prepared by expert teachers on the use of GeoGebra in primary schools; 1 was examples of activities created by teachers and students on volumes of diverse solids. The initial tutorial explained how to start to teach geometry with the program using the basic commands. For instance, the teachers are shown how to hide the algebraic part of the software and the Cartesian plane. They then learn how to insert the squared background. In this way, the children can recognize the squares of their squared copybook. Then they learn how to use the text function, to insert data, to change font size and color, to move geometric objects and also to use the function to insert points. The second tutorial shows the function of polygon: to insert a generic polygon and a regular polygon, to draw a polygon in an anti-clockwise way so as to be able to visualize the internal angles of the polygon. Regarding the regular polygon function, they are shown how to choose the length of the side and the number of sides. The third tutorial explains the area and perimeter considering the function of distance and length. Moreover, stretching the sides and reasoning on isoperimetries and equi-extensions were also included.

The link regarding the sample activities are open educational resources that can be modified and adapted depending on the needs of the teacher.

The student teachers were asked to prepare one or more lessons, on plane and space geometry, with particular reference to the area of plane figures and the volume of solids, integrating traditional methods with new methodologies, contextualizing problems, using software, in particular GeoGebra to check their results. The proposed didactic model suggests a synergistic use of pen-paper and software, (Farihah U., 2019), and combines the 5E instructional model and that of Revised Bloom's Taxonomy (RBT, Anderson et al. 2001). The 5E instructional model is a student-centered model for teaching, which consists of five stages: engagement, exploration, explanation, elaboration and evaluation (Bybee, 1997). Regarding the definition of learning outcomes, we follow Bloom's taxonomy, reviewed by Anderson et al., in 2001, which provides a list of six categories of skills, in ascending order from the simplest to the most complex,

1. Knowledge: ability to memorize information;
2. Comprehension: being able to translate and interpret the stored information;
3. Application: knowing how to extend the acquired concepts to unusual situations;
4. Analysis: ability to distinguish elements of information and qualitatively separate data;
5. Evaluation: critically examine a situation;
6. Creating: put elements together to form a new coherent or functional whole; reorganize elements into a new pattern or structure (design a new set for a theater production, write a thesis, develop an alternative hypothesis based on criteria, invent a product, compose a piece of music, write a play).

In the first phase of the project, the objectives to be achieved according to the model were termed "S.M.A.R.T." conceived by G. T. Doran and published in the November 1981 issue of Management Review. With the acronym S.M.A.R.T., the characteristics of the objectives to be pursued are indicated:

- Specific: Results should not be too general
- Measurable: the results must be able to be measured, it is therefore necessary to devise methods to quantify progress
- Achievable: the results must be achievable
- Relevant: the desired outcomes must be relevant to the course
- Time-limited: It must be clear within how long the goals should be achieved.

Tutorials for installing and using the software for primary school students were indicated and some examples were provided.

Through a questionnaire, future primary school teachers were analyzed regarding the time taken to carry out the assigned activity, the positive aspects of using the GeoGebra software and any negative aspects.

4. RESULTS

The degree of acquisition of GeoGebra skills was divided into three levels: low, medium and high.

Level	GeoGebra skills	Percentage
High	Full mastery of the tools, functions and commands of the software relating to the 2D and 3D shapes	22%
Medium	Good mastery of the tools, functions and commands of the software relating to the 2D and 3D shapes	48%
Low	Limited mastery of the tools, functions and commands of the software relating to 2D and 3D shapes	30%

Figure 2. Degree of acquisition of GeoGebra skills

In the first level of skills acquired with the GeoGebra software, students describe, name and classify plane figures. For example, imagine asking students to count the number of squares and triangles present in a plane figure or to use different colors based on the different geometric shapes, as illustrated in the following figures.

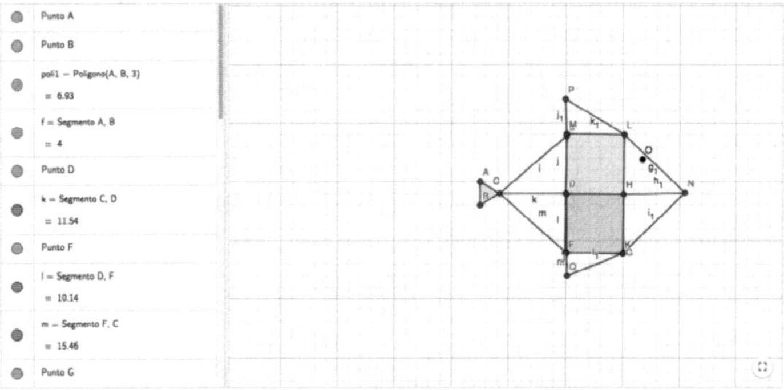

Figure 3. On counting squares and triangles

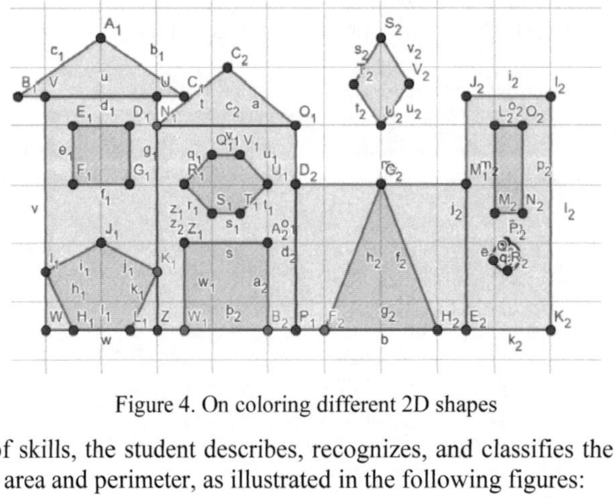

Figure 4. On coloring different 2D shapes

In the second level of skills, the student describes, recognizes, and classifies the 2D shapes and uses the functions for calculating area and perimeter, as illustrated in the following figures:

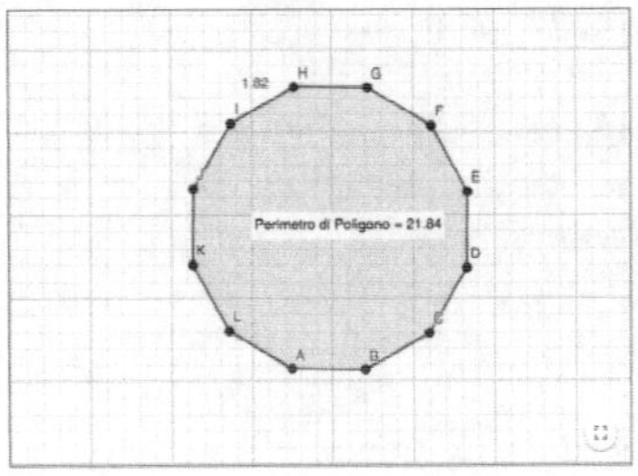

Figure 5. Perimeter of a dodecagon

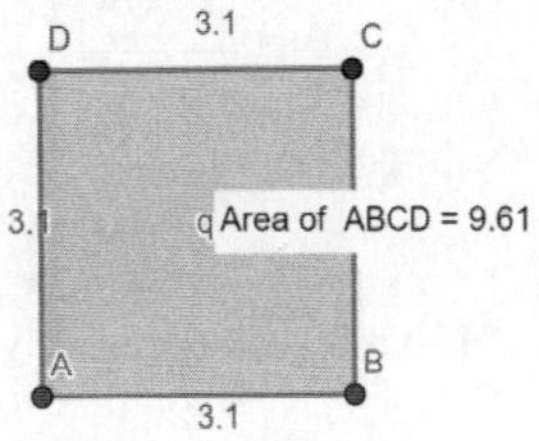

Figure 6. Area of a square

In the third level of skills, the student describes, recognizes, and classifies 2D and 3D shapes and uses the functions for calculating area, perimeter, volume and the slider to change the size of a side as illustrated in the following figures:

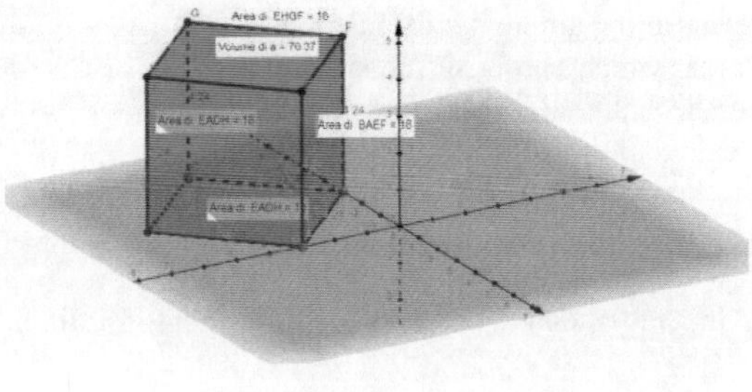

Figure 7. Area of the faces and volume of a cube

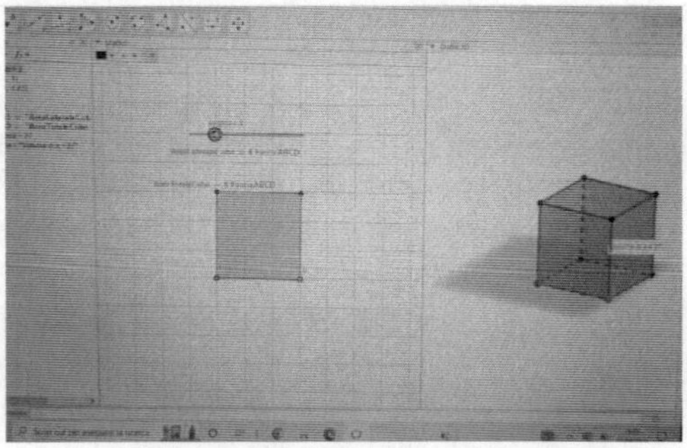

Figure 8. Slider to change the size of an edge

In the following tables, we summarize the positive and critical aspects perceived by the students on the proposed activity:

Table 1. Students' positive considerations on the proposed activity

Fast, simple and fun, it stimulates the pupil's interest, adapts to the educational needs of the pupil and to the very different levels of preparation.
It improves learning processes, helps keep pupils' attention high, helps pupils actively participate, is useful for consolidating pupils' knowledge, presents an alternative way of teaching, increasing pupils' skills.
Makes learning challenging; helps fill the gaps and achieve goals more easily and effectively.
The use of educational software in primary school stimulates pupils' reasoning and allows the teacher to create personalized educational paths.
The use of technology in education improves the quality of learning, facilitates access to resources and in some cases as group work; it also favors collaboration and the exchange of ideas.
The proposed methodologies help pupils' learning
The proposed methodologies help pupils develop skills.
The positive aspects are the simplicity of use for introducing topics. Pupils can use it to calculate perimeter and area of polygons and draw various geometric shapes.
GeoGebra is very intuitive, but with advanced features.

Table 2. Students' critical considerations on the proposed activity

Excessive use of GeoGebra software could become a source of distraction and not everyone has the possibility of accessing technological means outside the school walls.
Pupils may underestimate the lesson, cooperation with peers may be lacking, pupils may not be familiar with computers.
Limit creativity and dexterity in writing or drawing; can be distracting during the lesson.
Pupils may consider the lesson and/or explanation as if it were a game.
Educational software could lead to less interaction between pupils, since each child, having his own electronic device at his disposal, could isolate himself from the rest of the class.
Constraint on the use of digital resources.
A negative aspect may relate to the automatic calculation of the perimeter and of the area, which can push those who use it to rely completely on the automatic system without engaging individually in the calculation.
Time necessary to acquire knowledge of the commands, functions and potential of the software.

The different levels of acquisition of skills with GeoGebra also depend on the time spent in training and experimentation.

Hours/Days	Number of students	Percentage
1-2 hours	8	35%
3-5 hours	5	22%
1 day	3	13%
2-3 days	7	30%

Figure 10. Time taken to carry out the requested activity

The students were asked on a scale of 1 (not at all) to 5 (very much) about their interest in GeoGebra.

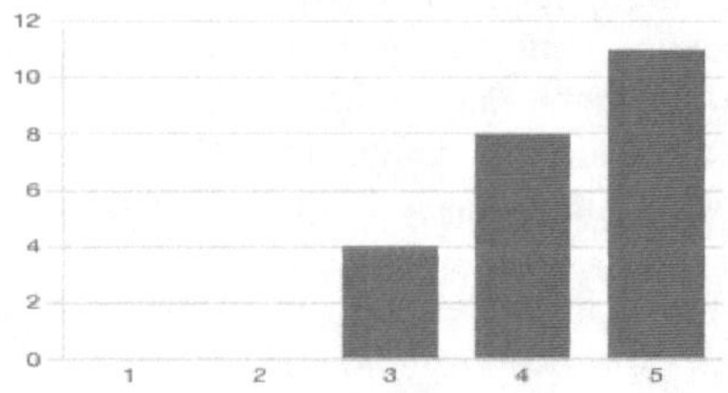

Figure 11. Interest in software GeoGebra

5. DISCUSSION

Feiman-Nemser, in 2001, explains specific principles and strategies on how to apply educational mentoring and some notions on how to carry them out. In Kokol-Voljc 2007, pedagogical advantages of the use of mathematical software in teaching are outlined. In Jirotková et al. 2019 the research focused on the issue of the relationship between perimeter and area and the presence of a misconception "same area, same perimeter". Our study is in line with the above mentioned works, whereby tutorials given by experts are seen as helpful, student teachers appreciate the software which helps them to understand the concepts, and finally the students' own misconceptions, often not trivial, are clarified, thus helping them teach better.

6. CONCLUSION

The case study highlights significant positive aspects of using the GeoGebra software, it is interesting, intuitive, useful, simple to use, fun, effective and stimulates curiosity. From analysis of the activities carried out by the future elementary school teachers, it emerges that only 22% of the students have acquired full mastery of the tool, 48% of the students have good knowledge of the software and 30% have limited knowledge. Therefore, a guided training course is certainly essential for its more effective and widespread use. Furthermore, the need for a complementary, but not exclusive, use was underlined. The risks of exclusive use of software would lead to a loss of dexterity in design and measurement, and in calculation ability.

ACKNOWLEDGEMENT

The author would like to thank the anonymous referees for useful and detailed comments on the previous version of the manuscript.

REFERENCES

Adair, J., (2009). *Leadership and Motivation: The Fifty-Fifty Rule and the Eight Key Principles of Motivating others*, London Kogan Page Ldt.

Alkhateeb, M. A. et al. (2019). The Effect of Using Mobile Applications (GeoGebra and Sketchpad) on the Students' Achievement, *Int. Electron. J. Math. Educ.*, vol. 14, no. 3.

Anderson, L.W., et al. (2001). *A taxonomy for learning, teaching, and assessing. A revision of Bloom's taxonomy of educational objectives*, New York, Addison Wesley Longman.

Atiyah, M. (2001). Mathematics in the 20th Century: geometry versus algebra. *Amer. Math. Mon.,* Vol.108, No. 7, pp. 654–666.

Bonanzinga, V. (2023), Some applications of linear algebra and geometry in real life, submitted.

Bonanzinga, V. et al (2020). Blended learning an experience with use of mobile devices for the course "Fundamentals of mathematics for Basic Education", *19th Conference on Applied Mathematics, Aplimat 2020, Proceedings*, pp. 119-125.

Bonanzinga, V. et al (2022). University training for future primary school teachers on automatic formative assessment, *Proceedings of the International Conference e-learning 2022*, IADIS press, pp. 119-126.

Bruner, J. S. (1957). *Going beyond the information given*. New York: Norton.

Bruner, J. S. (1961). The act of discovery. *Harvard Educational Review*, 31, pp. 21–32.

Bruner, J. S., (1973), *The relevance of education*, New York, NY: Norton.

Bruner, J. S., (1996). *The Culture of Education*. Cambridge, MA Harvard University Press.

Bulut, M. et al. Akçakın, Kaya, and Akçakın, (2015). The effects of GeoGebra on third grade primary students' academic achievement in fractions, *Math. Educ.,* vol. 11, no. 2, pp. 327–335.

Bybee, R. W. (1997). *Achieving Scientific Literacy: From Purposes to Practices*. Heinemann, Portsmouth, NH.

Dewey, J. (1933). *How We Think: A Restatement of the Relation of Reflective Thinking to the Educative Process*. Boston, MA: D.C. Heath & Co Publishers.

Dewey, J. (1938). Experience and Education. New York: Macmillan Company.

Doran, J. T. (1981). There's a S.M.A.R.T. Way to Write Management Goals and Objectives, *Management Review* (AMA Forum), November 1981, pp. 35-36.

Farihah, U. (2019). The synergy of students' use of paper-pencil techniques and GeoGebra in solving analytical geometry problems. *J. Phys.: Conf. Ser.* 1211.

Feiman-Nemser, S. (2001). Helping Novices Learn to Teach: Lessons from an Exemplary Support Teacher. *Journal of Teacher Education*, 52(1), pp. 17–30.

Feiman-Nemser, S., and Parker, M. B. (1990). Making Subject Matter Part of the Conversation in Learning to Teach. *Journal of Teacher Education*, 41(3), pp. 32–43.

Gunčaga, J., and Žilková, K. (2019). Visualisation as a Method for the Development of the Term Rectangle for Pupils in Primary School. *Eur. J. Contemp. Educ.* ,8, 52–68.

Hohenwarter, M. (2002). GeoGebra - Ein Software system für dynamische Geometrie und Algebra der Ebene. Master thesis, University of Salzburg.

Hohenwarter, M. and Jones, K. (2007). Ways of linking geometry and algebra: the case of GeoGebra. *Proceedings of the British Society for Research into Learning Mathematics*, 27, 3, 126-131.

Hohenwarter, M. and Preiner, J. (2007) Dynamic Mathematics with GeoGebra. *The Journal of Online Mathematics and Its Applications*, 7, Article ID: 1448.

Jirotková et al. (2019). Misconceptions about the relationship between perimeter and area. *Proceedings SEMT'19*, Prague: Faculty of Education, Charles University in Prague, pp. 221-231.

Kokol-Voljc, V. (2007). Use of mathematical software in pre-service teacher training: the case of GeoGebra, *Proceedings of the British Society for Research into Learning Mathematics*, 27(3), pp55-60.

MIUR, 2012. National Guidelines for the nursery school curriculum and the first cycle of education.

Munna, A.S. and Kalam, M.A. (2021). Teaching and Learning Process to Enhance Teaching Effectiveness: A Literature Review, *International Journal of Humanities and Innovation (IJHI)*, vol. 4, No. 1, pp. 1–4.

Putra, Z. H., ed al. (2021). GeoGebra Integration in Elementary Initial Teacher Training: The Case of 3-D Shapes. *International Journal of Interactive Mobile Technologies* (iJIM), 15(19), pp. 21–32.

Tatar, E. and Zengin, Y. (2016). Conceptual Understanding of Definite Integral with GeoGebra, *Comput. Sch.*, vol. 33, no. 2, pp. 120–132.

Tay, M. K. and Wonkyi, T. M. (2018). Effect of using GeoGebra on senior high school students performance in circle theorems, *African J. Educ. Stud. Math. Sci.*, vol. 14, pp. 1–18.

Velichova, D. (2011). Interactive Maths with GeoGebra. International Journal of Emerging Technologies in Learning (iJET), 6(S1), pp. 31-35.

Žilinskienė, I. and Demirbilek, M. (2015). Use of GeoGebra in Primary Math Education in Lithuania: An Exploratory Study from Teachers' Perspective, *Informatics Educ.*, vol. 14, no. 1, pp. 127–142.

Zulnaidi, H. et al. (2020). Effect of use of GeoGebra on achievement of high school mathematics students, *Educ. Inf. Technol.*, vol. 25, no. 1, pp. 51–72.

DESIGN DIDACTIC ACTIVITIES USING GAMIFICATION: THE PERSPECTIVE OF TEACHERS

Cecilia Fissore[1], Valeria Fradiante[1], Marina Marchisio[1] and Claudio Pardini[2]
[1]*Dipartimento di Biotecnologie Molecolari e Scienze per la Salute, Università di Torino*
Via Nizza, 52, 10126 Torino TO, Italy
[2]*Comitato Scientifico PP@S, Italy*

ABSTRACT

In recent years, gamification has raised considerable interest in various fields, including industry, healthcare, business and education. Moreover, the spread of technological and IT tools has contributed to the claim of new methodologies, approaches and tools, such as games, in the education field. Teachers are often attracted by gamification but they do not know how to put it into practice. It is therefore necessary to train teachers not only on the theoretical contents, but also and above all on the planning of didactic activities to adopt innovative educational approaches. The context of this research is the immersive 3-hour workshop on the theme "Gamification and education: innovative approaches to facilitate learning" which involved 54 teachers from all over Italy from primary to secondary school. After a short theoretical introduction, the teachers were guided through the individual step-by-step creation of a gamification activity. The research questions are: Which are the most frequent characteristics, factors and elements of gaming employed in the planning of didactic activities by school teachers? What were the most adopted gamification strategies? To answer the research questions, we analyzed teachers' responses to the initial questionnaire before the workshop and to the final questionnaire at the end. We considered also the design forms filled by teachers in order to understand how teachers employed the proposed methodology within their educational activity. The results show that teachers really appreciated the proposed methodologies and they tried immediately to put them into practice.

KEYWORDS

Gamification, Gaming and edutainment, Mathematics, Primary school, Secondary school, Teacher training

1. INTRODUCTION

According to the opinion of several researchers, if properly planned, gamification can favor and improve the learning and teaching of the subject to be studied (Behl et al., 2022). The term gamification refers to the use of the typical mechanisms of the game, such as the challenge, the use of points, levels and prizes, in a context that is essentially not a game (Deterding et al., 2011). It also represents a multidisciplinary approach that includes theories, practices and technologies. In education, teachers are often attracted by gamification but they do not know how to put it into practice within their teaching practices. To this purpose, it is necessary to train teachers not only on the theoretical contents inherent to these themes, but also and above all on the planning of didactic activities to adopt innovative educational approaches. In this sense, the Problem Posing and Solving (PP&S) project of the Ministry of Education (www.progettopps.it) promotes the training of Italian teachers of primary and secondary schools on innovative teaching methods, such as Gamification, with the use of ICT, and gives continuous support to teachers of all types and disciplines (Barana et al., 2019a; Fissore et al., 2020).

The context of this research is the immersive 3-hour workshop on the theme "Gamification and education: innovative approaches to facilitate learning", organized within the PP&S project. The workshop involved 54 teachers from all over Italy from primary to secondary school and it took place face-to-face in March 2023 with a desire to learn and innovate their teaching practices. The first hour of the workshop was focused on a theoretical introduction on the topic with group discussion and interactions between teachers and trainers. Then, several examples of educational activities of different levels using gamification approach were given. The rest of the workshop was spent to group planning of a didactic activities using gamification. In this stage teachers reflected on gamification strategies to be used in their activity, as well as on the theoretical contents, objectives and prerequisites. Trainers have provided teachers with a design form to guide them in planning the activity, which asked to explain the main characteristics of their work (e.g. synchronous/asynchronous, individual/in groups), the gaming factors (e.g. involvement, control, rewards, fun, progress, accumulation, personalization and adaptability) and the gaming elements (e.g. challenge, leaderboard, points/coins/treasures, make choices, variation depending on choices, avatars, storytelling). At the end and at the beginning of the workshop participants were asked to fill an initial and a final questionnaire. The research questions are:

- Which are the most frequent characteristics, factors and elements of gaming employed in the planning of didactic activities by school teachers?
- Which were the most adopted gamification strategies?

To answer the research questions, we analyzed teachers' responses to the initial questionnaire and to the final questionnaire. We considered also the 36 design forms developed by teachers in groups in order to understand how teachers employed the proposed methodology within their didactic planning.

The paper is structured as follows: in Section 2 the State of the art is outlined, in Section 3 research Methodology is presented together with workshop modalities and type of data collected, in Section 4 the Results are showed while in the final sections results are discussed and Conclusions are drawn.

2. STATE OF THE ART

2.1 Gamification

The spread of technological and IT tools in recent decades such as computers, smartphones, tablets, through which it is possible to play everywhere, has contributed to the claim of new methodologies, approaches and tools, such us games, in the education field. At the beginning there was a strong skepticism towards the use of games in education, since it was believed that games had the only purpose of entertaining the participants (De Freitas, 2006). Although, today gaming and gamification represent new ways of interaction, learning and exploration and contribute to the achievement of a defined purpose, different from pure entertainment (Dimoulas et al., 2021). The term gamification refers to the use of the typical mechanisms of the game, such as the challenge, the use of points, levels and prizes, in a context that is essentially not a game (Deterding et al., 2011). Unlike serious games, which are tools designed and developed as game tools from the very beginning, gamification represents an approach that involves the application of characteristic elements of the game within modules or learning units that were not necessarily born as games (Deterding et al., 2011). Many studies have shown how gamification can be successfully adopted in higher education contexts, (Čeponienė et al., 2019) and how in the case of distance learning it can stimulate interaction and engagement (De la Peña et al., 2021). It can be adopted as a methodology to simplify students' approach to disciplines, especially when combined with the use of adaptive learning, formative assessment and interactive feedback (Barana et al., 2022, Corino et al., 2022a). Furthermore, gamification enables the user to experience situations that would otherwise not be accessible and provides a safe place where multiple trials can be performed. For example, it is used in the health and safety field to educate people to behave responsibly (Dimoulas et al., 2021). One of the most difficult parts of using game mechanisms in non-game context consists in knowing how to wisely balance learning and game aspects (Bente & Breuer, 2010). On the one hand, game elements are important to engage and motivate students, but on the other hand it is necessary not to lose sight of the learning goals.

People who play are motivated by certain factors that characterize games:

- Involvement: players like to feel they have an active role in the game and they are therefore encouraged to participate dynamically. The challenge is one of the key elements to engage users within the game.
- Control: players like to feel they have power and control over their actions.
- Rewards: prizes, even small ones (e.g. badges or achievements) encourage players to keep going, especially when the rewards are regular.
- Fun: users enjoy the game because it is interesting and they will continue to play as long as there is interest.

- Progress: players are stimulated to continue the game if they have the impression that they are moving forward, getting better, gaining skills. The use of levels in the game is an incentive factor for the user who can be motivated by the progression in the challenge. Another element used in games is the leaderboard to stimulate users to play and get better and better results.
- Accumulation: those who play appreciate the possibility of being able to accumulate rewards (e.g. money, treasure, points).
- Personalization: players like to customize the game, for example by choosing an avatar to develop an own identity.
- Adaptability: players like the possibility of making the game vary depending on their decisions, whose path is the result of their own choices.

These are real needs of the players that the game must be able to satisfy. Gamification exploits the most commonly appreciated advantages and aspects of games to keep students' participation, involvement and motivation high in order to generate positive attitudes that can promote learning (Deterding et al., 2011). Storytelling represents another gamification strategy: telling a story, event, myth, legend or mission is one of the most used ways to involve the user (Chorianopoulos & Giannakos, 2014). A gamification activity can also be an opportunity to favor collaborative learning among students and for the achievement of specific common goals, for example by taking part in a challenge that involves teamwork. In fact, many games make available chats, forums or other tools to allow communication and interaction between users that helps to create a community that plays (Chorianopoulos & Giannakos, 2014).

2.2 The Digital Learning Environment of PP&S Project

The PP&S - "Problem Posing and Solving" - project (available at www.progettopps.it), headed by the Italian Ministry of Education, since 2012 has been promoting the training of teachers from primary to secondary schools on innovative teaching methods through the use of digital technologies. Teachers involved in the project learn how to use different kinds of digital tools and new methodologies, such as gamification, in order to enhance their daily didactic (Fissore et al., 2020). By enrolling in the project, totally free of charge, teachers have the possibility of having an integrated Digital Learning Environment (DLE) for all the classes of students they need. The use of a DLE, for online or hybrid teaching, can support the implementation of gamification teaching activities (Barana et al., 2022). Within the DLE, students are provided with multiple resources (interactive materials, links, videos, theoretical explanations, etc.) and numerous synchronous and asynchronous other online activities. At the same time, teachers develop gamification teaching activities for their students at a distance, within a DLE dedicated to the community of PP&S teachers, who also collaborate with each other exchanging ideas, teaching strategies and materials reviewed and tested in their classes. They also develop activities in collaboration with the tutors at a distance during the online training sessions (Barana et al., 2020).

Table 1 shows an example of designing mathematical activity using gamification, illustrated to teachers during online training activities and during this workshop. The activity also offered an example of compilation of the guided form that teachers have to fill in the planning of their activity.

Table 1. Battleship: example of a mathematical activity using gamification used to train teachers

Title	Battleship
Discipline	Mathematics
School grade	Grade 9/10
Duration	3 hours
Topic	Representation of points on the Cartesian plane
Goals	Familiarize with the Cartesian plane Use the Cartesian plane to locate points Recognize points, segments and figures on the Cartesian plane
General characteristics o the activity	Synchronous and asynchronous moments Groups of 2/3 students Self-assessment
	Sink the enemy ship which lies in the rectangle [-2,-1]x[1,2] (see Fig.1). In turn, a student chooses a point on the plane and communicates it to his mate, who selects an abscissa and an ordinate and shoot

Brief description of th activity	to hit the goal. More rounds of the game are foreseen if teacher varies the position of the goal and it is possible to introduce a partial and total leaderboard of the participants. A score based on the number of attempts is assigned (high number of attempts = low score). More levels could be inserted depending on the score obtained. The prize for each round could be the chance of choosing the goal position, instead of the teacher, for the next round during which the other teams will compete. The activity continues with a common mission for the class: sink the enemy ship which is located in an unknown area of the Cartesian plane. The exchange of information among groups is fundamental in identifying the winning area and can take place in asynchronous ways, such as through a forum or a specific chat. The game can be repeated several times by varying the winning area
Gaming factors	Involvement, control, rewards, fun, progress, accumulation, and adaptability
Gaming elemen	Challenge, leaderboard, points, make choices, variation depending on choices, storytelling

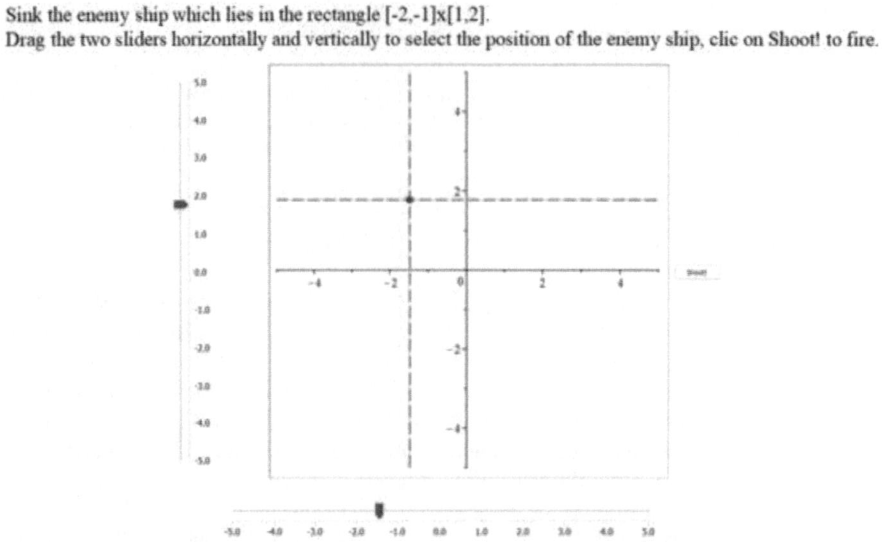

Figure 1. Interactive educational material used for the “Battleship” activity in Table 1

The DLE of the PP&S support gamification strategies. For example the forum activity supports collaboration and discussion between users. There are several tools to show students the level achieved, the position in the ranking and the progress. Moreover, it is possible to access to interactive worksheets, that we used to build the “Battleship” activity. The platform also allows users to receive badges depending on their goals.

3. METHODOLOGY

The context of this research is the immersive 3-hour workshop on the theme "Gamification and education: innovative approaches to facilitate learning" which involved 54 teachers from all over Italy from primary to secondary school. To answer the research questions, we analyzed the design forms provided by trainers to guide teachers in planning the gamification activity and their responses to the initial and final questionnaires. The first one investigated teachers’ data (age, level of education, school, discipline, etc.) and prior experience and knowledge about gamification and the perception of their own teaching practices. Using Likert scale questions (1=Not at all, 2=Very little, 3=Somewhat, 4=Much, 5=Very much) we asked teachers how much they usually pay attention to the following aspects in their teaching practices: motivation, engagement, cooperation between students, adaptability, inclusion and involvement. In the final questionnaire teachers reflected on methodologies proposed to support innovative teaching and how these favor the understanding of different aspects: catching students' attention, increasing motivation for the subject to study, involving students, developing their autonomy and responsibility and personalizing learning activities. Furthermore, they expressed their appreciation for the workshop in various aspects and for the proposed methodologies to support innovative teaching. The first hour provided a brief introduction to the topic and then a group discussion and interactions between teachers and tutors. The example of a Mathematics activity using gamification in Table 1 was also shown. The rest of the workshop was spent group planning a didactic activity through the compilation of a guided form to help teachers in designing a didactic activity using gamification. This form was created on the basis of the illustrated theoretical framework. The requests of the form are shown in Table 1. The activity presented to the teachers, in fact, served both to show an example of educational gamification activity and to explain how to design the activity using the form. The first part of the form required general information about their activity, such us title, disciplines involved, target, duration, topics and goals. The second part, focused on gamification, asked teachers to explain the main characteristics (synchronous/asynchronous, individual/in groups), the gaming factors (involvement, control, rewards, fun, progress, accumulation, personalization and adaptability), and the gaming elements (e.g. challenge, leaderboard, points/coins/treasures, make choices, variation depending on choices, avatars, storytelling etc.) of their gamification activities.

4. RESULTS

Our sample is composed by 47 teachers for the initial questionnaire and 36 for the final questionnaire. 36 activities were designed by teachers through the guided form, less than the number of participants since some of them worked in groups and submitted the same activity. From the initial questionnaire it was found that only one teacher was under 30 years old, 9 were between 30 and 40 years old, 21 between 41 and 50 years old, 15 between 51 and 60 and only one was more than 60 years old. Only one was primary school teacher, 18 were lower secondary school teacher, 27 were upper secondary school teacher and one was a head teacher. As shown in Fig. 2, half of them already had prior knowledge of gamification before the workshop. Most of them stated that they usually pay attention to those aspects in their teaching practices: motivation and engagement of students (98%), collaboration between peers (98%), adaptability of learning (85%), inclusion (98%) and involvement (100%). Therefore, the sample considered is composed by teachers who are willing to actively involve and motivate their students in the learning process; perhaps it is only necessary to make them reflect with a greater awareness on the methodologies they already employ for this purpose and to train them on innovative educational approaches and practices. In this sense, they attended the workshop to learn how to put in practice those aspects through gamification, and it will be easier for them to approach this methodology because the key elements will not be new.

Figure 2. Teachers' prior experience and knowledge about gamification

Even if some teachers answered only the initial questionnaire (11 teachers), since they could not stay until the end of the workshop, we consider all their responses as an expression of their pre-workshop perspectives on didactics and gamification. Those who filled the final questionnaire also designed the gamification activity. 86% of teachers stated that they gained interesting ideas and stimuli from their teaching practices from the workshop. About half of them identified the workshop as a useful occasion to reflect and rethink on their teaching methodologies in order to achieve greater clarity and awareness for didactic design. Teachers created 36 different activities, less than the number of the workshop participants, since some of them had collaborated and designed the same activity. 22 of them are addressed to upper secondary school, 13 for lower secondary school and one for primary school. Regarding the disciplines involved, 15 activities are focused on Mathematics topics, 5 on Science, 3 on Computer Science, 3 on Philosophy, History and Italian, 1 on Music and 9 are multidisciplinary activities which cover simultaneously different subjects, such as Civics,

Art, Geography. All activities are designed to be performed synchronously and 17 also include asynchronous moments. 35 out of 36 activities provide collaboration between students in small or big groups. We have analyzed also the most frequent gaming factors (involvement, choice, control, rewards, fun, progress, accumulation, personalization and adaptability) and the gaming elements (e.g. challenge, leaderboard, points/coins/treasures, make choices, variation depending on choices, avatars, storytelling etc.) employed in their designed activities. As shown in Fig. 3, teachers preferred to include some gaming factors and elements instead of others. In particular, all the activities are designed to actively involve students and they provide rewards (75%), personalization (61%) and make students' progress (78%). Teachers employed other factors less: control (42%), fun (36%), accumulation (36%), adaptability (25%). Perhaps they found these aspects more difficult to put in practice in a didactic activity. It is worth noting that the adaptability of learning was the indicator that obtained the lowest score in the initial questionnaire: its implementation in teaching practices turned out to be complicated. Regarding gaming factors, teachers employed different strategies: the most part inserted challenges (86%), levels (67%) and leaderboards (53%). About half of activities enable students to accumulate points/coins/treasures (50%), to win prices (44%) and make choice (44%). Only few teachers inserted avatar (28%) or invented stories (33%) or provided variations depending on choices (33%) within their activities. These aspects are also more complex to design and implement in a teaching activity.

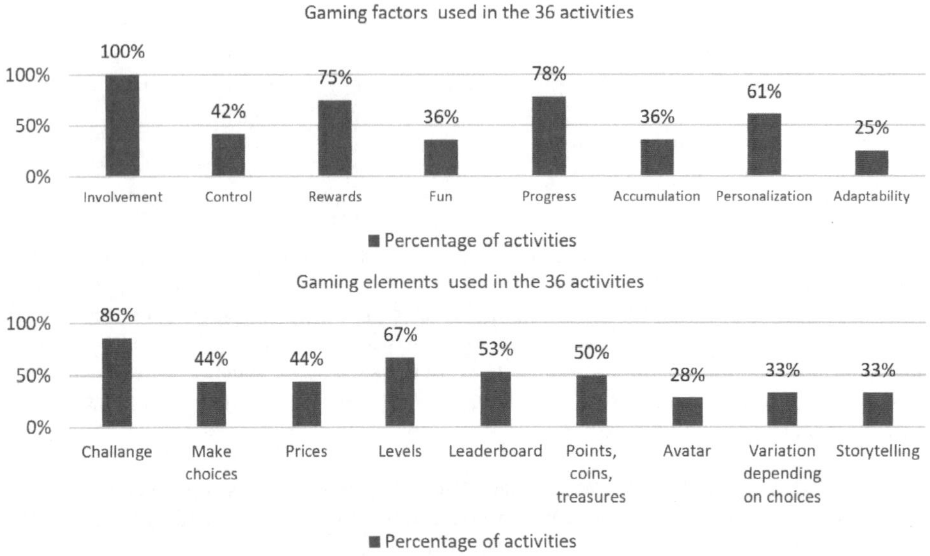

Figure 3. Teachers' choices of gaming factors and elements.

One of the most successful activities, showed in Table 2, is entitled "Crossword-Euclid". The activity was designed by a teacher of Mathematics of lower secondary school on Euclidean geometry topics. The goals of the activity are: to familiarize with fundamental geometric entities; to understand the main axioms of Euclidean geometry, and to distinguish and classify fundamental geometric entities. The activity is very interesting because it uses a well-known gamification element such as the crossword puzzle, but in a completely new context. The activity also involves both designing and solving a crossword puzzle. Since there is a score, the students know that they must not formulate the definitions too easily but neither too difficult or incomprehensible, because then it would not be mathematically correct. The most interesting aspect of the activity is that it links linguistic aspects to mathematical aspects. The activity could lead to a collaboration among Mathematics and Italian teachers, since students do not always know how to speak and write about Mathematics topics and often students' mathematical difficulties depend on linguistic difficulties (Corino et al., 2022b). It would be interesting to implement this task using encrypted crossword puzzles. Linguists consider these question types effective to develop language competences and have already used them profitably in traditional teaching and language learning (Barana et al., 2019b).

Table 2. Crossword-Euclid: example of activity designed by Mathematics teacher

Title	Crossword-Euclid
Discipline	Geometry
School grade	Grade 6
Duration	3 hours
Topic	Fundamental geometric entities: segments, angles, lines and polygons
Goals	Familiarize with fundamental geometric entities Understand the main axioms of Euclidean geometry Distinguish and classify fundamental geometric entities
General characteristics the activity	Synchronous Groups 3/4 students, Collaboration between students
Brief description of the activity	Students create a crossword puzzle using an open-source software, where they have to insert terms of the fundamental geometric entities and their respective definitions. Each group must solve the crosswords made by the other groups which will be randomly assigned in various rounds. A leaderboard of the groups that manages to achieve the highest number of correct definitions in the shortest time is drawn up for each round. There is a time penalty: in each round those who lie in the last 3 positions of the leaderboard must wait 1 minute to solve the next crossword. The first group to complete all crosswords correctly wins a badge.
Gaming factors	Involvement, rewards, fun, adaptability
Gaming elements	Challenge, leaderboard, points, prize

To engage the learners and increase motivation, crossword puzzles can be implemented using an automatic grading system in a DLE (Barana et al., 2019b). This is why it is important to train teachers on innovative teaching methodologies but also on the development of digital skills. Table 3 shows another example of activity designed by teachers on conscious daily use of water. The activity not only allows to develop curricular skills, but it is aimed at practicing skills that can be used in everyday life: to understand the amount of water used in daily life and to learn to read and use different kinds of graphical representations.

Table 3. Use or waste: example of activity designed by a Technical Education teacher and a Science teacher

Title	Use or waste
Discipline	Technical Education and Science
School grade	Grade 7
Duration	2 hours
Topic	Estimation, reading and using graphs
Goals	Understand the amount of water used in daily life Learn to read and use different kinds of graphical representations
General characteristics the activity	Synchronous, Asynchronous, Groups 5 students, Collaboration between students, Discussion between teachers and students
Brief description of the activity	The teacher divides the class into teams (5 students per team). Each team has got an amount of water and has to manage and use it for daily uses for 3 days. First rule is not to give up basic functions (washing and cooking). Whoever manages to survive up to 3 days with the amount of water given wins, perhaps with still water available.
Gaming factors	Involvement, rewards, fun, adaptability accumulation
Gaming elements	Challenge, make choices, variation depending on choices, levels, prize, leaderboard

5. CONCLUSION

This paper presents the results of a workshop on gamification and education that involved 54 Italian teachers of different levels. From the initial questionnaire we found that the sample considered is composed by teachers who are willing to actively involve and motivate their students in the learning process and that they were open to discover and learn new teaching methodologies to use in their teaching practices for these purposes. This fact highlights the need to train teachers who are very often attracted by innovative methodologies such as gamification, but they do not know how to put them into practice in their daily teaching practices. As the final questionnaire shows, thanks to the workshop they were able to discover interesting and clear methodologies and strategies to support innovative teaching, useful to better engage and motivate students in learning processes and to help them develop competencies. Through the analysis of the guided forms filled by teachers during designing stage, it was possible to study which are the most frequent gamification strategies used by teachers to create activities that allow students to achieve goals and skills in the disciplines they teach. All the didactic activities entail collaboration among students, for example to achieve a common mission or goal. Regarding gaming factors, the most frequent are involvement, rewards, personalization and progress, which have been implemented mainly through the use of challenge, levels, leaderboards and points/coins/treasure. From the initial questionnaire and from the guided designing forms, teachers' difficulties in achieving an adaptability of learning emerges. A future challenge could be to help teachers enhance their teaching practices, though practical tools and advices, to achieve a greater adaptability. The experience led to the production of valid gamification educational activities and was full of very interesting ideas. The same experience can also be offered to the students. Since teachers really appreciated activities and methodologies showed during the workshop, we think that a further step can be the creation of a database of didactic activities using gamification approach on different topics and disciplines that can be shared among a wide community of teachers, from primary to upper secondary school, in a vertical perspective, as well as the PP&S community. The results show that the experience can be presented at all levels and facilitate dialogue between teachers who teach different subjects at different schools.

REFERENCES

Barana, A. et al, 2022. Teaching online EMI Mathematics courses: a proposal to combine gamification and adaptive learning. *Book of Abstracts International Conference on Gamifying Mathematics in CLIL Contexts: Approaches and Good Practices.* Online, pp. 304-324.

Barana, A. et al, 2020. Automatic Formative Assessment Strategies for the Adaptive Teaching of Mathematics. *Proceedings of International Conference on Computer Supported Education*. Online, pp. 341-365.

Barana, A. et al, 2019a. Immersive teacher training experience on the methodology of problem posing and solving in Mathematics. *Proceedings of 5th International Conference on Higher Education Advances*. Valencia, pp. 667-675.

Barana, A. et al, 2019b. Adapting STEM automated assessment system to enhance language skills. *Book of Abstracts* Elearning and Software For Education, Vol. 2, pp. 403-410.

Behl, A. et al, 2022. Gamification and e-learning for young learners: A systematic literature review, bibliometric analysis, and future research agenda. *In Technological Forecasting & Social Change*, Vol. 176.

Bente, J. and Breuer, J., 2010. Why so serious? On the relation of serious games and learning. *In Journal for Computer Game Culture*, Vol. 4, No. 1, pp. 7-24.

Čeponienė, L. et al, 2018. Implementing gamification in a university-level UML modeling course: A case study. *In Computer Applications in Engineering Education,* Vol. 27, No. 2, pp. 332-343.

Chorianopoulos, K. and Giannakos, M., 2014. Design Principles for Serious Video Games in Mathematics Education: From Theory to Practice. *In International Journal of Serious Games*, Vol. 1, No. 3.

Corino, E. et al, 2022a. Adaptive Exercises and Formative Assessment for English Remedial Action. *Orchestration of Learning Environments in the Digital World.* pp. 3-19.

Corino, E. et al, 2022b. Data Driven Learning activities within a Digital Learning Environment to study the specialized language of Mathematics. *Proceedings of IEEE 46th Annual Computers, Software, and Applications Conference.* Online, pp. 167-176.

Deterding, S. et al, 2011. Gamification: Toward a Definition. *CHI 2011 Gamification Workshop Proceedings*. Vancouver, pp. 12-15.

De Freitas, S., 2006. *Learning in immersive worlds: a review of game-based learning*. JISC, London.

De la Peña, D. et al, 2021. Learning through play: Gamification model in university-level distance learning. *In Entertainment Computing*, Vol. 39, No. 2, pp. 1-24.

Dimoulas, C. et al, 2021. A Serious Game for Mediated Education on Traffic Behavior and Safety Awareness. *In Educational Sciences*, Vol. 11, No. 3.

Fissore, C. et al, 2020. Secondary school teacher support and training for online teaching during the covid-19 pandemic. *In EDEN Conference Proceedings*. Online, pp. 311-320.

VIDEO-PLAYING LOGS FOR ANALOGOUS VIDEOS REVEAL LEARNERS' PROACTIVE LEARNING STRATEGIES

Satoshi Yazawa[1], Masashi Inutsuka[2], Terutaka Komahashi[3], Kento Kanemura[1] and Kazuo Hiraki[1]
[1]*Graduate School of Arts and Sciences, The University of Tokyo, Meguro-ku, Tokyo, Japan*
[2]*Samuraikyoiku Co., Bunkyo-ku, Tokyo, Japan*
[3]*E Cubed (Online English School), Yokohama City, Kanagawa, Japan*

ABSTRACT

On-demand lecture videos play a major role in the networked classroom environment. However, there have not been many studies on the methods for optimal lecture video production as the mainstream approach is to statistically analyze indicators such as video length, playback speed, and the number of operations based on viewing records of existing lecture videos like MOOCs. We collaborated with an English lecture specialist to perform a controlled experiment by creating two videos with the same topic, lecturer, and studio, but different editing policies. By analyzing the detailed logs, we found that the number of points increased with learning time for the learners who viewed the lecture videos edited in the form of explicit chapters, and that learners changed their behavior such as seeking and changing the playback speed, depending on the video editing policy and their own ability. This suggests that it is important for lecture video creators to organize chapters and design user interfaces based on the assumption that learners will actively seek to optimize their viewing behavior according to their abilities.

KEYWORDS

Learning analytics, On-demand lecture video

1. INTRODUCTION

On-demand lecture videos are increasingly being used as an auxiliary tool in situations where face-to-face teaching is no longer possible due to the COVID-19 pandemic (Impey, 2021). Lectures using on-demand video are the primary form of educational material for online educational services, such as MOOCs (Massive Open Online Courses) (Baturay, 2015). Since MOOCs have been reported to have high dropout rates (Onah, 2014), ways to improve lecture videos and obtain high retention rates has become an important issue. One approach to improving such lecture videos and other learning materials is learning analytics. Learning analytics is the measurement, collection, analysis, and use of learner data and their situations for the purpose of understanding and optimizing learning content and its environment (Long, 2011). It has been mentioned that most learning analytics studies have analyzed learner behavior, but only a few have addressed the improvement of learning materials (Zhu, 2022).

A study focusing on learners' video viewing behavior reported that learners who played the video at 1.25x speed received more certificates than those who played it at 1x speed when an intervention was used to change default playback speed (Lang, 2020). A playback speed of 1.25x was found to be optimal for the group with low prior mastery regarding lecture videos, but that of 1.5x helps the group with higher mastery obtain better learning effects (Mo, 2022). It is unclear, however, how these trends change with the content and style of lecture videos. A study that focused on improving lecture videos analyzed the relationship between lecture videos and engagement from MOOC logs (Guo, 2014). The researchers used video playback time normalized by video length as a measure of engagement, and evaluated videos shorter than three minutes as having high engagement. However, it is unclear how the seek bar and playback speed change functionality of the video player affects engagement and learning effectiveness. The lack of studies on the characteristics of lecture videos could be due to the lack of variety in teaching materials relative to the differences in learners' learning abilities. In other words, it is difficult to estimate how factors other than the length of the video, such as the speaker's appearance, speaking style, presentation technique, filming technique, and editing technique, affect the learning log if there are two video materials on the same subject.

Therefore, in cooperation with a professional instructor who creates videos, we conducted a comparative experiment using two types of video materials on the same subject with partially different editing policies. In particular, we focus on the presence or absence of clear chapter division in lecture videos as an editing policy. We analyze how learners' behavior changes with and without chapter division according to their own level of understanding, how learning effects change, and what elements are necessary for an effective lecture video to be usable by learners.

2. MATERIALS AND METHODS

Subjects were asked to take a pre-test prior to the experiment and a post-test immediately after watching a lecture video. The two types of lecture videos were randomly assigned to the subjects. The task sequence presented to the subjects and the type of lecture video are shown in Figure 1.

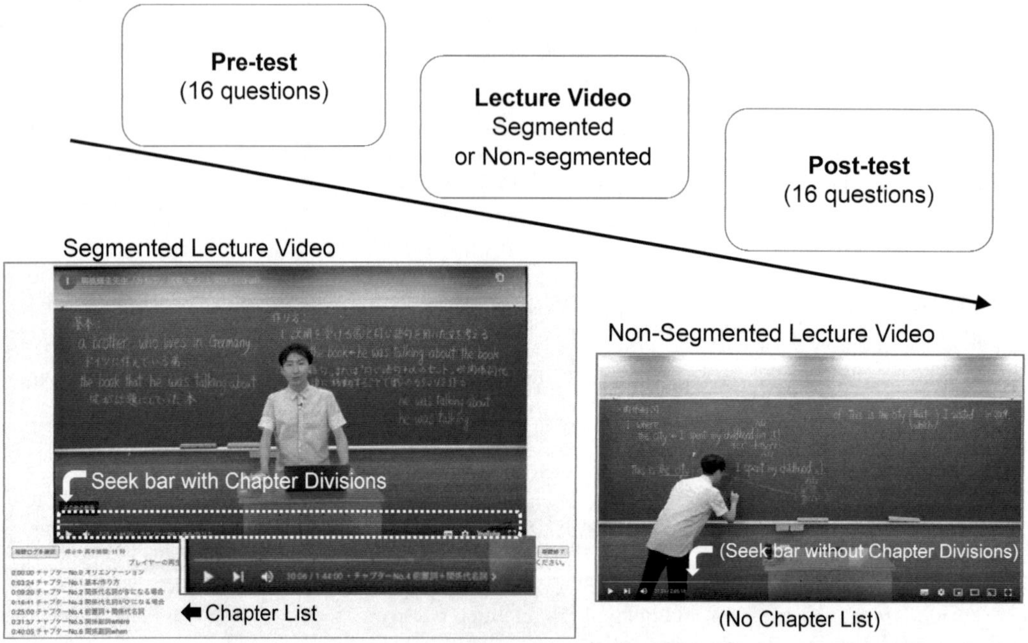

Figure 1. Task flow and prepared lecture videos

Two types of lecture videos (segmented and non-segmented) on the theme of relative pronouns in English grammar were prepared. These videos were created by co-author Komahashi, a professional cram school teacher. Each video showed a lecture by Komahashi himself, performed in the same studio, and filmed in the same location. The segmented video is divided into chapters explaining each topic, and chapter information was added for easy headings. In contrast, the non-segmented video is a video recording of a lecture, which includes the writing on the blackboard. The subjects' behavior toward these videos were compared. Komahashi also created two types of quizzes to check comprehension: a pre- and post-test with four choices. The content of the pre-test and post-test was the same for all subjects. Each test displayed the correct and incorrect answers and total score for each question immediately after the answers.

2.1 Design of Lecture Video

The lecture video consists of 15 topics for both the segmented and non-segmented videos. The list of topics is shown in Table 1.

Table 1. List of video topics. Each type of video has different lengths of topics

No	Title	Duration (Non-segmented; board writing; sec)	Duration (Non-segmented; without board writing; sec)	Duration (Segmented; sec)
0	Orientation	0	50	204
1	Basics/How to make	0	426	356
2	When the relative pronoun is S	169	92	441
3	When the relative pronoun is O	0	110	499
4	Prepositions + Relative pronouns	0	722	417
5	Relative adverb: where	171	355	488
6	Relative adverb: when	145	236	293
7	Relative adverb: why	154	360	379
8	When the relational adverbs where, when, and why create [noun group]	280	525	547
9	Restrictive and Non-Restrictive Usage	162	696	529
10	some of whom/all of which	234	380	384
11	which + noun and the noun of which	227	409	374
12	Relative pronoun: what	179	298	526
13	Idiomatic expressions using the relative pronoun what	85	637	366
14	Chained relational clauses	165	311	437
	(Total amount of time)	-	5607 (including time to write board = 7578)	6240

The segmented video was divided into chapters for each of the topics in Table 1. Information on the topic and starting time can be found on the seek bar and in an area outside the video. Although the non-segmented video does not contain explicit chapters like the segmented video, its topic content and order of presentation are the same as in the segmented video. Each type of video was taken independently. The segmented video was divided into chapters on editing points, while the non-segmented video was a single shot video. The total time for the non-segmented video is longer than that for the segmented video because it includes the time for board writing. In addition, no oral explanations were given while the instructor was writing on the board. Note that the orientation clarified that the learner is free to change the playback speed and seek, especially in the segmented video.

2.2 Implementation of Log Collection System

The experiment was implemented online via the experimental platform GO-E-MON (Yazawa, 2020). Video playback was implemented using the YouTube embedded player, and the JavaScript code defined on GO-E-MON was used to obtain YouTube IFrame API (Google Developers, 2021) events for video viewing logs. As a video viewing log, YouTube IFrame API records player state change events (Ended, Playing, Paused, Buffering, Cued), playback speed change events, and playback quality change events. When these events are detected, it records the player's playback status (NotStarted, Ended, Playing, Paused, Buffering, Cued), playback time on the video, playback speed, and playback quality together. Since YouTube IFrame API does not record seek operation events, they are detected by polling the player status every 100msec, checking the playback position on the video, and detecting discontinuous playback position changes.

3. RESULTS

The participants were recruited from first-year high school students. We asked the English teachers who teach the first-year students to introduce the experiment. Data were analyzed on 78 participants (41 females and 37 males) who voluntarily agreed to participate in the study based on ethical form 571-15, and completed the post-test. The segmented and non-segmented videos were randomly assigned to each participant, with each type of video being viewed by 39 participants. The participants then completed the post-test.

Figure 2 shows the increase in post-test scores relative to pre-test scores after viewing the segmented and non-segmented videos.

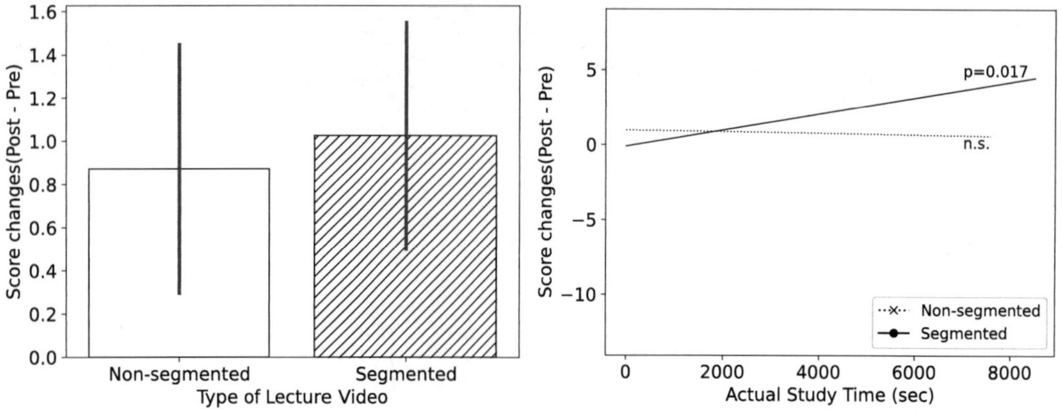

(a)Score changes per lecture video type

(b) Score changes for actual study time

Figure 2. Change in score between the segmented and non-segmented video

The average change in score per video type (a) showed an increase of 1.03 points for the segmented video and an increase of 0.87 points for the non-segmented video, indicating that the segmented video tended to improve performance, but there was no significant difference based on the t-test. However, for score improvement (b) relative to actual learning time, there was a significant correlation in the Pearson's correlation coefficient (with checking Q-Q distribution of score improvements and actual learning times, and calculating correlation coefficients) in the segmented video. The score improvement trend was observed for up to about one hour of study time for the non-segmented video, but no increase in score was observed for longer study time. A four-year study on HarvardX and MITx (Chuang, 2016) also suggested the link between actual study time and certificate rate, so the trend in the segmented video is to be expected.

In both the segmented and non-segmented videos, the score increase tends to be higher for lower pre-test scores. Pre-test scores were divided into three score ranges (low, middle, and high), and the changes in scores are shown in Figure 3.

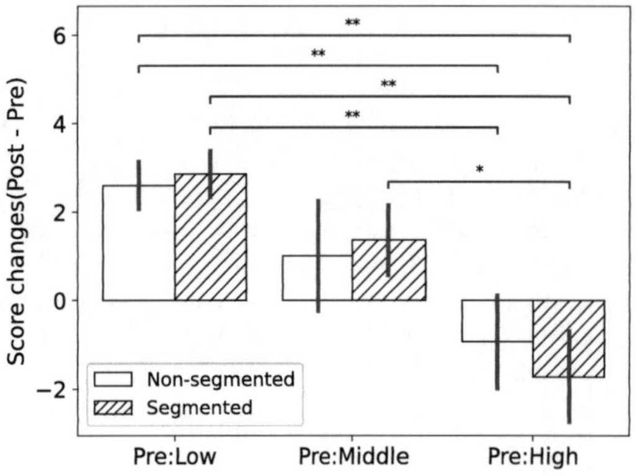

Number of learners per Pre-test

Level	Pre-test(S)	Non-segmented	Segmented
Low	S≦3	15	14
Middle	3<S<6	9	14
High	6≦S	15	11

Figure 3. Differences in increasing scores by pre-test score range

The threshold for each score range was determined by sorting all participants' pre-test scores in ascending order, with the score corresponding to one-third (decimal rounded down) of the total number of participants being the upper limit of the low level and that corresponding to two-thirds (decimal rounded down) of the total number of participants being the lower limit of the high level. The change in scores by pre-test score range shows that the lower the pre-test (Pre:Low), the greater the increase in scores, and the higher the pre-test

(Pre:High), the smaller the increase. It is possible that the basic parts of the lecture videos were easy to understand and contributed to the increase in scores, but the applied parts could not be fully mastered just by watching the videos. In addition, when it comes to the high level, the average score is negative for both segmented and non-segmented videos. It is possible that the difficulty level of the pre- and post-tests was not sufficiently adjusted, and that a more accurate consideration of the increase in scores could be made by interchanging the pre- and post-tests and taking counterbalance into account.

3.1 Zapping and Playback Speed While Video Viewing

The differences in behavior between the segmented and non-segmented videos are shown in Figure 4, which presents the average video playback speed and number of zapping times for each pre-test score range.

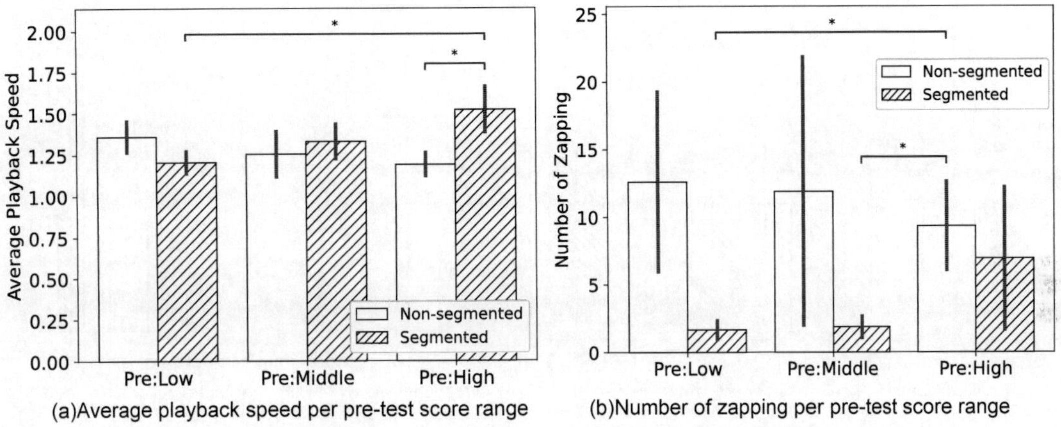

Figure 4. Relationship between Pre-test score range, zapping, and playback speed

The average playback speed is the total time watched continuously (in the video) divided by the actual total time spent on that playback. The number of zapping times is the count of the number of seek operation events that did not exceed 100 msec when immediately moving to another playback position. The average playback speed (a) shows that learners in the high-performance group of the pre-test who referred to the segmented video sped up the playback time, whereas those in the low performance group of the pre-test referred to the video at a speed close to the actual (1x) speed. Although the trend is not significant in the t-test, there is also a tendency for the low-performing group to play the non-segmented video at a faster speed and the high-performing group to play it at a speed closer to the actual speed.

The number of zapping times (b) shows that the segmented videos tended to have significantly fewer zapping times than the non-segmented videos in the low and middle pre-test score range. Thus, it is possible that many learners in the pre-test low-to-medium score range who watched the segmented video were able to use their viewing time more effectively without operating the seek bar as frequently as the non-segmented one. These results clearly show that learners behaved differently depending on the pre-test score range and the type of video they were watching.

3.2 Selection of Playback Points

The video viewing log allows a visualization of which topics in the video are referenced and in what order. Figure 5 shows a state transition diagram of the order and quantity in which topics are referenced.

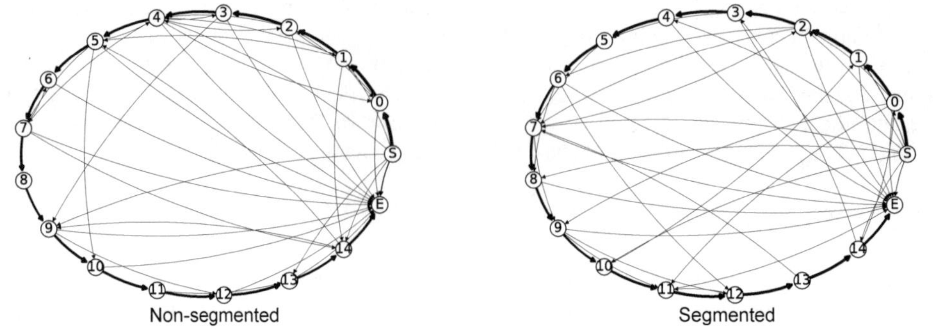

(a)State transition diagram by video type

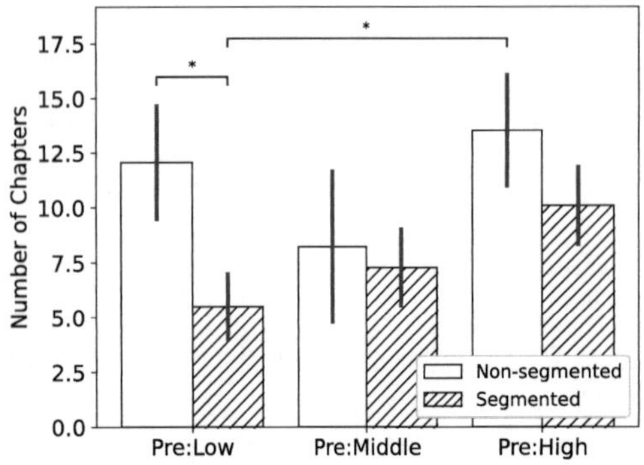

(b)Number of chapters per pre-test score range

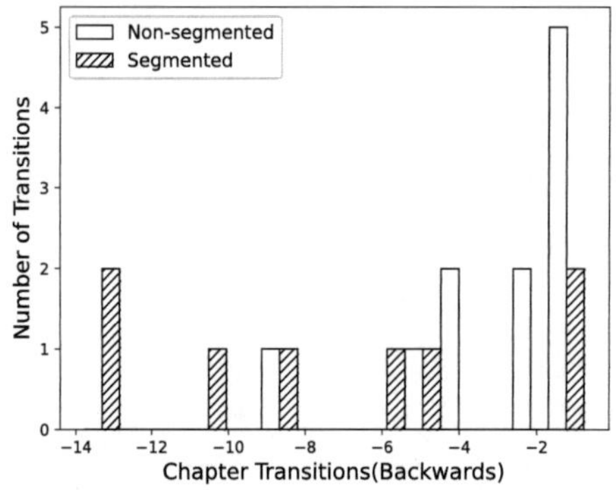

(c)Frequency of backward chapter transitions by video type

Figure 5. State transition diagram for two videos and chapter reference and transition trends

In state transition (a), S indicates the start of playback, E indicates the end of playback, and the numbers indicate the topic number (referenced for more than 15 seconds; including board time). The thickness of the arrows is the total number of transitions for all users for each video type. The majority of users for both video types refer to the videos in order starting with the orientation. The trend (b) in the total number of topics viewed by learners in each pre-test score range indicates that the low score group in the segmented type referred to fewer chapters than the high and the low score groups in the non-segmented type. In addition, there are a small number of transitions (c) that go back to the previous topic. While many transitions in the non-segmented video return to the previous topic, transitions in the segmented video go back to a few topics earlier. By clearly indicating chapters, learners can learn even a part of a video in a short time, and those who study for a long time can refer back and review points they did not understand, which is thought to contribute to the linear effect on actual learning time.

4. DISCUSSION

This experiment showed that learners change their video viewing behavior depending on their pre-test scores, that is, their level of understanding of the topics covered by the lecture video, and the structure of the target lecture video. Although previous studies have examined learner characteristics in a video-based learning environment, such as that by Yoon (Yoon, 2021) who investigated the active and passive characteristics of learners based on their learning behaviors, few have examined learner behavior based on the structure of the video itself. This study is unique in that it showed that learners adapt their video viewing behavior according to their own abilities. The process of creating lecture videos is a hard task that requires a great deal of know-how, including experience in conducting lectures and video editing skills. However, we believe that this is an important approach in that it makes it possible to evaluate not only quantitative aspects such as the length and playback speed of existing videos, but also qualitative aspects such as composition methods, through controlled experiments.

4.1 Optimal Video Playback Speed

In this experiment, it was confirmed that the higher the pre-test score, the more likely the viewer was to watch the segmented video at a faster speed. In the study that explored the appropriate playback speed by intervention, it was shown that the low mastery level group achieved a good learning effect at 1.25 times the playback speed and the high mastery level group at 1.5 times the playback speed (Mo, 2022). This result indicates that learners themselves may be adjusting the playback speed to their own cognitive load according to their own level without external control. Conversely, the non-segmented video might show the opposite trend, which might be influenced by the writing time on the board. Assuming that the learner is also writing down notes while watching the instructor write on the board, as in an actual lecture, the cognitive load in such a case would be higher than that of just listening, and the learner may tend to slow down the playback speed. In contrast, for learners who do not engage in such behavior, board time is a time that is not cognitively loaded and can be considered to have encouraged the behavior of speeding up the playback speed change. In the future, we would like to collect behaviors other than video playback during video viewing to clarify the details of learners' adaptive behaviors to lecture videos.

4.2 Optimal Video Length

Several previous studies have concluded that the optimal length of a lecture video is a few minutes (Guo, 2014, Zhu, 2022), and increasing video engagement with short videos may be an important approach for learning services that are primarily evaluated on a course credit. On the other hand, in this experiment, the segmented videos, in which learners can easily select viewing points and methods based on their own judgment, showed score improvement in proportion to the actual time spent, which may be effective for learning services that emphasize whether the learner has completed the course or not. We believe that for learning services that emphasize mastery-based learning, preparing videos of a certain length that are systematically and appropriately divided into chapters may encourage optimal video viewing behavior according to the learner's own level of understanding and allow the learner to efficiently obtain learning results.

5. CONCLUSION

We conducted a controlled experiment with the help of a lecture specialist by creating two videos on the same topic but with different editing policies. The detailed analysis of the logs revealed that the learners adaptively changed their behavior, such as seeking and changing the playback speed of videos, according to their own abilities. This experiment demonstrated the possibility of promoting efficient learning by designing chapters based on the assumption that learners actively try to optimize their viewing behavior according to their abilities when creating lecture behaviors, and by exposing the playback control function to them.

ACKNOWLEDGEMENT

We thank Nozakura S., Satomi Y. and Maeda Y. in Miura Gakuen High School and Iwasawa T. for their support during our experiments. We would like to thank Editage (www.editage.jp) for English language editing.
This research was funded by CREST of JST, grant number JPMJCR18A4. Also supported by Cross-ministerial Strategic Innovation Promotion Program (SIP).

REFERENCES

Baturay, M.H., (2015). An Overview of the World of MOOCs. *Procedia - Social and Behavioral Sciences*, Vol. 174, pp. 427–433.

Chuang, I., & Ho, A. D., (2016). HarvardX and MITx: Four years of open online courses--Fall 2012-Summer 2016. Available online: https://papers.ssrn.com/sol3/papers.cfm?abstract_id=2889436

Google Developers, accessed (2021). Youtube IFrame Player API. Available online: https://developers.google.com/youtube/iframe_api_reference (accessed on 28 Jul 2021)

Guo, P.J., et al., 2014. How video production affects student engagement: An empirical study of MOOC videos. *In Proceedings of the first ACM conference on Learning @ scale conference (L@S '14). Association for Computing Machinery*, New York, NY, USA, 41–50.

Impey, C., and Formanek, M., (2021). MOOCS and 100 Days of COVID: Enrollment surges in massive open online astronomy classes during the coronavirus pandemic. *Social Sciences & Humanities Open*, 4(1), 100177.

Lang, D., et al, (2020). Is faster better? A study of video playback speed. *In Proceedings of the Tenth International Conference on Learning Analytics & Knowledge (LAK '20). Association for Computing Machinery*, New York, NY, USA, 260–269.

Mo, C-Y., et al., (2022), Video playback speed influence on learning effect from the perspective of personalized adaptive learning: A study based on cognitive load theory. *Frontiers in Psychology*, 13, 839982.

Onah, D.F., Sinclair, J., and Boyatt, R. (2014). Dropout rates of massive open online courses: behavioural patterns. *EDULEARN'14*, 5825–5834.

Siemens, G., & Long, P., (2011). Penetrating the fog: analytics in learning and education. *EDUCAUSE review*, 46(5), 30.

Yazawa, S., et al., (2021). GO-E-MON: A new online platform for decentralized cognitive science. *Big Data and Cognitive Computing*, 5, 76.

Yoon, M., et al., (2021). Video learning analytics: Investigating behavioral patterns and learner clusters in video-based online learning. *The Internet and Higher Education*, 50.

Zhu, J., et al., (2022). The impact of short videos on student performance in an online-flipped college engineering course. *Humanities and Social Sciences Communications*, 9, 327.

Zhu, M., et al. (2022). Trends and issues in MOOC learning analytics empirical research: A systematic literature review (2011–2021). *Education and Information Technologies*, 27, 10135–10160.

HIGH-PERFORMING ONLINE STUDENT BEHAVIOURS

Rhodora Abadia, Sisi Liu and Qiaoling Sun
University of South Australia, Australia

ABSTRACT

Research on understanding student high academic performance focused on studying students in a traditional classroom or blended learning environment. This paper presents key online engagement behaviours of students that contribute to achieving high academic performance. The study experimented on three years of data derived from online students' academic performance and online course engagement in a series of six computing courses. Patterns discovered in this study indicate that most of the high-performing students in the online introductory programming courses continue to be high-achieving in their succeeding online computing courses compared to the non-programming courses. The results also show that the more programming experience, the students' performances improve. In addition, students with the highest academic performance engage at least 100% more in online formative learning activities compared to non-high performing students. Their highest activity engagement behaviours were on forum views and quiz activities. Findings of this study will assist educators in identifying critical elements in their content design to help all types of students increase their scholastic performance, better engage online students, and elevate non-high performing students to narrow the gap in students' performance in online courses. Finally, this study can help educational institutions identify what needs to be improved in the current learning analytics tool to better model student behaviours and patterns.

KEYWORDS

Descriptive Learning Analytics, e-Learning, Online Engagement Behaviours, Online Academic Performance

1. INTRODUCTION

Self-motivation and regulation are important for students to achieve high academic performance in a fully online learning and teaching environment. Numerous education research and studies have focused on exploring the factors that differentiate academically high-performing students from low-performing students. To date, studies on understanding students' high academic performances have concentrated on the traditional or blended classroom using self-report instruments (Guo et al., 2019; Kaplan, 2018; Wang & Liou, 2018). However, self-report instruments have been criticized for being limited due to subjectivity. Inaccuracies in self-assessment can be resulted from ingenuine assessments, such as participants altering answers to make them more socially acceptable (Duckworth & Yeager, 2015) being unable to assess themselves accurately (Araujo et al., 2017) exhibiting response bias and having difficulty mapping their answers to the rating scales, and being unable to fully retrieve information from memory (Rosen et al., 2017). Students tend to retrieve distinctive information that is time-bound, but self-reported questions are seldom distinct. With the popularity of online learning and the advancement of data mining techniques, Learning Management Systems (LMS) have been developed and implemented by various online education providers in online courses to automatically record students' engagement and performance data. It is reported that using LMS data to support learning analytics and educational data mining provides a more objective picture of students' learning through data-driven approaches (Liu et al., 2017).

Understanding the online engagement behaviours that contribute to online students achieving high performance will assist online facilitators in continually supporting high-performing students and identifying critical factors to help other students improve their academic performance. Online facilitators can set a guide to online engagement to uplift the performance of non-high performing students. This reasoning is consistent with Dweck (2006)'s work on growth mindset, which is now a synonymous of high expectation. It is hypothesized that students' achievement is strongly affected by what the teacher expects of them, and this has been justified by many education researchers (Campbell et al., 2020; Johnston et al., 2019; Robinson, 2017).

1.1 The Study

The engagement data used in this study were collected from first- and second-year students who were enrolled in a 100% asynchronous online IT degree. Demographics of online students include those who are already working, parents who are unable to go to campus, have disabilities or live or work in a remote area. For them, online learning is more flexible and accessible. Identifying online engagement behaviours of high-performing students using the engagement and performance data recorded through the LMS not only provides a more objective view, but also gives the instructors an insight into how these students' online behaviours differ from the rest of the online students. Identifying these online behaviours can also help instructors highlight these behaviours and encourage changes in practice. Specifically, instructors can suggest specific strategies and provide guidance to non-high performing students on how to adopt these online engagement behaviours. Moreover, personalized learning interventions can be motivated from these behaviours and designed such that students are guided in the development of these positive online behaviours. Results can also help students reflect on their behaviours and learn to self-regulate. Finally, this study will also help educational institutions identify what needs to be improved in the current learning analytics tool to better understand student behaviours and patterns.

This paper presents a longitudinal study investigating online student engagement and academic performance from LMS data of high-performing students enrolled in fully online introductory computing courses. It aims to answer the following research questions:

1. Do high-performing students in introductory online computing courses (programming and non-programming) continue to have high performance in their succeeding computing courses?
2. What are the associations between students' academic performance and engagement behaviours in fully online programming courses?

2. RELATED STUDIES

This section discussed related literature on high-performing students, self-regulated learning, and student engagement data from LMS are used in learning analytics.

2.1 High-Performing Students

Studies have shown a positive relationship between student motivation and academic achievement (Pintrich & De Groot, 1990). It is also indicated that high-achieving students show a refined ability to select and modify study behaviours (Mai et al., 2021), manage their time and use more effective study strategies (Dunlosky et al., 2013) and routinely engage in and adapt skills to pursue these behaviours (Zimmerman & Schunk, 2001). Despite evidence of the association between high performance and effective study behaviours as a primary study strategy, Persky (2018) found in his longitudinal study that high-achieving students relied on re-reading texts and re-watching videos. All these studies revolved around the idea that students perform better if they can fully understand concepts when they self-regulate their learning behaviours. This idea is especially true in an online learning environment where self-regulated learning is critical to students' academic success due to them having limited interactions with peer learners and assistance from the instructors.

A study undertaken by Alqurashi (2022) compared seven aspects of student engagement (i.e., higher-order learning, reflective and integrative learning, learning strategies, quantitative reasoning, and collaborative learning, student–faculty interaction, and effective teaching practices) using survey data collected from senior-level undergraduate students who studied online courses. Their findings showed that low achieving students had significant higher student-faculty interactions than high-performing students. Although this study looked at students' engagement in an online learning environment, the data used were from student survey, which may include inaccuracies in self-assessment of their engagement. Also, the study was not for students who studied fully asynchronous online course and did not explore the engagement behaviours when interacting with online course activities. In this paper, the focus is the monitoring of students' learning process manifested through their online engagement behaviours using engagement and academic performance data captured in the LMS.

2.1.1 Student Engagement and Performance through LMS Data

In an online learning environment, data are often collected from an LMS. An LMS is a critical tool in a fully online learning environment to facilitate the teaching and learning process. In addition to distributing and managing course materials, it also captures students' engagement and performance data that can be further utilized to support the learning analytics and educational data mining. Recent LMS-related research has shifted the focus from exploring interactive and creative functionalities of an LMS to analyzing the LMS data, such as the log data (Henrie et al., 2018) and activity data (Simanullang & Rajagukguk, 2020) to discover patterns of student engagement, evaluate students' academic performance and improve instructors' teaching pedagogical practices. Recent studies reveal that learning analytics (Conole et al., 2011) can be applied to LMS data to visualize student engagement patterns, derive insights from student engagement data for better learning design and improve student learning experience (Henrie et al., 2018; Toro-Troconis et al., 2019). However, empirical data were limited in these studies due to an LMS implemented as a supplement to the traditional classroom teaching or very few fully online courses available for research purpose.

To date, research on understanding student high academic performance has focused on studying students in a traditional classroom or blended learning environment. Nevertheless, few studies have attempted to look at the longitudinal data in online learning and identify the online engagement behaviours of students achieving high academic performance in a fully online learning environment. In this paper, three-year data from six courses were collected from an LMS that stores various online courses having frequent engagement data of different online activities to conduct a longitudinal study to further analyze the online students' engagement behaviours and their association with students' academic performance.

3. WORKFLOW MODEL

The analysis used students' academic performance and online student engagement data. Online engagement behaviours included are the following formative online data activities: forum, quiz, URL, lesson, file, and folder. For empirical experiments, six courses were analyzed, among which three were a series of programming courses and three were a series of non-programming IT courses. The data were divided into two case scenarios, including student academic performance alone, and student academic performance with online student engagement. These scenarios were subsequently used to identify the high-performing students in each course, if they continue to perform well in the succeeding courses, and the relationship of their engagement to the online course to their high performance. For example, Programming 1 is a prerequisite for the Programming 2 course, thus in the experiment, the number of high-performing students in the pre-requisite course that continue to be high-performing in the subsequent courses were identified. The workflow model used in this study comprises of three main phases: data collection, data processing, data exploration and visualizations

3.1 Data Collection

The data were collected from student engagement and academic performance reports from the LMS. A total of 804 student academic and engagement data were used in the study. These reports were taken from a series of three online programming courses and a series of three online non-programming courses. For programming courses, three years (2018-2020) of reports were taken from two first year courses and one second year course. For non-programming courses, three years (2019-2021) of reports were taken from two first year courses and one second year course. The data contained a list of students who were enrolled in the courses with de-identified student IDs, their online engagement data, including the number and percentage of forum views and contributions, quizzes attempted and scores (if recorded), files, URLs, lessons, files, folders accessed, and students' academic performance involving summative assessment grades and final course grades. The forums are discussion forums where students can post questions and contribute to answering the questions; quizzes are formative assessments for students to self-check what they've learned; files are content related materials (instructions, additional materials); URLs are links to optional content materials; and folders contain a group of content material files usually containing instructions and files associated to learning activities or assessments. Demographic data of the students were not gathered as a result of previous study (Bretana, Kaushal, Nair, & Cheung, 2020) which discovered that the available demographic characteristics (e.g., age, location, grade point average, gender, degree, full-time/part time study) were not effective indicators of achievement in online programming courses.

3.2 Data Preprocessing

The outcome variable is named as *IsHighPerformance*, which is calculated from *Grade*. A student is considered high-performing if his/her *Grade* has a final raw score of 75 and above. Students can engage with seven types of activities in a course: forum, quiz, assignment, URL, lesson, file, and folder. Assignment is a summative activity while the rest are formative activities. We will only analyze formative activities because these are not required activities. URL, Lesson, File and Folder are formative activities which contains the content materials and does not require contributions from the students while forums and quizzes require students to participate. Each activity has three information types: total engagements of each student in an activity, views of a student, and contributions of a student in an activity as shown in Table 1.

Table 1. Generalized data structure of each activity engagement

Column Name	Data Type	Description
SP, Year	Nominal	Study Period and Year
Course	Nominal	Course name
Sum<activity>Views	Continuous	Total view count of a student in an activity
SumPercent<activity>Views	Continuous	Total percentage of views of a student in an activity
Sum<activity>Contributions	Continuous	Total contributions of a student inan activity.
SumPercent<activity >Contributions	Continuous	Total percentage of contributions ofa student in an activity
IsHighPerformance	Binary	Target variable

3.3 Data Exploration and Visualizations

Initially, the data sets for all courses were explored by studying the students' academic performance and then by studying the details of the students' online engagement and combined academic performance and engagement. These analyses aided in the identification of patterns and descriptive analytics of the data. Next, the academic performance of continuing students, who enrolled in the series of programming courses (2018-2020) and students who enrolled in the series of non-programming courses (2019-2021), was explored. These students' views and contributions of the formative activities were also explored. Since the focus was on formative activities, summative activities were not included in the analysis of engagement behaviours, that is, viewing of assignment and file submissions were not included in the total count. The reason is that these are required activities and all high-performing students, by default, engage in these summative activities to get a high grade. We are interested in exploring the behaviour patterns of formative activities.

Patterns were explored to identify whether the high-performing students enrolled in the prerequisite programming courses continue to have high performance in the succeeding programming courses, and their engagement behaviours in each of these courses. Results of the data exploration not only led to the understanding of the data, but also reveal the patterns showing whether high-achieving students continue to perform well in succeeding courses. It is important to note that Programming 1 is the prerequisite to both Programming 2 and Programming 3, and it is recommended that students enroll in Programming 1, followed by Programming 2, then Programming 3 in different study periods. However, if there are schedule conflicts, students can take Programming 1, followed by Programming 2 and 3 at the same time. IT Fundamentals is the prerequisite of the System Analysis course, which in turn is the prerequisite of the System Design course. The results and analysis of the data exploration and presented in the next section.

4. RESULTS AND ANALYSIS

The results and analysis in the following subsections explored and visualized the consistency of high-performing students in their series of introductory computing courses, and the associations between students' high academic performance and engagement behaviours in a fully asynchronous online degree.

4.1 Exploring Consistent High Performance of Students in Programming and Non-programming Courses

Table 2 suggests that 52.00% of the high-performing students in Programming 1 continue to obtain high grades in the succeeding programming course. 12.5% of the students who were non-high performing in Programming 1 and enrolled in Programming 2 had a better grade outcome and become high performing in Programming 2. The significant difference between the two percentages (52.00% - 12.50% = 39.50%) indicates that students are more likely to perform high in the successive courses (e.g., Programming 2) if they performed high in prerequisite courses (e.g., Programming 1). Similar observations have existed for students who are high performing in Programming 1 that enrolled in Programming 3 (71.43% continued to be high performing), and Programming 2 and then enrolled in Programming 3 (75% continued to be high performing). Note that 53.33% students who were in the non-high performing category in Programming 2 have performed well in Programming 3. This result is consistent with the experience factor that has been widely studied (Wilcox & Lionel, 2018). Although these studies looked programming experience prior to taking the introductory course, the experience factor can be applied in the study, that is programming experience influence better performance.

Table 2. Academic performances of students in the series of courses

	Prog 1 to Prog 2	Prog1 to Prog 3	Prog 2 to Prog3	IT Fund to Sys Analysis	Sys Analysis to Sys Design	IT Fund to Sys Design
Continued to be High performing	52%	71.43%	75%	41.38%	68.75%	25%
Non-high-performing to high-performing	12.50%	10%	53.33%	19.15%	22.73%	44.44%

The academic performances of students in the non-programming courses (Table 1) show the percentage of students who have high-performance in pre-requisite that continue to have high performance in succeeding non-programming courses. However, compared to the programming courses, the percentage of high-performing students are lower. For high-performing students in IT Fundamentals that enrolled in System Analysis, 41.83% continued to perform well. For IT Fundamentals to System Design, only 25% continue to perform well. However, from Systems Analysis to System Design, the percentage of students who continually perform high is significant (68.75%). The above results showed that out of six pairs of prerequisites and succeeding courses, five courses were observed that high-performing students in prerequisite courses had a much higher chance of being a high-performing student in the succeeding courses. Only one pair of courses (IT Fundamentals to System Design) showed the opposite trend. Compare to System Analysis and System Design where the contents are more related in the context of software development. IT Fundamentals covers a broad range of IT topics and may not be directly connected to System Design.

Students who are high-performing in the introductory programming courses have a higher chance of getting a high-performing grade in the succeeding programming courses In non-programming courses, it is observed that there is a lower percentage of students continuing to have high-performance if the pre-requisite course contents are not significantly related to the contents of the succeeding courses.

4.2 Online Engagement

Online activities are used to help students progress their learning and prepare them for assessment tasks. The total views of formative activities in programming courses are 108,668 and non-programming courses is

126107. The next highest views are the URL and Quiz. The total views in forum posts are 91% higher than URL and 92% higher than quizzes. For non-programming courses, the total views in forum posts are 99% higher than URL and 86% higher in quizzes. This shows that high performing students significantly views the forum posts. Comparing to non-high performing students, the online engagement data of high-performing students (Figure 1) suggest that the distribution of forum views where the average views of high-performing students for both programming and non-programming courses is more than two-fold (112 % higher) than that of non-high performing students. The active use of forums by high-performing students is consistent with studies from Moström (2011) and Pedrosa et al. (2016) of successful programming students' behaviour in tradition classroom learning that they apply different strategies when they get stuck in programming. One common strategy is through social interaction with peers and teachers. In online learning, one equivalent form of interaction through peers and teachers is through engagement in forums. High-performing students' behaviour of having almost twice activity views than non-high performing is consistent with the self-regulated learner's behaviour of reflection. Viewing helps students to reflect and think about what they are seeing, which help develop their skills and knowledge to analyze what they have just viewed. It is also consistent to studies that indicated that students' viewing activities have direct positive influence on their completing learning tasks (Ma et al., 2015).

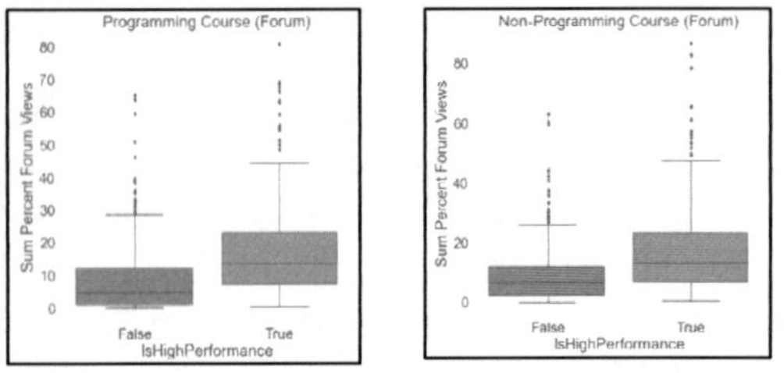

Figure 1. Boxplot of percentage forum views

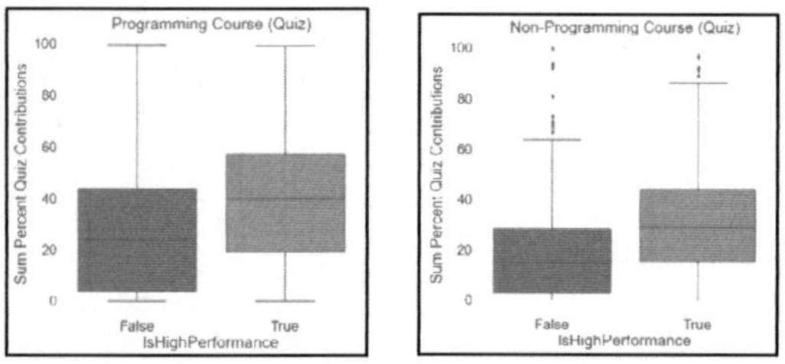

Figure 2. Boxplot of percentage quiz contributions.

The quiz is significantly higher in contributions for both programming and non-programming courses (57,393 total contributions for programming courses and 76,211 for non-programming). Forum contributions is significantly lower (90% lower for programming and 92% lower in non-programming courses). Other activities have zero contribution because students these formative activities provide content to the online student but was not designed for interaction with them. For quiz contributions (Figure 2), the percentage difference in the average quiz contributions in programming courses of high-performing students is 44% higher compared to non-high performing students, while for non-programming courses the percentage difference is 112%. We've also seen that students who were high performing for all programming courses (Programming 1, Programming 2, and Programming 3) have a 159% difference in contribution and 198% difference in views. For non-programing courses (IT Fundamentals, System Analysis and System Design), students who were consistently high performing in these three courses have a 170% difference in views and 105% in contribution.

The quiz formative activity is the online activity where high-performing students contribute the most in programming courses and non-programing courses. This behaviour of having high contributions to formative activities such as quizzes in coding is consistent with the studies that high-performing students in programming education have been consistently active in practice as their study progresses (Hassinen & Mäyrä, 2006; Mai et al., 2021; Pedrosa et al., 2016).

4.2.1 Statistical Significance

Student's t-test has been performed to evaluate statistically if the mean values of online engagement between high performance and non-high-performance students are equal. Table 3 shows that all p-values of two online engagement data sets of both programming (8.99e-11 for forum views and 2.69e-9 for quiz contributions) and non-programming courses (9.83e-16 for forum views and 5.33e-15 for quiz contributions) are much lower than 0.05, indicating a rejection of the null hypothesis, i.e., the online engagement is statistically significantly different between high-performing and non-high performing students.

Table 3. P-values of the students' t-test on two different online engagement

	Programming Courses	Non-Programming Courses
Sum Percentage Forum Views	8.99e-11	9.83e-16
Sum Percentage Quiz Contributions	2.69e-9	5.33e-15

5. CONCLUSION AND RECOMMENDATION

The study presented in this paper analyzed three-year engagement and academic performance data of students enrolled in a 100% asynchronous online IT courses. Students' academic performance and online engagement data were collected from the LMS to provide a more objective data and explore the relationship between online engagement behaviours and academic performance. The data were further analyzed to discover the patterns of online engagement behaviours of high-performing students. The study revealed that an average of 54.5% of high-performing students in the introductory programming courses continued to have high performance in the succeeding programming courses. High-performing students in the first programming courses have a higher chance (61.43%) of maintaining the high performance in Programming 3 than Programming 2. The study also showed that as students get more programming experience, that is after they have completed Programming 2, more than fifty percent of the non-high-performing students became high-performing in Programming 3. For non-programming courses, high-performing students in their first course (IT Fundamentals) has a lower percentage of achieving a high performance in the succeeding courses. Students who are high performance in System Analysis has a higher chance (46.025) of having consistent high performance in System Design. As the results indicated a continuity of high-performing students in an introductory course and the succeeding courses especially for programming courses, their online engagement behaviours were further explored.

The online engagement data of high-performing students is consistent with self-regulated learner behaviours. The average forum views of high-performing students for both programming and non-programming courses was 112 % higher than that of non-high performing students. In the quiz activity contributions, the average contributions of high-performing students for programming courses were 44% higher while 112% higher for non-programming courses comparing to non-high performing students. Online students use the online equivalent medium or tool to manifest the face-to-face student behaviours (e.g., social interaction with peers and teachers in face-to-face and the equivalent discussion forum tool in online learning as a form of social interaction).

Though this study utilized a data-driven approach based on learning analytics to understand the online engagement behaviours of high-performing students studying at a 100% asynchronous online IT courses, the study was limited to the data available through the LMS. Viewing of content videos is an activity that has been excluded due to the transition from one video tool to another which resulted in the inability to access the older data. For future directions, we aim at further analysing the data collected in this study to identify the predictive factors that contribute to students' high performance and the continuity of their high performance and extending this study by understanding the mindset of these high-performing students when engaging and learning online.

REFERENCES

Alqurashi, E. (2022). What do students engage with the most? A comparative study between high and low achieving students within online learning environments. *Open Learning: The Journal of Open, Distance and e-Learning*, *37*(3), 219-234.

Araujo, T., Wonneberger, A., Neijens, P., & de Vreese, C. (2017). How much time do you spend online? Understanding and improving the accuracy of self-reported measures of internet use. *Communication Methods and Measures*, *11*(3), 173-190.

Bretana, NA., Robati, M., Rawat, A., Pandey, A., Khatri, S., Kaushal, K., Nair, S., Cheung,G. & Abadia,R (2020). Predicting Student Success for Programming Courses in a Fully Online Learning Environment. H. So et al. (Eds.) (2020). Proceedings of the 28th International Conference on Computers in Education. Asia-Pacific Society for Computers in Education.

Campbell, A., Craig, T., & Collier-Reed, B. (2020). A framework for using learning theories to inform 'growth mindset'activities. *International Journal of Mathematical Education in Science and Technology*, *51*(1), 26-43.

Conole, G., Gašević, D., Long, P., & Siemens, G. (2011). Message from the LAK 2011 general & program chairs. International Learning Analytics & Knowledge Conference 2011,

Duckworth, A. L., & Yeager, D. S. (2015). Measurement matters: Assessing personal qualities other than cognitive ability for educational purposes. *Educational researcher*, *44*(4), 237-251.

Dunlosky, J., Rawson, K. A., Marsh, E. J., Nathan, M. J., & Willingham, D. T. (2013). Improving students' learning with effective learning techniques: Promising directions from cognitive and educational psychology. *Psychological Science in the Public interest*, *14*(1), 4-58.

Dweck, C. S. (2006). *Mindset: The new psychology of success*. Random house.

Guo, W., Lau, K. L., & Wei, J. (2019). Teacher feedback and students' self-regulated learning in mathematics: A comparison between a high-achieving and a low-achieving secondary schools. *Studies in Educational Evaluation*, *63*, 48-58.

Hassinen, M., & Mäyrä, H. (2006). Learning programming by programming: a case study. Proceedings of the 6th Baltic Sea conference on Computing education research: Koli Calling 2006,

Henrie, C. R., Bodily, R., Larsen, R., & Graham, C. R. (2018). Exploring the potential of LMS log data as a proxy measure of student engagement. *Journal of Computing in Higher Education*, *30*, 344-362.

Johnston, O., Wildy, H., & Shand, J. (2019). A decade of teacher expectations research 2008–2018: Historical foundations, new developments, and future pathways. *Australian Journal of Education*, *63*(1), 44-73.

Kaplan, H. (2018). Teachers' autonomy support, autonomy suppression and conditional negative regard as predictors of optimal learning experience among high-achieving Bedouin students. *Social Psychology of Education*, *21*, 223-255.

Liu, D. Y.-T., Bartimote-Aufflick, K., Pardo, A., & Bridgeman, A. J. (2017). Data-driven personalization of student learning support in higher education. *Learning analytics: Fundaments, applications, and trends: A view of the current state of the art to enhance e-learning*, 143-169.

Ma, J., Han, X., Yang, J., & Cheng, J. (2015). Examining the necessary condition for engagement in an online learning environment based on learning analytics approach: The role of the instructor. *The internet and higher education*, *24*, 26-34.

Mai, T. T., Crane, M., & Bezbradica, M. (2021). Student behaviours in using learning resources in higher education: How do behaviours reflect success in programming education?

Moström, J. E. (2011). *A study of student problems in learning to program* Umeå University, Department of Computing Science].

Pedrosa, D., Cravino, J., Morgado, L., & Barreira, C. (2016). Self-regulated learning in computer programming: strategies students adopted during an assignment. Immersive Learning Research Network: Second International Conference, iLRN 2016 Santa Barbara, CA, USA, June 27–July 1, 2016 Proceedings 2,

Persky, A. M. (2018). A four year longitudinal study of student learning strategies. *Currents in Pharmacy Teaching and Learning*, *10*(11), 1496-1500.

Pintrich, P. R., & De Groot, E. V. (1990). Motivational and self-regulated learning components of classroom academic performance. *Journal of educational psychology*, *82*(1), 33.

Robinson, C. (2017). Growth mindset in the classroom. *Science Scope*, *41*(2), 18.

Rosen, J. A., Porter, S. R., & Rogers, J. (2017). Understanding student self-reports of academic performance and course-taking behaviour. *AERA Open*, *3*(2), 2332858417711427.

Simanullang, N., & Rajagukguk, J. (2020). Learning Management System (LMS) based on moodle to improve students learning activity. Journal of Physics: Conference Series,

Toro-Troconis, M., Alexander, J., & Frutos-Perez, M. (2019). Assessing student engagement in online programmes: Using learning design and learning analytics. *International Journal of Higher Education*, *8*(6).

Wang, C.-L., & Liou, P.-Y. (2018). Patterns of motivational beliefs in the science learning of total, high-, and low-achieving students: Evidence of Taiwanese TIMSS 2011 data. *International Journal of Science and Mathematics Education*, *16*, 603-618.

Zimmerman, B. J., & Schunk, D. H. (2001). *Self-regulated learning and academic achievement: Theoretical perspectives*. Routledge.

EU SECURITY AND DEFENCE STUDENTS' PERCEPTION AND USE OF DIGITAL COMPETENCIES IN HIGHER EDUCATION

Marina Marchisio[1], Fabio Roman[1], Matteo Sacchet[1], Enrico Spinello[2],
Linko Nikolov[3], Malgorzata Grzelak[4] and Alin-Constantin Sava[5]
[1]Department of Molecular Biotechnology and Health Sciences, University of Turin, Via Nizza 52, 10126 Torino, Italy
[2]Education and Training Command and School of Applied Military Studies, Via Arsenale 22, 10121 Torino, Italy
[3]Department of Communication Networks, National Military University, Karel Shkorpil 1 str., 9700 Shumen, Bulgaria
[4]Faculty of Safety, Logistics and Management, Military University of Technology, 00-908 Warszawa 49, Poland
[5]Military Technical Academy "Ferdinand I", 39-49 George Cosbuc, Sector 5, Bucharest, Romania

ABSTRACT

The use of digital tools in teaching and learning has become increasingly common in recent years. Even prior to 2020, their usage was steadily growing, but the outbreak of the COVID-19 pandemic in the early part of the current decade significantly accelerated their adoption. This trend has persisted even as the virus is no longer a pressing emergency, indicating that certain changes will have a lasting impact. Utilizing digital tools in education offers advantages, but it also presents challenges. Transitioning from using computers and other devices for everyday purposes to incorporating them into educational contexts is not as straightforward. In this research, we examine how students perceive higher education during the transition prompted by the pandemic. We conducted a survey among military and civilian students enrolled in security and defence studies, focusing on their experiences and perceptions as they undergo significant changes in their educational routines. Analysis of the questionnaire revealed that students encountered difficulties in various aspects, including establishing connections with teachers and experiencing the same level of engagement when unable to interact with them in person. An increase in the students' perception of digital competencies was detected. This research is part of the activities undertaken in the European project DIGICODE, which aims to enhance the quality of education in security and defence by promoting the proper use of digital tools and fostering the development of digital competencies among both students and teachers.

KEYWORDS

Digital Education, Distance Learning, Online Education, Security and Defence, Student Training

1. INTRODUCTION

Security and defence education needs suitable tools that facilitate learning in various ways, such as blended learning (Marchisio et al., 2022c), hybrid learning (Marchisio et al., 2022d), and other modalities that enable strategic internationalization of educational settings without mandating purely online courses (Mihalova, 2006). To foster effective and long-term collaboration, it is beneficial for military officers and civilians involved in security and defence to cooperate early in their educational journey, such as during their student or training years (Marchisio & Spinello, 2021), also through digital media: in fact, e-learning widespread adoption across the globe took a significant leap forward during the COVID-19 pandemic (Hodges et al., 2020). It is important to provide specific training to teachers, students, and all stakeholders, as it helps them understand the unique application of digital tools in education (Marchisio, M. et al, 2022a). The DIGICODE project, funded by the European Union as an Erasmus+ Key Action 2 Strategic Partnership project, is facing these challenges at the European level. It involves several nations such as Bulgaria, Italy, Poland, and Romania. The project, which stands for DIGItal COmpetencies for improving security and Defence Education, aims to promote effective utilization of digital tools for teaching in the military context. This paper aims to examine the digital competencies of students both before and during the COVID-19 pandemic. To achieve this, we conducted a questionnaire survey among students, which assessed various aspects including engagement, communication, development of digital competencies, and achievement of learning outcomes. Additionally, we compared students' time management when using computers and electronic devices. Finally, open-ended questions explore some complementary insights. The study focuses on European students enrolled in security and defence courses, interdisciplinary programs with a strong emphasis on international cooperation. A similar survey has already been conducted and analyzed among teachers (Marchisio et al., 2022a), enabling us to establish comparisons between the perspectives of students and teachers. The paper is organized as follows: Section 2 provides the theoretical framework in which this research takes place, while Section 3 discusses the research question and methodology. Section 4 presents the findings, and Section 5 provides a comprehensive discussion. Finally, Section 6 offers concluding remarks.

2. THEORETICAL FRAMEWORK

Numerous recent studies have examined the significance of acquiring digital competencies in the context of security and defence. Firstly, it is evident that digital tools form the foundation of e-learning, necessitating all stakeholders, particularly those involved in education, to possess digital competencies. Attitude, motivation, self-efficacy, and use of technology play a significant role in the cognitive engagement and academic performance of students (Patricia Aguilera-Hermida, 2020). However, a critical issue arises when individuals overestimate their skills, leading them to believe that their existing knowledge is sufficient and that they can rely on others with better tool proficiency if assistance is needed. This misconception is prevalent among both students (Buffardi & Taddeo, 2017) and teachers (Tomczyk, 2021), resulting in a diminished emphasis on the importance of acquiring adequate digital competencies. Pinchuk and Prokopenko (2021) analyzed the experiences of various countries, including the United States, Australia, China, Britain, Israel, Korea, and Singapore, in implementing modern educational approaches to STEM subjects. They identified the potential for transdisciplinary integration in providing advanced training for the Armed Forces of Ukraine, their home country, and the need for effective ownership of tools for planning and organizing project work. Barron and Rowles (2021) emphasized the significance of digital literacy, specifically in the Air Force and other military branches. The optimization of technology for educational purposes requires technical tools to serve as means rather than ends (Goldin & Katz, 2009), thereby involving various other skills (Van Laar et al., 2017). One well-known concern is that despite digital competencies aligning with future skills (Ehlers, 2020, and references therein), nearly half of Europeans lack even basic digital skills, with a gender gap and digital divide exacerbating the situation based on 2017 data (European Education Area, 2020). The European Union's publication of the Digital Education Action Plan (DEAP) in 2020 underscores the strategic importance of digital competencies in education, inspiring the initiation of the DIGICODE project. Overall, these skills are crucial for addressing the challenges of sustainable education (Mentsiev et al., 2022) and align with the United Nations' Sustainable Development Goals (United Nations, 2015).

3. RESEARCH QUESTION AND METHODOLOGY

Fixing the context of security and defence higher education in a transitional scenario caused by the COVID-19 pandemic, our study was driven by the need to address the following research questions: (RQ1) what are the perceptions of military and civilian students and (RQ2) what are the perceptions of digital competencies? To assess these factors, we submitted, measured and analyzed a survey that examined both quantitative and qualitative aspects about these perceptions of students. The research in this paper considers 6 pairs of questions, a subset of the whole questionnaire. The quantitative aspects allowed for a comparison of the situations before and during the COVID-19 pandemic. We utilized Likert scales for Pairs 1-4 and categorical levels for Pairs 5-6. For ratings, we established a five-level Likert scale, with 1 representing the lowest score and 5 indicating the highest. Regarding time spent, we categorized it based on the actual number of daily hours in one case, and in the other case, we utilized reasonable ranges of weekly hours (e.g., "from 4 to 10 hours"). We performed both descriptive and inferential analyses on the resulting numerical data in particular we performed the paired Wilcoxon signed-rank test and the paired t-test. This approach enabled us to effectively demonstrate any existing differences and establish their statistical significance. Similar questions provided to students and teachers allow us to compare the perceptions, highlighting similarities and differences in how both groups experienced the scenario. The qualitative aspects gave us some perspectives about students' perceived problems and threats resulting from remote education, and how to deal with them, along with some practices they deemed as good to address the situation both during and after the COVID-19. The data was collected from 832 European students in Security and Defence studies, categorized by age and gender as follows:

Table 1. Distribution of the students by age and gender

Age range	Females	Males	Did not specify
18-22 years old	103	239	4
23-25 years old	106	309	5
Over 25 years old	11	51	4

Approximately 83% of the participants are military students, the remaining 17% are civilians. They are nearly evenly divided between bachelor's and master's students, with a slight majority of the former. The questionnaire was proposed to students who experienced security and defence education in the transition scenario, thus from last year of bachelor's students upwards (almost 95% of the sample). We considered even other responses because they represent a very little part of the sample and some students attended military secondary schools, thus experiencing security and defence education even prior to university. Considering that the COVID-19 pandemic compelled the use of digital tools that were not previously mandatory, it is reasonable to anticipate that ratings and time spent are important indicators that have undergone changes.

4. RESULTS

In the subsequent tables, the term "before" refers to the period "before the pandemic, in a face-to-face context," while "during" refers to the period "during the pandemic, in an emergency or online context."

Pair 1: how do you perceive your personal engagement?

Table 2. Rating of students' personal engagement

Engagement	Before	During	Difference
Very low (1)	9	13	+4
Low (2)	21	97	+76
Average (3)	169	240	+71
Good (4)	404	356	-48
Very good (5)	229	126	-103

Table 2 demonstrates a decrease in students' perception of their engagement during the pandemic. Out of the total number of scores, 344 decreased, 131 increased, and 357 remained the same. This decline is also evident in the average score, which decreased from 3.99 (standard deviation: 0.82) to 3.58 (standard deviation: 0.93). The analysis of the paired data reveals a mean difference of -0.41 (standard deviation: 1.10). The paired Wilcoxon signed-rank test and the paired t-test (both approximated normally), indicate the significance of this deterioration with respective z-scores of 9.77 and 10.69. Generally, a value higher than 3 is considered significant, which is clearly surpassed in this case. The practical implication is that students faced difficulty in maintaining the same level of engagement remotely as they did in the classroom.

Pair 2: how do you rate the/your communication with teachers?

Table 3. Rating of communication with teachers

Communication	Before	During	Difference
Very low (1)	6	16	+10
Low (2)	22	69	+47
Average (3)	168	268	+100
Good (4)	432	339	-93
Very good (5)	204	140	-64

Table 3 presents a decrease in the quality of communication between students and teachers. Out of the total scores, 301 decreased, 119 increased, and 412 remained unchanged. This decline is evident in the average score, which decreased from 3.97 (standard deviation: 0.79) to 3.62 (standard deviation: 0.92). The mean difference is -0.35 (standard deviation: 1.06). The z-scores for the Wilcoxon signed-rank test and the t-test indicate statistical significance, with values of 8.57 and 9.45 respectively. This suggests that students faced difficulties in feeling equally comfortable communicating with teachers remotely compared to face-to-face interactions, despite the general tendency of young people to communicate through virtual means.

Pair 3: how do you rate your own development of digital competencies?

Table 4. Rating of development of digital competencies

Digital competencies	Before	During	Difference
Very low (1)	7	5	-2
Low (2)	32	16	-16
Average (3)	219	148	-71
Good (4)	405	444	+39
Very good (5)	169	219	+50

In contrast to Tables 2 and 3, Table 4 reveals an increase in scores. Out of the total, 234 scores increased, 97 decreased, and 501 remained unchanged. The average score rose from 3.84 (standard deviation: 0.82) to 4.03 (standard deviation: 0.76), with a mean difference of 0.19 (standard deviation: 0.79). The z-scores for the Wilcoxon signed-rank test and the t-test are 6.43 and 7.00 respectively, indicating statistical significance. However, the significance is lower compared to Pairs 1 and 2. This suggests that students developed some digital competencies because of the situation, but not uniformly. In fact, more than 60% of students did not improve these competencies as stated.

Pair 4: how do you rate your achievement of the learning outcomes?

Table 5. Rating of achievement of the learning outcomes

Learning outcomes	Before	During	Difference
Very low (1)	6	10	+4
Low (2)	18	29	+11
Average (3)	206	243	+37
Good (4)	465	413	-52
Very good (5)	137	137	0

In this instance, Table 5 presents a more balanced situation. Out of the total scores, 145 increased, 193 decreased, and 494 remained unchanged. The average score shifted from 3.85 (standard deviation: 0.74) to 3.77 (standard deviation: 0.81), resulting in a variation of 0.09 (with a standard deviation of 0.89 for the mean difference). These changes appear to be less pronounced compared to previous cases, as indicated by the z-scores of 2.80 for the Wilcoxon signed-rank test and 2.78 for the t-test. These scores imply statistical significance, but with a higher p-value, close to 0.01. This implies that some students may have felt satisfied with their achievements as learners in the transitional scenario. The uncertainty surrounding the situation could have potentially led to even greater difficulties, leading to their perception of relative success.

Pair 5: how many hours per day do you spend on the PC for learning purposes?

Table 6. Daily time spent on a PC

Daily time in front of PC	Before	During	Difference
Less than 1 hour (1)	172	35	-137
About 2 hours (2)	281	95	-186
About 3 hours (3)	146	93	-53
About 4 hours (4)	84	100	+16
About 5 hours (5)	30	98	+68
6 hours or more (6)	39	331	+292

In this case, we excluded 80 students who responded "It's hard to say" regarding the pre-pandemic period, resulting in a slightly smaller sample size of 752. The increments observed in the analysis are significant: the amount of time increased 468 times, while it decreased only 64 times (remaining the same in 220 instances). The average number of hours increased from 2.52 (standard deviation: 1.34) to 4.49 (standard deviation: 1.64), indicating a substantial jump of 1.98 (with a standard deviation of 1.95 for the mean difference).

The z-scores provide strong confirmation, with the highest scores among all the pairs considered: 19.13 for the Wilcoxon signed-rank test and 27.82 for the t-test. This implies that most students required more time on a/their PC, with significant differences observed for many students. However, a minority of students did not require additional time and some even needed less time. The students who required less or the same amount of time likely already used the PC sufficiently before the transition, as indicated by the evidence from the following pair of questions. It is important to note that these students had a reasonable level of PC usage before the pandemic, as there is substantial uniformity in the reported number of hours.

Pair 6: how much time per week do you spend studying for classes?

Table 7. Weekly time spent studying for classes

Learning outcomes	Before	During	Difference
Less than 1 hour (1)	46	59	+13
From 1 to 4 hours (2)	279	247	-32
From 4 to 10 hours (3)	313	286	-27
From 10 to 20 hours (4)	131	154	+23
More than 20 hours (5)	63	86	+23

In the final pair of questions, the situation is once again more balanced. The category increased 226 times, decreased 172 times, and remained unchanged 434 times. The average score increased from 2.86 (standard deviation: 1.00) to 2.95 (standard deviation: 1.09), resulting in a mean difference of 0.09 (standard deviation: 1.05). Like Pair 4, these modifications hold less significance, as indicated by the z-scores of 2.50 for the Wilcoxon signed-rank test and 2.47 for the t-test. This implies a p-value higher than 0.01, indicating the inability to reject the null hypothesis of insignificance in the changes with a 99% confidence level (although it is possible to reject it with a 95% confidence level). Consequently, students feel less compelled to significantly increase the time they dedicate to studying, and it can be assumed that a portion of them already devoted enough time to studying before the COVID-19 pandemic. It could be argued that the subdivision of the scale may influence these considerations, but the proximity of the numbers in relation to the increments and decrements reasonably suggests the absence of a clear trend regardless.

Table 8. Pairs of questions versus differences before/during (percentages)

Pairs	-5	-4	-3	-2	-1	0	+1	+2	+3	+4	+5
1		0.0	3.7	11.7	26.0	42.9	12.6	2.4	0.5	0.2	
2		0.7	2.4	9.9	23.2	49.5	10.8	2.9	0.6	0.0	
3		0.1	0.2	1.7	9.6	60.2	23.6	4.1	0.4	0.1	
4		0.1	0.6	5.5	16.9	59.4	14.1	2.5	0.7	0.1	
5	0.0	0.1	0.9	2.5	4.9	19.8	13.4	14.2	16.2	18.4	9.4
6		0.6	1.0	3.8	15.3	52.2	19.5	5.6	1.7	0.4	

Table 8 summarizes the situation of increase and decrease of students' answers in all pairs of questions. We see that the differences relative to Pairs 1 and 2 are asymmetrically distributed toward the negative values (left), while those concerning Pairs 3 and 5 have an asymmetric distribution favoring the positive values (right), which is particularly marked in the latter case. Pairs 4 and 6 exhibit a more equilibrated scattering. Let us present some qualitative aspects now. We asked students if during the pandemic they had experienced problems with activities such as laboratories and projects, and if so, which ones. Only about 20% of them answered "yes", but we found some interesting responses:

- "Many times the assessment was not as fair as possible, because it was more difficult to develop connections between the teacher and the student, making it difficult to evaluate in the best way the quality of the knowledge depending on the progress of the student during the semester."
- "I failed an exam because the professor did not hear my correct answer due to audio buffering. Projects were difficult to manage, since there was no face-to-face interaction between students."

- "Very little time to take the test, in face-to-face classes a test would have never been administered under the same conditions; sometimes I also experienced lack of understanding for problems."

From these answers, we can see that sometimes technology causes problems (such as in the case of buffering), but more frequently it is the methodological approach that is perfectible, requiring for example an adaptation of the assessment to the changed conditions. In another question, we asked students if they had good practices in learning during and after the pandemic, and if so, to share them. Again, not more than 20% of students answered "yes", but several noteworthy answers occurred:

- "Discipline had a more important role during the pandemic. Before students lived closer one to the other and could hear news from their school easily. During the pandemic, it was necessary to check your e-mail periodically and to evaluate our level of knowledge before an exam ourselves."
- "Organizing the acquired materials from classes on an ongoing basis, controlling and verifying emerging activities in teams so as not to get lost in the course of teaching, subject scope, material..."
- "Reading a lot of scientific papers and books and learning by myself from extra topics than the ones the teachers were presenting."

The general impression has been that a certain degree of autonomy, higher than the one required before the pandemic, can help in learning under the new conditions. Finally, a question regarding students' perception of the biggest threats that could result from remote education was asked. In this case the number of answers was considerably higher, here some examples:

- "From an educational point of view, I believe that my institution has adapted and countered the threats of exam fraud. I believe that the biggest threat is not the online or hybrid learning system, but the isolation of the individual from the collective (individual and not collective work). The military system is based on teamwork, and the exclusive use of online courses can damage the student's integration in the future workplace. I believe that the hybrid system my institution adopted is ideal."
- "At a military level, we need to stay in contact with people and with the military context. In my opinion there are no threats for university studies, so we should study remotely but we should also have some hours where we can do (in presence) only military things."
- "We may risk shifting the focus from deeply understanding the subject to being able to answer quizzes and online questions."

We see that remote learning itself cannot be blamed for generating threats, but rather problems arise when it implies a significant reduction of social interactions, especially in the security and defence context where working in a team is pivotal. It is also recognized how solutions like hybrid and blended learning, which contemplate the coexistence of a part in presence with one at a distance, can deal with this issue. From last perception, we can see that students may be more tempted to perform specific training fine-tuned on the tests' format rather than really study the subject.

5. DISCUSSION

These findings can be connected to the theoretical framework and research question. Starting from the quantitative ones, the decrease in averages for Pairs 1-2 (engagement and communication) and the increasing standard deviations indicate that students react differently, to some extent, to the challenges posed by the pandemic, likely influenced by their initial circumstances. Like what was found by Hodges et al. (2020), flexibility with learning activities, course policies, and institutional policies should be considered. Personalized learning paths could help address this diversity. Additionally, since students' digital competencies vary, targeted training to develop these skills (Pair 3) could be beneficial, since it was evidenced by Patricia Aguilera-Hermida (2020), students with no previous experience with online learning may think that online delivery is not desirable. It would have been expected to observe increments in Pair 5, but trained students tend to require less time for technical tasks, thereby reducing the additional time spent using PCs. Regarding Pair 4 (learning outcomes) and Pair 6 (weekly study time), the results indicate a more stable situation. However, improving students' engagement could enhance the quality of study time. Another crucial aspect is the need for accommodation: it takes time to adjust to changes, even when they may bring benefits. This is in line with other findings which showed that students struggled to adapt to online learning (Patricia Aguilera-Hermida, 2020). In our context, this adjustment depends on the extent to which students use technology in education. It is evident that teachers should lead by example and utilize digital tools themselves.

We can also compare the recently obtained results with the corresponding survey conducted on teachers, as depicted in (Marchisio et al., 2022a). Pairs 1-2 are directly related to the questions posed to teachers, where they were asked to rate student engagement (corresponding to Pair 1) and communication with students (corresponding to Pair 2). In both cases, we observe a statistically supported decline in the ratings. This suggests that both students and teachers shared the perception that students faced greater difficulty in engaging, and that communication became more challenging in both directions—teachers found it harder to communicate with students, and vice versa. There is also a similar agreement regarding Pair 5, as we asked teachers about the number of hours per day they spent on the computer for teaching or preparing teaching materials. Both students and teachers experienced a significant increase in the daily time spent on educational purposes using computers, confirming the necessity for all stakeholders to engage with digital tools to a much greater extent than before. On the other hand, Pair 3 is less directly related to the questions asked of teachers regarding competencies. Teachers were asked to rate the development of students' competencies, while students were asked to rate their own development of digital competencies. This resulted in different outcomes: teachers perceived a significant decline in the development of students' general competencies, while students themselves perceived an increase in their digital competencies. Pair 4 also has only partial relevance to the question posed to teachers regarding learning outcomes, where they rated the implementation of learning outcomes, while students rated their own achievement. While teachers perceived a strong difficulty in implementing learning outcomes like pre-pandemic times, students had a less negative perception, with difficulties still present but not so marked. Lastly, in Pair 6, we return to a higher level of similarity. However, while teachers strongly reported needing more time to fulfill their commitments, the statistical evidence suggesting the same for students was relevantly weaker. This can be explained by the fact that adapting learning to new scenarios may be easier than adapting teaching methods.

The discussion on the qualitative findings can start from this last observation: about experiencing problems while a practical part is involved, the proportion of teachers answering “yes” was about one third, higher than the 20% of students, thus supporting the greater ease in the adaptation of learning with respect to teaching. However, the two sides did not perceive the issues exactly in the same way. Several students’ answers focused on the assessment, highlighting the difficulties in guaranteeing the same fairness a more traditional evaluation may assure; teachers spoke instead mostly about didactic methodologies and practices (Marchisio et al., 2022b). If on the one hand this is understandably related to their different goals, on the other hand this does not mean that students are interested in the assessment alone, since they recognized the risks in specifically preparing for tests rather than for gaining knowledge on the subjects. Concerning good practices, the similarities between students’ and teachers’ answers are more marked, since both groups of stakeholders agree on the fact that autonomous organization plays a more important role in these forms of didactics. Finally, yet importantly, students and teachers concur also on having a teaching approach comprehensive of blended or hybrid elements to highly consider the human factor. Practically, educators and policymakers should address the challenges by means of proper strategies, including the consideration of digital tools to foster collaboration across disciplines and implement creative approaches.

6. CONCLUSION

This research provided us with insights into how students' perception of higher education changed during the transition caused by the COVID-19 pandemic, providing a general response to (RQ1). We found students’ difficulties in various aspects, including establishing connections with teachers, and increase in the perception of digital competencies (RQ2). These perceptions and changes have practical implications. Digitally skilled teachers and organizations, enabling students to develop more specific competencies beyond their existing ones, can overcome the experienced difficulties. Furthermore, incorporating interdisciplinary activities and innovative methodologies through digital tools can broaden students' understanding of the subject matter. The impact of design and activities on teachers and students is under examination, specifically in the context of the DIGICODE Learning, Teaching and Training Activities like the school "Systems for Command and Control in Security and Defence Field". A large part of the adopted approach is not limited to security and defence education, but the international nature of the project offers a noteworthy context for studying and implementing suitable interventions in education. As future work, each partner of the DIGICODE project is interested in collaborating more to study specific competencies related to security and defence education, like mathematical competencies, critical thinking and problem-solving skills. Moreover, further research will include surveys and interviews to measure other variables such as cultural differences, academic results, and degree of satisfaction.

ACKNOWLEDGEMENT

The research activity described in this paper has been carried out inside the Erasmus+ Key Action 2 Strategic Partnership N. 2020-1-PL01-KA226-096192 "Digital Competences for Improving Security and Defence Education – DIGICODE".

REFERENCES

Barron, A. and Rowles, P., 2021. Evaluation of digital literacy as an Air Force foundational competency. https://apps.dtic.mil/sti/pdfs/AD1143616.pdf

Buffardi, A. and Taddeo, G., 2017. The Web 2.0 Skills of Italian Students: An Empirical Study in Southern Italy. *Italian Journal of Sociology of Education*, vol. 9, no. 1, pp. 45-76.

Ehlers, U.D., 2020. *Future Skills – Future Learning, Future Higher Education.* Springer, Karlsruhe, Germany,

European Education Area, 2020. Digital Education Action Plan. https://education.ec.europa.eu/document/digital-education-action-plan

Goldin, C.D. and Katz, L.F., 2009. *The Race between Education and Technology.* Harvard University Press, Cambridge, UK.

Hodges, C., et al., 2020. The Difference Between Emergency Remote Teaching and Online Learning. *EDUCAUSE Review*, vol. 27, no. 1, pp. 1-12.

Marchisio, M. and Spinello, E., 2021. Internationalization for enhancing the European Security and Defence Higher Education. *15th International Conference on e-Learning (EL2021) – Held at the 15th Multi-Conference on Computer Science and Information Systems (MCCSIS2021)*, pp. 99-106.

Marchisio, M. et al, 2022a. Teachers' digital competences before and during the COVID-19 pandemic for the improvement of security and defence higher education. *16th International Conference on e-Learning (EL2022) – Held at the 16th Multi-Conference on Computer Science and Information Systems (MCCSIS2022)*, pp. 68-75.

Marchisio, M. et al, 2022b. Teachers' perception of higher education in a transition scenario. *Proceedings of IEEE 46th COMPSAC Conference*, pp. 139-144.

Marchisio, M. et al, 2022c. Teaching Mathematics to non-Mathematics majors through Problem Solving and new technologies. *Education Sciences*, vol. 12, paper 34.

Marchisio, M. et al, 2022d. Valuable features of hybrid teaching in a higher education context. *Communications in Computer and Information Science*, vol. 1639, pp. 16-21.

Mentsiev, A. U. et al, 2022. Digital skills development as a basis for sustainable education development. *AIP Conference Proceedings*, 2647, 040085.

Mihalova, G., 2006. E-learning as internationalization strategy in higher education. *Baltic Journal of Management*, vol.1, no. 3, pp. 270-284.

Patricia Aguilera-Hermida, A. , 2020. College students' use and acceptance of emergency online learning due to COVID-19. *International Journal of Educational Research Open*, vol. 1, 2020, 100011, ISSN 2666-3740.

Pinchuk, O. and Prokopenko, A., 2021. Actual areas of development of digital competence of officers of the Armed Forces of Ukraine. *Educational Dimension*, vol. 5, issue 57, pp. 89-108.

Tomczyk, L, 2021. Declared and Real Level of Digital Skills of Future Teaching Staff. *Education Sciences*, vol. 11, paper 619.

United Nations, 2015. The 2030 Agenda for Sustainable Development. https://sdgs.un.org/2030agenda

Van Laar, E., et al, 2017. The relation between 21st-century skills and digital skills: A systematic literature review, *Computers in Human Behavior*, vol. 72, pp. 577-588.

PANDEMIC IMPACT ON ARGENTINIAN HIGHER EDUCATION. THE CASE OF THE NATIONAL UNIVERSITY OF LA PLATA

Francisco Javier Díaz, María Alejandra Osorio, Ana Paola Amadeo and Isabel Kimura
LINTI – Research Laboratory in New Computer Technologies
School of Information Technology – National University of La Plata
La Plata, Argentina

ABSTRACT

This article analyzes the impact of the pandemic in Argentina, considering that the National Government and provincial governments issued isolation resolutions throughout the 2020 academic year and mandatory social distancing measures in 2021, which prevented in-person activities at National Universities, such as classes and exams. The analysis focuses on the use of virtual means (such as Moodle, Zoom, Webex, etc.) by students and the graduation rate. Official statistics from the Secretary of University Policies of the Ministry of National Education are used, comparing them with those of the National University of La Plata. The study includes a gender analysis to determine if the pandemic's impact in our country maintained or modified the proportions of incoming students, enrolled students, and graduates. The results show that while the pandemic changed habits, it did not have such a negative impact as expected on the overall university population.

KEYWORDS

Higher Education, Pandemic, Public Policies, Impact, Students

1. INTRODUCTION

The pandemic caused by the coronavirus disease (COVID-19), resulting from the SARS-CoV-2 virus, provoked different governmental reactions in each country around the world to mitigate the negative social and economic impact. According to data from the report by CEPAL, OREALC, and UNESCO on education during the COVID-19 pandemic (UNESCOa, 2020), it is estimated that by May 2020, more than 190 countries had closed schools, affecting over 1.2 billion students. This physical closure of institutions, particularly in higher education, led to the adoption of emergency remote education models.

In higher education, many institutions had different virtual tools that aided in the transition, unlike other educational systems. In Latin America and the Caribbean, the impact of the pandemic led university institutions in the region to shift to virtual teaching. The leading institutions called for a virtual modality to address the immediate situation and consider hybrid education models for a post-COVID-19 scenario, especially in private universities (Salto, 2021, p. 63).

According to Francesc Pedró, director of IESALC, from the perspective of UNESCO's Chairs' students worldwide (UNESCOb, 2020), the following points stand out as the most crucial issues in the region concerning the COVID-19 problem:

- Internet connectivity difficulties: only 51% of households in Latin America have connectivity with quality standards.
- The negative perception of social isolation is not as significant for students in the region.
- Maintaining a regular study schedule, self-regulation skills, and discipline for distance learning at the undergraduate level. This highlights the need to rethink how to learn.
- Economic concerns.

Regarding the first point, internet access in the region varies widely. It ranges from 3 out of 10 people in Nicaragua to over 70% of the population in Argentina and Chile. In Argentina, according to INDEC, 42.6% of the population uses a computer, and 77.7% of the population uses the internet. 92.1% has internet access from home (mobile or fixed network), with 62.6% having a computer in urban households (INDEC, 2023). However, quality is also a factor to consider. In addition to the connectivity issues caused by the pandemic, it is important to analyze pedagogical continuity, emotional support, and the balance between family life and personal projects at home, particularly for women. The support and repatriation of international students are also important. Concerning the economic situation, the pandemic in Argentina coincides with an economic crisis preceding it, reflected in high levels of indebtedness and inflation (Miranda, 2020)

On March 12, 2020, Argentina decreed the ASPO (Preventive and Mandatory Social Isolation), which remained in effect until November 8, 2020. From this date onwards, the DISPO (Social, Preventive, and Mandatory Distancing) came into effect, which lasted throughout 2022, varying according to the different provinces of our country. This decree underwent modifications, transitioning from strict confinement to health recommendations, capacity limits, etc. This date is a few days before the start of the academic year in universities, which begins on April 1 and continues until March 30 of the following year. Argentina has 134 universities and university institutes, 70 of which are public (CIN, 2023) and 64 are private (CRUP, 2023). These institutions have been impacted by the pandemic, including their teachers, students, and non-teaching staff. Over 1.8 million students enrolled in degree programs in 70 public universities shifted from face-to-face to distance learning, without a break. In-person classes moved to virtual formats, many of them through video conferences. All universities chose to immediately transition to distance learning, with some initial reservations, and it is still a modality offered by very few public institutions, even today (Salto, 2021, p. 63).

Despite the health and economic crisis, public universities developed institutional policies to mitigate the impact of the pandemic on students from low-income backgrounds, such as connectivity grants, access to free platforms, and economic assistance scholarships, among other initiatives. As the situation progressed in all educational levels, the government reached agreements with mobile and internet companies to provide free access to educational platforms (García de Fanelli et al., 2020).

The Argentine government also implemented a set of public policies for research, such as the Coronavirus Unit of the Ministry of Science, Technology, and Innovation of the Nation and the "PISAC COVID-19: Argentine Society in the Post-pandemic" program, aimed at funding research projects on pandemic and post-pandemic societal studies.

The PlanVES (Higher Education Virtualization Plan) of 2020 was an initiative of the Secretariat of University Policies of the Ministry of Education of the Nation, which sought to improve technological tools for inclusion, teaching and learning development, evaluation in virtual environments, monitoring and tutoring, connectivity, and infrastructure adjustments. The PlanVES II, the Higher Education Virtualization Plan II, was developed in 2021 and aimed to provide university institutions with resources to strengthen face-to-face and remote training through the development and implementation of hybrid classrooms. Hybrid classrooms seek to combine virtuality with face-to-face interactions, offering a pedagogical experience that combines both modalities. They represent a response from the national university system to the global pandemic situation experienced recently. The specific objectives include developing and implementing hybrid classrooms that guarantee access to innovative and quality education, strengthening virtual training through teacher training, developing innovative proposals through project development, problem-solving, and addressing local needs, strengthening institutions with the necessary software and hardware, and improving the connectivity of students and teachers. For this purpose, approximately $1,490,000,000 (around 6,000,000 USD) was allocated to national and provincial universities and the University Information System (PlanVES, 2021) (Rojas, 2021).

In 2022, the PlanVES III was launched to consolidate hybrid classrooms in the post-pandemic context through specific funds allocated to higher education for updating the equipment of hybrid classrooms, networks, storage, and evaluation for virtualization strategies (Institutional Strengthening, 2022).

As mentioned earlier, the Argentine university system consists of a total of 134 university institutions, including state universities (63), private universities (50), and university institutes (21). University statistics are managed by the University Information Department, under the Secretariat of University Policies of the Ministry of Education of the Nation. Information for each university institution is collected twice a year through the SIU-Araucano system (DIU, 2023).

The UNLP (National University of La Plata, UNLP, 2023) is a public university, ranking third in the country in terms of student enrollment, with over 117,000 undergraduate students, 17,600 postgraduate students, and more than 5,000 students from pre-university schools and colleges (SIU Araucano, 2022). It offers a wide range of academic programs, from Aerospace Engineering to Popular Music Education and English and French Translation, including Medical Sciences and Computer Science. These programs are offered in 17 faculties and 5 pre-university colleges. During the pandemic, the UNLP, like all national universities, suspended face-to-face classes and transitioned from in-person to distance learning. With varying levels of digital literacy among teachers and students and available technological resources, the UNLP implemented different programs to support and retain students, such as the "Your PC for Studying" program (UNESCO, 2023) (UNLP, 2022). The UNLP's Institutional Distance Education System (SIED UNLP) reported in its extraordinary technical report for 2020 that there were 4,832 courses created in virtual environments, involving over 106,000 enrolled students and 16,935 teachers. In 2019, the report indicated 2,643 active classrooms and 81,561 active participants (SIED, 2019) (SIED, 2020).

Next, an analysis of new enrollments, students, and graduates from 2016 to 2021 at the UNLP will be presented, comparing them with the national university system in Argentina and available regional statistics on higher education.

2. IMPACT OF THE PANDEMIC ON UNLP

In Argentina, university statistics are managed by the University Information Department, under the Secretariat of University Policies of the Ministry of Education of the Nation. The information is collected annually using the information system called SIU-Araucano. This system allows reporting on admission, enrollment, and graduation statistics for undergraduate and postgraduate programs in both public and private Argentine universities (SPU, 2016, 2017, 2018, 2019, 2020, 2021).

This study uses national statistics on undergraduate and pre-degree university students from public institutions between the years 2016 and 2021 to make a comparison with the numbers from the National University of La Plata (UNLP).

Figure 1 (or Table 1) shows the evolution of total students, new enrollments, and graduates in Argentina and at UNLP between the years 2016 and 2021. The choice of these years is to analyze the impact of the pandemic on university statistics. The data includes the number of undergraduate and pre-degree students from public institutions.

Table 1. Evolution of Students, New Enrollments, and Graduates of Undergraduate and Pre-degree Programs in Public Institutions in Argentina and UNLP. Years 2016, 2017, 2018, 2019, 2020, and 2021

Year	Total Students		New Enrollments		Graduates	
	Argentine	UNLP	Argentine	UNLP	Argentine	UNLP
2016	1,519,797	105,551	372,467	25,959	82,731	6,897
2017	1,584,392	104,615	396,303	23,270	86,174	7,039
2018	1,640,405	108,047	424,959	25,908	86,958	6,546
2019	1,751,607	109,874	471,827	25,941	88,080	6,383
2020	1,872,591	116,824	513,151	31,725	73,832	5,690
2021	2,065,115	117,277	556,628	31,385	93,917	7,825

In terms of the number of students and new enrollments, an increase is observed both at the national level and at UNLP. However, there is a decrease in the number of graduates in 2020. This trend could be attributed to the decision made by the majority of Argentine public universities to sustain virtual classes as a means to ensure academic activity during the quarantine period, as reflected in a joint declaration by the National Interuniversity Council (Fanelli et al., 2022). The UNESCO report on higher education in Latin America and the Caribbean (UNESCO, 2023) also indicates that student enrollment remained stable at 28.9 million between 2019 and 2020.

Regarding the evolution of approved courses at UNLP, it can be observed that during the pandemic, the proportion of students who passed 1 course experienced a decrease in 2020 and a rebound in 2021. Those who passed 6 courses or more increased in 2020 and slightly decreased in 2021, while the number of students who passed 2, 3, 4, and 5 courses remained stable.

Table 2. Evolution of the percentage of students according to the number of approved courses at UNLP between the years 2016 and 2021

Year	0 subject	1 subject	2 subjects	3 subjects	4 subjects	5 subjects	6 or more subjects
2016	21,024	14,462	10,367	8,437	6,857	5,587	12,858
2017	31,784	17,858	13,068	10,763	8,814	7,251	15,077
2018	32,808	19,142	13,658	11,214	8,900	6,753	15,572
2019	30,722	19,789	13,969	11,764	9,423	7,235	16,972
2020	29,939	20,906	15,902	12,528	9,810	7,961	19,778
2021	27,872	24,075	16,161	12,868	10,044	7,812	18,445

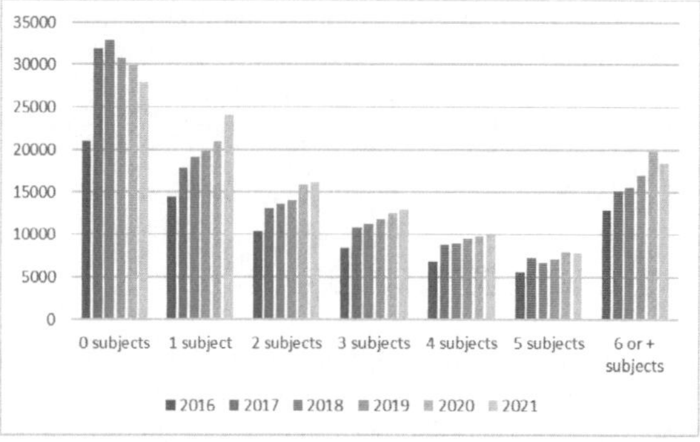

Figure 1. Evolution of the percentage of students according to the number of approved courses at UNLP between the years 2016 and 2021

2.1 New Enrollments by Field of Study

The SPU classifies Argentine university programs into 5 fields: Social Sciences, Applied Sciences, Humanities, Health Sciences, and Basic Sciences. Social Sciences include areas such as social communication, political science, law, geography, economics and management, sociology, anthropology, and social work. Applied Sciences encompass architecture and design, astronomy, biochemistry and pharmacy, agriculture and soil, meteorology and industries, computer science, and engineering. Humanities cover audiovisual arts, education, philosophy, history, languages, and psychology. Health Sciences include medicine and dentistry, paramedics and assistants, and veterinary studies. Basic Sciences involve biology, physics, mathematics, and chemistry.

Figures 2 and 3 show the evolution of enrollments in Argentine public universities according to the chosen field of study by students between the years 2016 and 2021. Figure 3 displays enrollments at UNLP by field of study in the years 2016, 2017, 2018, 2019, 2020, and 2021.

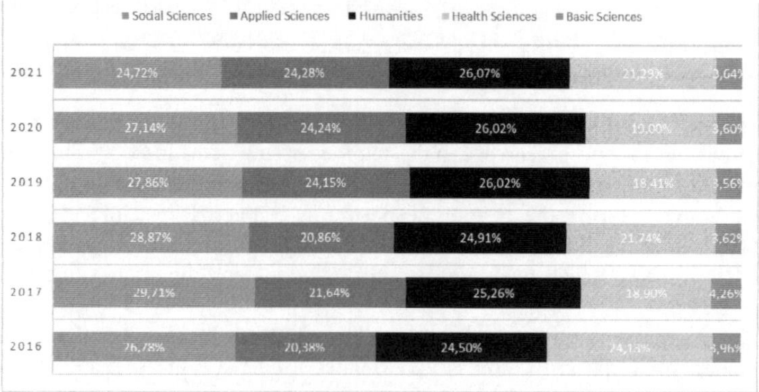

Figure 2. New enrollment to Argentinian public universities according to field of study

It can be observed that the majority of students choose Social Sciences, but there is a decrease from 37.50% in 2019 to 33.30% in 2021. On the other hand, two fields of study of great importance for the country, namely Health Sciences (essential during the pandemic) and Applied Sciences (strategic for development), have increased from 19.60% in 2019 to 20% in 2021, and from 20.40% in 2019 to 22.90% in 2021, respectively.

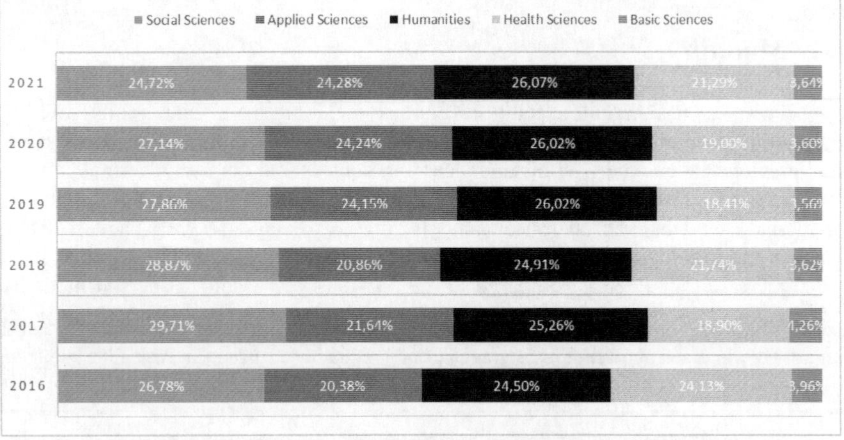

Figure 3. New enrollments to UNLP according to field of study

At UNLP, similar to the national level, there is a decrease in enrollments in Social Sciences programs, declining from 27.86% in 2019 to 24.72% in 2021. Notably, at UNLP, a higher proportion of students choose Applied Sciences and Health Sciences programs, representing near 50% of the institution's new enrollees.

In the UNESCO report analyzing enrollment in higher education by field of study and the participation of women in enrollment in each field of study in Latin American countries between 2015 and 2020, there is also an increase in women's participation in fields more associated with caregiving roles (health and well-being and education) and social sciences.

2.2 Women Participation

Another interesting aspect to analyze is the impact of the pandemic on women's participation in university enrollment. There is an increase in women's participation between the years 2016 and 2021, which were marked by the pandemic, as shown in Figure 4.

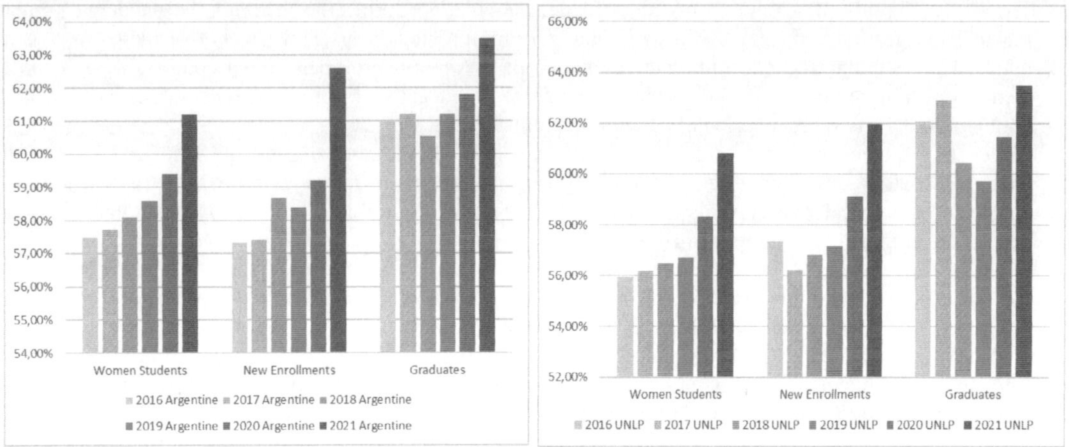

Figure 4. Women participation in Argentinian public universities and UNLP

The increased incorporation of women in the Argentine university system is evidenced by the rise in new enrollments from 58.40% in 2019 to 62.60% in 2021 nationwide. A similar trend is observed at UNLP, with figures of 57.15% in 2019 and 61.97% in 2021.

According to the UNESCO Institute of Statistics (UNESCO, 2023 page 11) between 2000 and 2020, the enrollment rate has increased by 36.6% for women and 25.6% for men. Over the past twenty years, there has been a growing participation of the female population.

2.2 Study Modality

During the pandemic, most Argentine state universities, including UNLP, incorporated virtual environments for a significant portion of their scheduled courses. While all universities adopted the same strategy, some already had previous experience with distance learning courses and the use of virtual campuses for teaching activities (Fanelli et al., 2020). Table 3 shows the evolution of the number of students in public universities according to study modality. Distance learning offerings recorded a 56.95% increase from 2020 to 2021, while in-person offerings saw a 27.53% increase during the same period. The number of distance learning students also increased by 17.82% compared to the previous year, while in-person modality experienced a 9.30% increase.

Table 3. Offerings and number of undergraduate and pre-graduate students in Argentine public universities according to study modality

Year	Offering		Students	
	Distance Learning	In-person Learning	Distance Learning	In-person Learning
2016	225	4,954	49,623	1,470,174
2017	221	5,065	60,580	1,523,812
2018	200	4,256	58,029	1,582,376
2019	381	6,272	172,078	2,015,214
2020	381	6,191	185,871	2,132,384
2021	598	7,896	219,005	2,330,784

The UNLP does not offer undergraduate and pregraduate programs in distance learning mode. It is interesting as a starting point to generate new study plans and curricula that adopt this modality. There is a growing availability of postgraduate programs in distance learning mode, accelerated by the pandemic, with an increase from only 5 offerings in 2019 to 30 offerings in 2023.

3. CONCLUSION

The study presents the evolution of the university population during the years of the pandemic's greatest impact in Argentina and compares it with data from the previous year. This shows that there have been no major changes in the trends regarding the evolution of university enrollment in our country, with the notable mention of a higher proportion of women. This shows no negative impact in the overall university student population of the two years period of Covid quarantine in Argentina

Another dimension that emerges from the analysis of the number of students who pass two or more courses is that students who were already studying at universities adapted better to virtual learning and approved more academic subjects compared to pre-pandemic values. It is likely that this will impact positively on an increase of the graduates.

As a future work we foresee that in the coming years, it will be necessary to analyze whether the increased virtualization in education and the creation of hybrid spaces/classrooms have a positive impact on the system and how they affect different disciplines and genders.

Lastly, another analysis of the pandemic's impact on universities will be to observe the progress of students who completed their secondary education in 2020 and 2021 in their chosen university programs, to compare their progress with those of previous years.

REFERENCES

CIN, 2023. Consejo Interuniversitario Nacional. Disponible en https://www.cin.edu.ar/instituciones-universitarias/

CRUP, 2023. Consejo de Rectores de Universidades Privadas. Disponible en http://www.crup.org.ar/universidades/

Fanelli, 2022, et al. ACCIÓN Y REACCIÓN EN ÉPOCA DE PANDEMIA: LA UNIVERSIDAD ARGENTINA ANTE LA COVID-19 Disponible en: https://ri.conicet.gov.ar/bitstream/handle/11336/109267/CONICET_Digital_Nro.947e5167-fdb0-4395-a8ce-dbbe0de57eea_A.pdf?sequence=2&isAllowed=y

Fortalecimiento institucional, 2022. Plan de Virtualización de la Educación Superior. Disponible en https://www.argentina.gob.ar/educacion/universidades/calidad-universitaria/fortalecimiento-institucional

INDEC, 2023. Acceso y uso de tecnologías de la información y la comunicación. EPH. Instituto Nacional de Estadísticas y Censo de Argentina. Disponible en https://www.indec.gob.ar/indec/web/Nivel3-Tema-4-26

Miranda, Estela. 2020. Políticas de educación superior en Argentina. Entre la COVID-19 y la deuda externa heredada. Universidades, 85, 194–213

PlanVES. Plan de Virtualización de la Educación Superior 2021. Secretaría de Políticas Universitarias. Ministerio de Educación de la Nación. Disponible en https://www.argentina.gob.ar/sites/default/files/2021/06/presentacion-planvesii.pdf

Rojas, 2021. Covid, Universidad y después. Revista de la Universidad Nacional de San Juan, Argentina. Disponible en http://www.revista.unsj.edu.ar/?p=4500

Salto, Dante. 2020. La pandemia y la educación superior privada en América Latina. Recopilación de trabajos del encuentro “La Universidad en la Era Exponencial”, Universidad Nacional de la Plata, Buenos Aires, Argentina. 8, 9 y 10 de noviembre de 2021.

SIED, 2019. Informe Técnico. Dirección General de Educación a Distancia 2019. Disponible en https://sied.ead.unlp.edu.ar/assets/files/Informe%20Tecnico%202019.pdf

SIED, 2020. Informe Técnico edición especial estraordinaria ASPO. 2020. Disponible en http://sedici.unlp.edu.ar/bitstream/handle/10915/120345/Documento_completo.pdf-PDFA.pdf?sequence=1&isAllowed=y

SPU, 2019, 2020, 2021, Secretaría de Políticas Universitarias del Ministerio de Educación de la Nación.. Estadísticas Universitarias Argentinas Disponible en: https://www.argentina.gob.ar/educacion/universidades/informacion/publicaciones

UNESCO, 2023. Avances hacia el ODS 4 en educación superior: desafíos y respuestas políticas en América Latina y el Caribe. Disponible en: https://unesdoc.unesco.org/ark:/48223/pf0000384366?posInSet=12&queryId=aa854886-74f7-4da5-9196-935f559e83f5

UNESCOa, 2020. Informe CEPAL, OREALC y UNESCO: “La educación en tiempos de la pandemia de COVID-19”. Disponible en. https://www.iesalc.unesco.org/2020/08/25/informe-cepal-y-unesco-la-educacion-en-tiempos-de-la-pandemia-de-covid-19/

UNESCOb, 2020. How do universities guarantee pedagogical continuity and plan for the future? Emerging contributions of the first forum of rectors in face of the COVID-19 pandemic. Disponible en https://www.iesalc.unesco.org/en/2020/06/09/how-do-universities-guarantee-pedagogical-continuity-and-plan-for-the-future-emerging-contributions-of-the-first-forum-of-rectors-in-face-of-the-covid-19-pandemic/

UNLPb, 2022. Beca TuPC para estudiar. Disponible en https://unlp.edu.ar/gestion/bienestar_universitario/estudiantes/beca-tu-pc-para-estudiar-49352/

BEYOND THE HYPE: CHALLENGES TO IMPLEMENTING INSTRUCTIONAL TECHNOLOGY IN HIGHER EDUCATION

Kathryn Cormican[1], Suzana Sampaio[1], Sagar Shinde[1], Özge Andiç Çakır[2], Firat Sarsar[2], Manon Van Leeuwen[3], Tsvetelina Georgieva[4], Seher Kardirova[4] and Nuno Pombo[5]
[1]Enterprise Research Centre and LERO—The Irish Software Research Centre, School of Engineering, University of Galway, Ireland, H91 TK33
[2]EGE University
35100 Izmir, Türkiye
[3]EOLAS SL,
Carretera de Sevilla KM 6 06080 Badajoz, Spain
[4]University of Ruse Angel Kanchev,
Studentska Street 8, 7017, Ruse, Bulgaria
[5]Universidade da Beira Interior,
Convento de Santo Antonio, 6202-001, Covilhã, Portugal

ABSTRACT

Educators in higher education must reinvent their learning environments so that digitalization augments and enhances instruction and learning. The widespread use of digital devices and applications makes this possible. We argue that educators must therefore develop their digital skills and adopt more innovative instructional strategies. Using available technological instruments for pedagogical purposes, instructional technology permits the development of engaging, effective learning experiences. However, the implementation of instructional technology in practice is difficult. It is complex, multifaceted and fraught with challenges, and consequently, educators struggle to realize the potential benefits. This exploratory study aims to uncover key challenges to implementing instructional technology faced by educators in order to better understand their needs and requirements. To do this, we conducted a comprehensive qualitative study involving 41 higher education educators via focus groups in Ireland, Turkey, Bulgaria and Portugal. Our analysis uncovers many challenges and we curate and discuss a prioritized list in this paper. The findings present pertinent challenges for educators and institutions across four countries and suggest how to overcome them. This study also offers insight into future research and development needed to expand instructional technologies to accommodate and address respondents' concerns.

KEYWORDS

Instructional technology, Challenges, Qualitative analysis, Higher education

1. INTRODUCTION

To remain relevant, universities must reinvent their learning environments so that digitalization expands and complements the student-educator relationship. The extensive use of digital devices and applications enables this to happen. Therefore, to respond to the challenges of pivoting towards online learning, changes in the student population and the increasingly pervasive digital environment, educators must develop their digital skills and adapt to more innovative teaching approaches. This is supported by the voice of the students. While it is important to support higher education educators to develop digital competencies, this is not sufficient. Educators must change their perspective and way of teaching to ensure that no student is left behind. Instructional technology may provide an answer to these challenges. Instructional technology refers to the creation of engaging, effective learning experiences using available technological tools for pedagogical purposes. It concerns the design, development, use, management, and evaluation of the process of learning mediated by technology applications. It goes beyond digital tools and fosters a more applied systems approach that accommodates successful learning technologies.

The concept of instructional technology has evolved in recent years, and the field has moved from a narrowly defined focus on tools towards a more holistic, systems approach (Spector, 2016). The benefits of using such technologies are many. For example, prior work has demonstrated that integrating and using technology in the classroom increases student engagement, collaboration, and achievement (Davies, Dean and Ball, 2013; Rasheed, Kamsin and Abdullah, 2020). However, many educators struggle to use these technologies and fail to reap their benefits (Ng and Park, 2021). An analysis of the state of the art indicates that educators are reluctant to adopt this technology for many reasons such as poor understanding of pedagogical issues and difficulties in integrating the technologies with their existing content (Mutambara, Bayaga and Others, 2020; Chaipidech *et al.*, 2021; Gandolfi, Ferdig and Kratcoski, 2021). Furthermore, deploying these technologies in practice is also a challenge and researchers have found that educators lack the required resources and support (Ishak, Din and Hasran, 2021). A lack of student engagement and poor inclusiveness of students has also emerged as challenges (Lee and Choi, 2011; Kayan Fadlelmula, Sellami and Le, 2022).

Consequently, this study aims to explore these challenges in more detail. We targeted STEM educators as STEM education has a strong practice and lab-based component in the pedagogical approach which adds more challenges compared to more theoretical courses. Qualitative focus groups were conducted with 41 higher education STEM educators in Ireland, Turkey, Bulgaria and Portugal to explore context-specific challenges and uncover deep insights. By doing so, the study hopes to contribute to existing knowledge about instructional technology and a broader discussion about instructional technology's challenges in higher education. This study offers suggestions for educators and administrators involved in developing STEM programs. The remainder of this paper is organized as follows. The next section discusses prior work relating to challenges to instructional technology, then we present the scientific method employed in the study. The results are discussed while the conclusions aim to draw out the salient learnings from our work.

2. ANALYSIS OF CHALLENGES

While the use of instructional technology has grown significantly in recent years, educators struggle to implement it to its full potential. An analysis of the extant literature reveals that the challenges to implementing instructional technology are complex and multifaceted. However, it seems that the most significant challenge is instructors' inability to seamlessly transfer face-to-face learning content to the online environment (Lee and Choi, 2011; Almazova, Krylova, *et al.*, 2020). Owston, York and Murtha, (2013) also assert that educators feel overwhelmed by the vast array of available instructional technologies, leading to confusion and uncertainty regarding which tools to use and how to use them. While Chawinga, (2017), notes that educators are apprehensive about integrating technology into their teaching practices due to concerns about the time and effort required to learn how to use it effectively. Students also encounter difficulties when utilizing technology, posing additional problems for faculty. This can lead to frustration among students attempting to use them and faculty attempting to support student technology use (Reid, 2014). Students may not know how to learn when confronted with new pedagogical approaches (Alexander, 2001; Johnson and Howell, 2006) or how to use resources and technology critically (Jackson *et al.*, 2008; Wang, Woo and Zhao, 2009). Several researchers, such as Lucas and Wright, (2009), Osika, Johnson and Buteau, (2009), and Eunjoo Oh and Suhong Park, (2009), have cited a lack of institutional support as a significant challenge. Scholars point out that faculty members and administrators lack a firm grasp of instructional technology adoption goals (Ross and Orr, 2009; Simpson, 2010). Others identify a lack of training and support (Picciano, Seaman and Allen, 2010), peer support (Margot and Kettler, 2019), and funding (Fairweather and Others, 2008; Baran and Keles, 2011).

Holotescu and Grosseck, (2013) found that the absence of a well-defined pedagogical framework for integrating technology into teaching poses a significant obstacle for educators. The authors argue that although there is an abundance of technological tools available to educators, there is a lack of clear guidance on how to use these tools to promote student learning. This is supported by other scholars (Bergstrom and Zhang, 2016; Calvin-Naylor *et al.*, 2017). Bergstrom and Zhang, (2016) assert that an in-depth understanding of pedagogical methodologies is imperative for educators to proficiently incorporate technology into the classroom. The authors suggest that an emphasis on technological tools rather than educational goals culminates in unfavourable academic consequences for students. Hrastinski, (2019) also warns that inadequate knowledge of instructional technology among educators can lead to poor academic outcomes and decreased student engagement. Table 1 synthesizes some of the literature relating to the challenges of implementing instructional technologies in practice. While this list is not exhaustive, it does indicate the level of complexity involved.

Table 1. Challenges to implementing instructional technologies

Challenge	References
Difficult to transfer face-to-face course materials to the online environment	Park and Choi, (2009); Tucker and Neely, (2010); Kyei-Blankson *et al.*, (2011); Almazova, Bernavskaya, *et al.*, (2020)
Lack of time to integrate instructional technologies into the learning process	Chawinga, (2017)
Lack of confidence and self-efficacy	Jones and Davis, (2008); Bingimlas, (2009); Buabeng-Andoh, (2012); Owston, York and Murtha, (2013); Stodel, Thompson and MacDonald, (2006)
Difficult to use technologies	Arnold and Paulus, (2010); Syverson and Slatin, (2010); Tucker and Neely, (2010); Hrastinski, (2019)
Lack of a well-defined pedagogical framework for integrating new technology	Holotescu and Grosseck, (2013); Bergstrom and Zhang, (2016); Calvin-Naylor *et al.*, (2017)
Lack of a student-centred approach	Alexander, (2001); Spodark, (2003); Xu and Morris, (2007); Park and Choi, (2009)
Lack of resources e.g. funding	Fairweather, (2008); Baran and Keles, (2011)
Poor institutional support	Tabata and Johnsrud, (2008); Lucas and Wright, (2009); Osika, Johnson and Buteau, (2009)
Lack of peer support for educators	Stodel, Thompson and MacDonald, (2006); Margot and Kettler, (2019)
Lack of student support e.g., training, peer learning and access to relevant technologies	Alexander, (2001); Johnson and Howell, (2006); Simelane, Blignaut and Van Ryneveld, (2007); Wang, Woo and Zhao, (2009); Hartman, Townsend and Jackson, (2019)

3. RESEARCH METHODOLOGY

This exploratory study utilized qualitative methods to identify the underlying challenges associated with instructional technology for higher education. The research process comprised four stages of development as illustrated below.

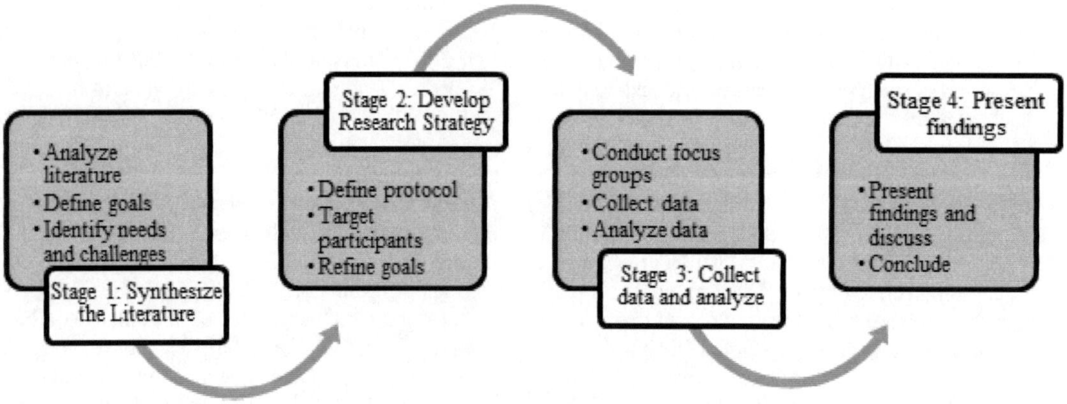

Figure 1. The scientific approach to data collection and analysis

The initial phase encompassed a thorough examination of existing academic literature and discourse pertaining to instructional technology in higher education. The study identified and narrowed down research gaps and issues, subsequently prioritizing and refining them to establish specific goals for the study. The most pertinent challenges were subsequently identified. The aforementioned constructs underwent expert verification to ensure their alignment with the specific context and were then refined based on feedback received during this process.

The second stage in the research process involved developing a research strategy. As this study is exploratory in nature qualitative methods were deemed appropriate to understand the issues and capture narratives, opinions and insights. The study used a purposive, high-variation sampling strategy (Flick, 2014) comprising experienced and non-experienced higher education teachers working in STEM from Ireland, Portugal, Bulgaria and Turkey. A profile of the respondents is presented in Table 2. Focus groups were used to explore the issues in detail and encourage discourse among the participants. This approach allowed us to elicit respondents' feelings, beliefs, and attitudes concerning the topic (Morgan and Krueger, 1993). Focus groups helped us discover new challenges, as the participants contributed collectively to a greater variety of perspectives, leading to further debate and discussion. An interview schedule was developed to guide the questions and to ensure consistency in the data collection process. The focus group protocol and supporting script were developed based on key themes and challenges derived from existing literature. The script was first pilot-tested, and amendments were made based on the feedback provided.

Table 2. Profile of participants

Category	Description	Number | Percentage
Country	Ireland (IE)	11 (27.0%)
	Turkey (TR)	12 (29.2%)
	Bulgaria (BGL)	13 (31.7%)
	Portugal (PT)	5 (12.1%)
Age	Less than 39	8 (20.7%)
	40 to 49	23 (55.2%)
	More than 50	10 (24.1%)
Gender	Male	13 (31%)
	Female	28 (69%)
Experience	10 years and below	6 (13.8%)
	11 years to 20 years	17 (41.4%)
	21 years and above	18 (44.8%)

The third stage of the research process involved collecting and analyzing the data. Data were collected in the local language to ensure a comprehensive understanding of the topics under discussion. Focus groups were recorded, and the data was transcribed and translated from each country. Two researchers independently analyzed the transcripts. Content analysis was used to generate themes and identify relationships for explaining the data. This was done iteratively, as revising and refining categories and themes was necessary. The study used Maxqda, a computer-based qualitative analysis program to facilitate this process. Miro boards were used to organize and validate the data. An example from one of the focus groups is illustrated in Figure 2. The data underwent an initial analysis at the national level, followed by a consolidation of data from four countries to identify shared and distinct challenges.

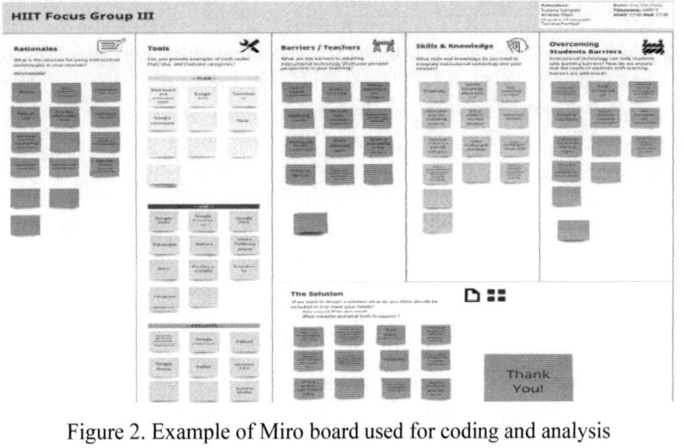

Figure 2. Example of Miro board used for coding and analysis

4. FINDINGS

Perceptions regarding the need for and use of instructional technologies were mixed in the higher education STEM context. According to one participant "*In the education of engineering students, more hands-on exercises are needed….*" while another highlighted the importance of this and emphasized the role instructional technology plays in this regard by stating that "*learning increases when I integrate real problems with learning objectives, instructional technology use within the classroom makes students more active and It also activates the student in data analysis*". However, it is evident that instructional technology cannon be seen as a panacea, educators also spoke about the negative effects of excessive technology use, with one saying that "*Quite often, their excessive use deprives students of adequate practical knowledge and skills in some of the disciplines.*" It is also evident that educators encounter various obstacles when utilizing instructional technology ranging from technical to social. We found that while there are some similarities between the countries, differences also exist. For instance, participants from Turkey expressed a more pronounced apprehension towards organizational process aspects, whereas participants from Ireland exhibited greater concern towards social aspects. While many important challenges were identified the following sections discuss a prioritized list of issues from our analysis.

4.1 Student engagement

Managing student engagement seems to be a major challenge to the effective implementation of instructional technologies. Respondents highlighted that the overuse of technologies can interfere with student engagement. For example, a participant from Turkey stated that "*I use instructional technologies, but lately, I have been hesitant to use them because it distracts students*". An educator from Bulgaria noted that "*Quite often, their excessive use does not stimulate students to make greater efforts for self-improvement,*" while another stated that "*The barrier of teaching online is the transforming the lab practical to an online-based platform and this has left many students disillusioned and under-skilled when entering their final year to work on projects*". This suggests that overusing technology may lead to student disengagement or lack of motivation. Our analysis revealed that students lose a sense of belonging when too much technology is used. In addition, educators lack sensory perception which makes it difficult to read and interpret cues which can also alienate students.

4.2 Educator self-efficacy

Some educators acknowledge that "*keeping up with new technologies is hard and overwhelming*". While another expressed feeling like a "*technophobe and having embarrassing moments*" due to this aspect. Findings from our analysis highlight the importance of stimulating teachers to use instructional technologies. One participant stated that using technologies may be daunting from the outset but it is possible to succeed with some practice. Another stated that "*Our habits and lack of knowledge get in the way. I met the pandemic with learning management systems; I tried to use Zoom, MS Teams, and whatever I had on my hand. However, I could use it effectively at the end of the pandemic*". Respondents acknowledge that implementing new technologies takes time and patience. Some participants cited time constraints and coordination as major obstacles to effective technology use. One teacher noted that adapting to new techniques can be time-consuming, especially in areas where curricula are constantly being updated due to technological advancements. Another stated that "*The short duration of the lessons requires a long time to teach what to do and practice. Therefore, using technology takes time*". It is evident that repurposing materials and adapting to new techniques takes time, effort and patience. These attitudes highlight educators' need to gain more confidence using various instructional technologies. Several educators in Bulgaria and Portugal also emphasized the need for adequate training on using educational technologies, stating that they require time to learn and apply. They also highlighted the importance of "*the hand to touch, the eye to see*" and the need for educators to invest time in researching and developing relevant materials and resources.

4.3 Integration and support

Many educators faced challenges with technology integration, with one teacher acknowledging the slow adoption of technology in their university. They stated, "*Our effective acquaintance was with the pandemic...the transition has been slow.*" An Irish educator expressed concerns about the overwhelming number of new technologies and the cost of licensing, saying, "*Too many tools and cost licensing is a concern.*" Another teacher emphasized the importance of coordination and cooperation among instructors, stating that "*coordination and cooperation...can make this a daunting task, especially if the preparation of the course needs to be done in a short time period.*" Several educators highlighted the challenges of integration. For example, an educator mentioned the need for coordination and cooperation among instructors to properly adapt to new technologies, saying, "*Coordination and cooperation, as one rarely gives classes alone, and the adoption of new technologies would require all instructors to be in sync in terms of what technologies to use, and how to properly adapt to them, which, combined with the time constraints that are common to all, can make this a daunting task, especially if the preparation of the course needs to be done in a short time period.*" These concerns highlight the need for institutional support, resources, and training to effectively integrate technology into teaching and address external factors and individual attitudes that can impede technology use. Several participants expressed concerns relating to organizational support and stated that there is a "*lack of institutional support for teacher training*". An educator from Bulgaria emphasized the importance of institutional support to encourage technology use, stating that "*policymakers are important.*"

5. CONCLUSIONS, LIMITATIONS AND FUTURE RESEARCH

The recent sudden shift to online and distance learning requires a different approach to instruction (especially when it comes to practice or lab-based courses). We argue that STEM instructors in higher education must change their perspective to ensure that they are reaping the benefits of new technologies to meet the very changing needs of their students. Higher education teachers are predominantly research-focused and specialists in their specific domain. They are currently ill-equipped with the skills, competencies and capacity required to absorb the principles of instructional technology needed to create bespoke student-focused learning experiences in short time frames. This exploratory study set out to identify the challenges of incorporating instructional technology into STEM courses and modules in higher education. The research identifies challenges to implementing instructional technologies in specific content. Due to the rapid evolution of technology, the takeaway from this research is that it is critical for both educators and those in decision-making positions at universities to adopt an integrated holistic perspective and ensure that all technologies align with the needs of the end user. Furthermore, educational institutions must provide necessary support systems for STEM teachers to ensure confidence in incorporating instructional technology into their classrooms and enhancing pedagogical competencies to ensure a dynamic and interactive learning environment for the student that can be accessed anywhere.

Transforming engineering education is not an easy task at a systems level. It requires changes in attitude, time and effort. Every stakeholder, including students, educators and institutions, must play a crucial role in improving teaching and learning experiences in higher education. Higher education engineering institutions must create an organizational culture that enables their educators to devote enough time to explore how to take maximum advantage of instructional technology and plan and conduct effective programs and modules. Cultivating strong leadership and effective management of people and systems is a necessary part of the process of teaching and learning. Similar to many exploratory studies, our research may be perceived to have limited generalizability beyond our targeted sample and the geographical areas included in this research. This particular study also focused solely on HE engineering education in Ireland, Turkey, Bulgaria and Portugal. Therefore, future research should probe into whether and to what extent our findings are also valid for other fields of study. We, therefore, encourage scholars to collect more data in different contexts. A multilevel analysis comprised of various hierarchical samples from different disciplines and countries may yield additional intriguing results.

ACKNOWLEDGEMENT

This research was funded by the European Commission and the Higher Education Authority, Ireland under the Erasmus+ program [HIIT: Higher Education Teachers and Institutions and Instructional Technology 2021-1-IE02-KA220-HED-000032025]. This paper reflects the authors' view; and the Commission is not responsible for any use that may be made of the information it contains.

REFERENCES

Alexander, S. (2001) 'E-learning developments and experiences', *Education + Training*, 43(4/5), pp. 240–248. Available at: https://doi.org/10.1108/00400910110399247

Almazova, N., Krylova, E., *et al.* (2020) 'Challenges and opportunities for Russian higher education amid COVID-19: Teachers' perspective', *Education in science: the bulletin of the Association for Science Education*, 10(12), p. 368. Available at: https://doi.org/10.3390/educsci10120368

Almazova, N., Bernavskaya, M., *et al.* (2020) Interactive learning technology for overcoming academic adaptation barriers. Integrating Engineering Education and Humanities for Global Intercultural Perspectives: Proceedings of the Conference '*Integrating Engineering Education and Humanities for Global Intercultural Perspectives*', 25-27 March 2020, St. Petersburg, Russia, Springer.

Arnold, N. and Paulus, T. (2010) 'Using a social networking site for experiential learning: Appropriating, lurking, modeling and community building', *The Internet and Higher Education*, 13(4), pp. 188–196. Available at: https://doi.org/10.1016/j.iheduc.2010.04.002

Baran, B. and Keles, E. (2011) 'Case study discussion experiences of Computer Education and instructional technologies students about Instructional Design on an asynchronous environment', *The Turkish Online Journal of Educational Technology*, 10(1), pp. 58–70. Available at: http://files.eric.ed.gov/fulltext/EJ926554.pdf (Accessed: 28 April 2023).

Bergstrom, C. and Zhang, D. (2016) 'Geometry interventions for K-12 students with and without disabilities: A research synthesis', *International journal of educational research*, 80, pp. 134–154. Available at: https://doi.org/10.1016/j.ijer.2016.04.004

Bingimlas, K.A. (2009) 'Barriers to the successful integration of ICT in teaching and learning environments: A review of the literature', *Eurasia journal of mathematics science and technology education*, 5(3), pp. 235–245. Available at: https://doi.org/10.12973/ejmste/75275

Buabeng-Andoh, C. (2012) 'Factors influencing teachersâ adoption and integration of information and communication technology into teaching: A review of the literature', *International journal of education and development using information and communication technology*, 8(1). Available at: https://www.learntechlib.org/p/188018/ (Accessed: 28 April 2023).

Calvin-Naylor, N.A. *et al.* (2017) 'Education and training of clinical and translational study investigators and research coordinators: A competency-based approach', *Journal of clinical and translational science*, 1(1), pp. 16–25. Available at: https://doi.org/10.1017/cts.2016.2

Chaipidech, P. *et al.* (2021) 'Implementation of an andragogical teacher professional development training program for boosting TPACK in STEM education', *Educational Technology & Society*, 24(4), pp. 220–239. Available at: https://www.jstor.org/stable/48629257

Chawinga, W.D. (2017) 'Taking social media to a university classroom: teaching and learning using Twitter and blogs', *International Journal of Educational Technology in Higher Education*, 14(1), p. 3. Available at: https://doi.org/10.1186/s41239-017-0041-6

Davies, R.S., Dean, D.L. and Ball, N. (2013) 'Flipping the classroom and instructional technology integration in a college-level information systems spreadsheet course', *Educational technology research and development: ETR & D*, 61(4), pp. 563–580. Available at: https://doi.org/10.1007/s11423-013-9305-6

Eunjoo Oh and Suhong Park (2009) 'How are universities involved in blended instruction?', *Journal of educational technology & society*, 12(3), pp. 327–342. Available at: http://www.jstor.org/stable/jeductechsoci.12.3.327

Fairweather, J. and Others (2008) 'Linking evidence and promising practices in science, technology, engineering, and mathematics (STEM) undergraduate education', *Board of Science Education, National Research Council, The National Academies, Washington, DC* [Preprint]. Available at: http://otl.wayne.edu/wider/linking_evidence--fairweather.pdf

Flick, U. (2014) 'Mapping the field', *The SAGE handbook of qualitative data analysis*, 1, pp. 3–18. Available at: https://books.google.com/books?hl=en&lr=&id=R-6GAwAAQBAJ&oi=fnd&pg=PA3&dq=flick+2014&ots=L69f10t3Qg&sig=EXAdupTNKsLV4pj04aE9Z2wE__w

Gandolfi, E., Ferdig, R.E. and Kratcoski, A. (2021) 'A new educational normal an intersectionality-led exploration of education, learning technologies, and diversity during COVID-19', *Technology in society*, 66, p. 101637. Available at: https://doi.org/10.1016/j.techsoc.2021.101637

Hartman, R.J., Townsend, M.B. and Jackson, M. (2019) 'Educators' perceptions of technology integration into the classroom: a descriptive case study', *Journal of Research in Innovative Teaching & Learning*, 12(3), pp. 236–249. Available at: https://doi.org/10.1108/JRIT-03-2019-0044

Holotescu, C. and Grosseck, G. (2013) 'An empirical analysis of the educational effects of social media in universities and colleges', *Internet learning* [Preprint]. Available at: https://doi.org/10.18278/il.2.1.3

Hrastinski, S. (2019) 'What do we mean by blended learning?', *TechTrends : for leaders in education & training*, 63(5), pp. 564–569. Available at: https://doi.org/10.1007/s11528-019-00375-5

Ishak, S.A., Din, R. and Hasran, U.A. (2021) 'Beyond Play: Conceptualising the Capability of a Good Digital Game to Stimulate Interest in STEM', *International Journal of Learning, Teaching and Educational Research*, 20(10). Available at: http://ijlter.myres.net/index.php/ijlter/article/view/13 (Accessed: 28 April 2023).

Jackson, L.A. *et al.* (2008) 'Race, gender, and information technology use: the new digital divide', *Cyberpsychology & behavior: the impact of the Internet, multimedia and virtual reality on behavior and society*, 11(4), pp. 437–442. Available at: https://doi.org/10.1089/cpb.2007.0157

Johnson, G.M. and Howell, G.J. (2006) 'The Impact of Internet Learning Technology: Experimental Methods of Determination', in *Selected Styles in Web-Based Educational Research*. IGI Global, pp. 282–301. Available at: https://doi.org/10.4018/978-1-59140-732-4.ch019

Jones, P. and Davis, R. (2008) 'Instructional Design Methods Integrating Instructional Technology', in *Handbook of Research on Instructional Systems and Technology*. IGI Global, pp. 15–27. Available at: https://doi.org/10.4018/978-1-59904-865-9.ch002

Kayan Fadlelmula, F., Sellami, A. and Le, K. (2022) 'STEM Learning during the COVID-19 Pandemic in Qatar: Secondary School Students' and Teachers' Perspectives', *EURASIA Journal of Mathematics, Science and Technology Education*, 18(6). Available at: https://eric.ed.gov/?id=EJ1345669

Kyei-Blankson, L. *et al.* (2011) 'The New Blend: When students are given the option to choose', in *Society for information technology & teacher education international conference*. Association for the Advancement of Computing in Education (AACE), pp. 433–436. Available at: https://www.learntechlib.org/p/36302/

Lee, Y. and Choi, J. (2011) 'A review of online course dropout research: implications for practice and future research', *Educational technology research and development: ETR & D*, 59(5), pp. 593–618. Available at: https://doi.org/10.1007/s11423-010-9177-y

Lucas, S.B. and Wright, V.H. (2009) 'Who am I? The influence of teacher beliefs on instructional technology incorporation', *Journal on excellence in college teaching*, 20(3), pp. 77–95. Available at: https://www.learntechlib.org/p/74544/ (Accessed: 28 April 2023).

Margot, K.C. and Kettler, T. (2019) 'Teachers' perception of STEM integration and education: a systematic literature review', *International journal of STEM education*, 6(1). Available at: https://doi.org/10.1186/s40594-018-0151-2

Morgan, D.L. and Krueger, R.A. (1993) 'When to use focus groups and why', *Successful focus groups: Advancing the state of the art.*, 271, pp. 3–19. Available at: https://doi.org/10.4135/9781483349008.n1

Mutambara, D., Bayaga, A. and Others (2020) 'Predicting rural STEM teachers acceptance of mobil elearning in the fourth industrial revolution', *Journal of Construction Project Management and Innovation*, 10(2), pp. 14–29. Available at: https://journals.uj.ac.za/index.php/JCPMI/article/view/404

Ng, O.-L. and Park, M. (2021) 'Using an enhanced video-engagement innovation to support STEM teachers' professional development in technology-based instruction', *Journal of educational technology & society*, 24(4), pp. 193–204. Available at: https://www.jstor.org/stable/48629255

Osika, E.R., Johnson, R.Y. and Buteau, R. (2009) 'Factors influencing faculty use of technology in online instruction: A case study', *Online Journal of Distance Learning Administration*, 12(1), pp. 1–14. Available at: https://www.academia.edu/download/79491238/osika121.pdf

Owston, R., York, D. and Murtha, S. (2013) 'Student perceptions and achievement in a university blended learning strategic initiative', *The Internet and Higher Education*, 18, pp. 38–46. Available at: https://doi.org/10.1016/j.iheduc.2012.12.003

Park, J.-H. and Choi, H.J. (2009) 'Factors influencing adult learners' decision to drop out or persist in online learning', *Journal of Educational Technology & Society*, 12(4), pp. 207–217. Available at: https://www.jstor.org/stable/pdf/jeductechsoci.12.4.207.pdf?_=1459005475270

Picciano, A.G., Seaman, J. and Allen, I.E. (2010) 'Educational transformation through online learning: To be or not to be', *Online learning*, 14(4). Available at: https://doi.org/10.24059/olj.v14i4.147

Rasheed, R.A., Kamsin, A. and Abdullah, N.A. (2020) 'Challenges in the online component of blended learning: A systematic review', *Computers & Education*, 144, p. 103701. Available at: https://doi.org/10.1016/j.compedu.2019.103701

Reid, P. (2014) 'Categories for barriers to adoption of instructional technologies', *Education and Information Technologies*, 19(2), pp. 383–407. Available at: https://doi.org/10.1007/s10639-012-9222-z

Ross, C. and Orr, R.R. (2009) 'Teaching structured troubleshooting: integrating a standard methodology into an information technology program', *Educational technology research and development: ETR & D*, 57(2), pp. 251–265. Available at: https://doi.org/10.1007/s11423-007-9047-4

Simelane, S., Blignaut, S. and Van Ryneveld, L. (2007) 'Preparing lecturers to integrate educational technology into their teaching and learning practices', *South African Journal of Higher Education*, 21(7), pp. 940–953. Available at: https://journals.co.za/doi/abs/10.10520/EJC37394

Simpson, C.M. (2010) 'Examining the relationship between institutional mission and faculty reward for teaching via distance', *Online Journal of distance learning administration*, 13(1), pp. 1–13. Available at: https://citeseerx.ist.psu.edu/document?repid=rep1&type=pdf&doi=7ea900ddd58353e1d860db2f6ec128a6ebb28b3b

Spector, J.M. (2016) 'Ethics in educational technology: towards a framework for ethical decision making in and for the discipline', *Educational technology research and development: ETR & D*, 64(5), pp. 1003–1011. Available at: https://doi.org/10.1007/s11423-016-9483-0.

Spodark, E. (2003) 'Five obstacles to technology integration at a small liberal arts university', *The Journal*, 30(8). Available at: https://www.learntechlib.org/p/96298/ (Accessed: 28 April 2023).

Stodel, E.J., Thompson, T.L. and MacDonald, C.J. (2006) 'Learners' Perspectives on what is Missing from Online Learning: Interpretations through the Community of Inquiry Framework', *The International Review of Research in Open and Distributed Learning*, 7(3). Available at: https://doi.org/10.19173/irrodl.v7i3.325

Syverson, M.A. and Slatin, J. (2010) 'Evaluating learning in virtual environments', *Learning Record* [Preprint]. Available at: http://www.learningrecord.org/caeti.html

Tabata, L.N. and Johnsrud, L.K. (2008) 'The Impact of Faculty Attitudes Toward Technology, Distance Education, and Innovation', *Research in higher education*, 49(7), pp. 625–646. Available at: https://doi.org/10.1007/s11162-008-9094-7

Tucker, J. and Neely, P. (2010) 'Unbundling faculty roles in online distance education programs', *The International Review of Research in Open and Distributed Learning*, 11(2), pp. 20–32. Available at: https://doi.org/10.19173/irrodl.v11i2.798

Wang, Q., Woo, H.L. and Zhao, J. (2009) 'Investigating critical thinking and knowledge construction in an interactive learning environment', *Interactive Learning Environments*, 17(1), pp. 95–104. Available at: https://doi.org/10.1080/10494820701706320

Xu, H. and Morris, L.V. (2007) 'Collaborative Course Development for Online Courses', *Innovative Higher Education*, 32(1), pp. 35–47. Available at: https://doi.org/10.1007/s10755-006-9033-5

GAMIFICATION APPROACH FOR TEACHING PROGRAMMING

Georgi Krastev, Valentina Voinohovska and Vania Dineva
University of Ruse 8 Studentska, str, Ruse, Bulgaria

ABSTRACT

The aim of the article is to present a study of two different approaches for using a digital game for improving problem-solving skills of the students through introductory computer programming learning. The game itself it is not predefined, it is developed as an engine, which shortly means that learners will develop their own logic. The research conducted an experiment with first year students taking introductory to programming course at University of Ruse. The results show a significant progress for using digital games when compared to not using it at all. The game was developed within the UNITe (Universities for Science, Informatics and Technologies in eSociety) project.

KEYWORDS

Digital Game, Programming, Problem-Solving

1. INTRODUCTION

Historically, programming education began in universities in the 1970s. In 1986, Bulgaria became the fourth country in the world to introduce, as a compulsory subject, Informatics at the high school level. At the beginning of the 21st century, in most European countries, computer programming was integrated into the lower secondary education stage, and in three countries – Estonia, Greece and England, it was introduced at the primary stage.

In the 21st century, the demand for the development of computational and digital skills, as elements of the fundamental literacy, will become increasingly intense (Papadakis, 2016). In the recent years, different studies have reported that novice programmers have difficulties in learning Object-Oriented Programming (OOP) concepts (Shoilekova, 2021 (a)). There has been a lot of critics of the traditional approaches for teaching computer programming stating that they don't support students to acquire and apply their knowledge, which leads to a decrease in students' motivation.

Teaching computer programming is a complex activity that requires a lot of practice and traditional teaching approaches have not been able to respond effectively (Queirós, 2020, Bosse, 2017). Moreover, Zhang (Zhang, 2013) reported that traditional and conventional teaching methods using static material such as books and slides no longer relevant and effective in teaching and learning of computer programming. Thus, Cheah (Cheah, 2019) elucidated that teaching material should be able to support spatial visualization by incorporating multimedia elements and interaction to explain the dynamic concept of programming (Assaf, 2022).

Technological advancements and their rapid development always create new and exciting ways to engage students learning and meet the growing needs of education (Kalogiannakis, 2021). Constantly emerging new technologies challenge the field of education. The proper use of digital educational technologies can improve learning and teaching (Shoilekova, 2021 (b)). However, to be helpful, these technologies need to be incorporated in advanced pedagogical creative teaching practices.

Studies in the field of learning through digital games and resources highlight their potential to achieve an effective learning process:

- Increasing the motivation to learn and providing interesting experiences that strengthen the interest of learners in the studied discipline.
- Encourage learning by doing.
- Encourage meaningful learning.

• Provide risk-free environments which enable learners to explore and experiment.
• Improving problem-solving skills.
• Improving students' understanding of computer programming concepts.

To establish the positive influence of digital games for conducting an effective educational process, it is first necessary to clarify the concept of the game and determine its main characteristic components.

A game is a rule-based software application with a variable and quantifiable consequence, where different outcomes are assigned different values. The player controls the course of action and exerts effort to influence the outcome. The game has a goal to which he aspires, and which he can achieve if he is sufficiently clever, subtle, or bold.

Serious games (SGs) have different terms in the literature. Smith (2013) pointed some of the popular terms, such as Educational Games, Digital Game-Based Learning, Edutainment, Virtual Reality, Immersive Learning Simulations, Simulation.

There are several computer games and programming tools that can be used to learn programming, including:

• CodeCombat – a multiplayer game that teaches students how to code in Python, JavaScript, and other programming languages.

• Screeps – a strategy game where students write code to control their own colony of creatures, using JavaScript as the programming language.

• Lightbot – a puzzle game that teaches programming concepts such as loops, conditionals, and functions.

• Human Resource Machine - A game where students use a visual programming language to solve puzzles and automate tasks.

• Robocode – a game where students program their own robots to compete against each other in a virtual arena.

2. DIGITAL EDUCATIONAL GAMES USAGE AND SUCCESS FACTORS

The interest in using computer games for educational purposes has also increased over the last decade, with researchers identifying key pedagogical factors that make serious games inherently strong learning tools. Several factors affect the success of using simulation software and serious games in teaching:

2.1 The Learning Environment Is Interactive, Flexible and Personalised to the Expectations of the Digital Generation Students

One of the grounded features of digital educational games is that **the training is tailored to the expectations of the digital generation**. Prensky (2001) highlights the gap between learners' expectations and what they learn in school. In fact, today's learners are a digital generation that has grown up in a world where interactive digital media (computers, the Internet, mobile phones or computer games) have always existed, and which differs from the pedagogical resources usually used in traditional classrooms (Prensky, 2001).

Digital games are becoming a major part of young people's lives as well as a significant educational force in their lives. From that point of view, digital educational games are related to students' interests, as well as to new ways of thinking and processing information. They provide a way of learning that continuously adapts to the cognitive level of the learners and their specific learning style.

2.2 Self-Study Opportunity

Digital educational games are interactive. They provide feedback to the learner and enable them to manage and take control of their own learning. According to Gee's (2003) principle of active learning, the learning process with digital games should include continuous reflection to establish the level of mastery of the intended knowledge and skills.

This way of learning is consistent with Vygotsky's "zone of proximal development." According to him, the zone of proximal development is the distance between the actual level of development, determined by what the learner can do on his own, and the level of potential development, determined by solving problems with the support of the teacher or in cooperation with peers. The Zone of Proximal Development suggests that interaction with experienced educator will help the students to reach their potential and specific goals in all areas of study. Digital educational resources provide this type of support to assist learners in dealing with the problem at hand.

2.3 Developing Cognitive Links

Digital educational games support learners in developing cognitive connections for isolated events. Learners must consider how a particular decision will affect their future actions and their consequences. Digital games can provide an authentic learning environment where learners work in a simulated reality and solve real-life problems. As a result, games provide an opportunity for learning in an established context.

2.4 Conducting Research Opportunity

Klopfer (2009) defines a game through five types of freedom, viz. freedom to fail, to experiment, to express one's identity and freedom to strive and create interpretations.

Digital educational games promote each of these characteristics. According to Gee's discovery principle, game-based learning encourages learners to experiment with their hypotheses by performing planned actions in a simulated environment. They may adopt different roles and identities (Gee, 2003).

2.5 Active Learning

Class activities are often considered boring. In contrast, digital games can stimulate students' engagement in learning tasks. The most discussed argument for the use of games in an educational context is their potential to enhance learners' intrinsic motivation and grasp their attention (Doncheva, 2017).

According to Gee's (2003) identity principle, digital educational games engage learners by providing an opportunity to adopt a different identity by creating different characters.

2.6 Providing a Safe Environment with Risk-Taking Opportunities

According to Gee (2003), learning with digital games provides opportunities for exploration and risk-taking in a virtual environment. Learners can experiment and develop their skills using different identities without fear of failure. Even so, however, failure is an integral part of learning and has a positive impact as participants will be motivated to choose new patterns to achieve the added goal. The use of simulation software can overcome the problems associated with traditional learning.

2.7 Flexible Assessment

Assessment in educational games happens in a natural way, as the actions of the learners are continuously analyzed. Therefore, games can serve as a non-invasive assessment tool by removing the pressure to continually prove achievement.

According to Gee (2003), learners can reflect on their actions, form hypotheses, test them and then accept, reject, or rethink them. Therefore, students could improve their knowledge, skills, and competencies through repetition.

2.8 Collaboration

Digital educational games provide an opportunity for collaborative learning to achieve common goals. Cooperation can take place during or outside of the game. Some of the games provide multi-user mode, use of the same physical space and synchronous online communication.

3. ORGANIZATION OF THE STUDY. ANALYSIS OF THE RESULTS

The methodology developed for the discipline Computer modelling offers a wide range of opportunities for integrating other disciplines studied by the students. The specifics of the learning material, the way it is structured, the methods of presenting it to the students, aim at improving the quality in the theoretical and practical preparation of students.

3.1 Organization of the Study and Research Questions

The organization of the students' practical activity in the classes in *Introduction to programming* is an important source of methodological skills, habits, competences and builds a creative attitude to the use of computational thinking. Through the inclusion of methodologically linking units in the content of the discipline *Introduction to programming* there are established a sustainable interest in the studied processes.

The new methodology was applied with first year students in the academic 2021/2022 at the Angel Kanchev University of Ruse. The experiment started after the teachers covered specific programming concepts and not at the beginning of the semester. The experiment design consists of three groups as follow:

Group 1 – students trained using the standard approach.

Group 2 – students trained based on the proposed new approach. They will take an induction lecture followed by a practical session and assignment, where learners will use the digital game to learn and apply their knowledge of programming. The students will play the game, pass the levels, achieve some goals and during game if they have a difficulty, they could use the hints.

Group 3 – students who are trained based on the proposed new approach. Students in this group will take lectures, where they will be guided to use the digital game to finish the same assignment as students in Group 1. The students will serve as administrators of the game for creating, deleting, and managing missions (generated from single XML file).

3.2 The Used Serious Game

Project is created with technology called JavaServer Pages (JSP). It helps software developers to create dynamically generated web pages based on HTML, XML or other document types. To deploy and run JavaServer Pages is required Apache Tomcat server.

Project is written in Java Programming Language because JSP requires it for creating so-called JavaBeans or models and for editor is selected Eclipse 4.4 Luna IDE.

Views or user interface (UI) of JSP are written in PrimeFaces, which is a component suit for JavaServer Faces based application.

Used database management system is My SQL For maintenance and administration of My SQL is used My SQL Workbench which is a visual design tool for My SQL databases. My SQL is one of the most popular database engines, also it is open source and reason for selecting My SQL is precisely this.

The digital game goal is to improve problem-solving skills of a students through introductory computer programming learning using a game. The game itself it is not predefined, it is developed as a game engine, which shortly means that users will develop their own logic. In many cases game engines provide a suite of visual development tools, such tools are provided in this web application too. These visual development tools are basic programming operator like "IF, THEN, ELSE".

In more details game represent a self-service snack bar with cook who ask clients what they want and take orders, prepare, and serve drinks and foods. For example one path of execution of a small program is provided in "Figure 1" below. Several of the exact commands that the cook could perform are shown in "Figure 2".

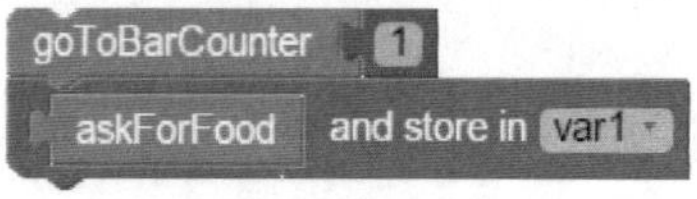

Figure 1. Simple program

"goToBarCounter [1]" – goes to client who seats in place 1

"Store operator" – purple block is an operator or function that executes functions inside it and stores result from that function in variable with name provided by user. In this example in variable var1.

"askForFood" – ask client which food he wants and returns it.

Main user interface of the web application is shown "Figure 3 Main user interface". User passes missions and try to achieve goal at next mission. For consistency an example with same algorithm is presented. Goal is achieved when user succeed to ask client what food he wants.

Different kinds of tips and hints, like one in the "Figure 3", "Drag a block to the workspace", are displayed to user during the game.

Figure 2. Cook commands

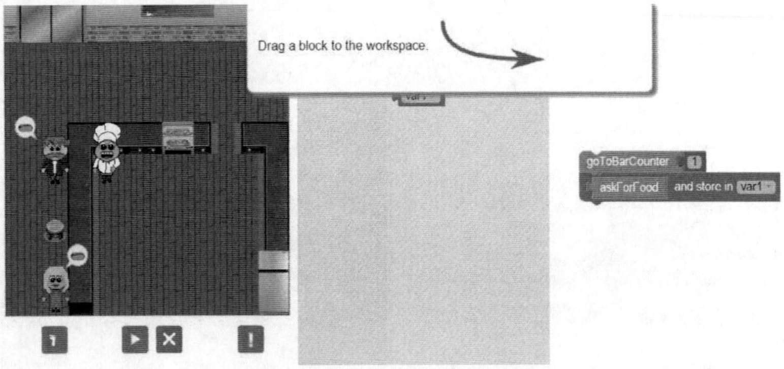

Figure 3. Main user interface

3.3 Analysis of the Results of the Study

3.3.1 General Characteristics of the Used Statistical Methods

In the statistical presentation of data there was made a comparative analysis between the groups at the entrance level for ascertaining the previous knowledge of the learners, and at the exit level for ascertaining the efficiency of the applied methodology in the three control groups.

For the achievements to be compared, dispersion analysis was used with three factors: METHODOLOGY – belonging to one of the three groups; FS – belonging to one of the two forms of study – full-time and distance and GENDER – male and female. When the results of the dispersion analyses were graphically represented, the average values were noted, together with 95% of confidential intervals. Details of the significance of the differences between the individual average values may be obtained with the aid of the Post-Hoc analysis which was carried out for the purpose of completeness of data after each dispersion analysis. The null hypotheses in this analysis, in applied form state, that the respective factor has a null effect (null result), and all verifying statistics have an F-distribution. The presence of a statistically significant effect on the methodology at specific location of the average values, is the main indicator for a positive result of the main model.

3.3.2 Analysis of the Results as Per the Individual Indicators

The test as a whole [alpha = 0.919], as well as the individual elements show very good reliability which is confirmed by the factor analysis based on the method of principal components. The factor analysis shows that there is a leading significance of 8.743650 with a next largest significance of 2.834316, which shows considerable one-dimensional state of the reviewed latent space.

To follow the effects of the different factors on the achievements as per the test indicators, it was used a dispersion analysis for recurring measurements with the following factors: METHODOLOGY – belonging to one of the three groups; FS – FORM OF STUDY - belonging to one of the two forms of study – full-time and distance; and GENDER – male and female. The results of the analysis are presented below.

From Figure 4 we can see that Group 2 and Group 3 show significantly high results as compared to Group 1 (the standard approach). There can also be noted slightly higher achievements in Group 3 in comparison to Group 2, which are statistically significant.

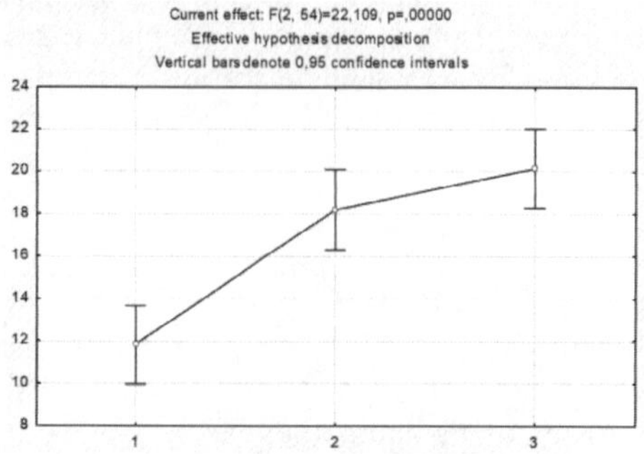

Figure 4. Diagram of the effects of the factor Methodology (test-indicators)

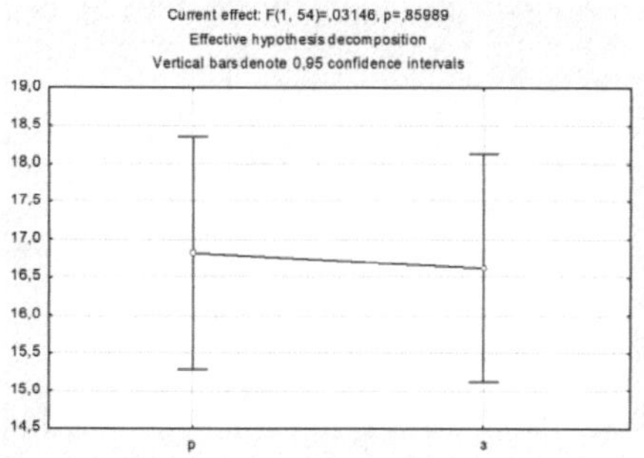

Figure 5. Diagram of the effects of the factor FS (form of study) (test-indicators)

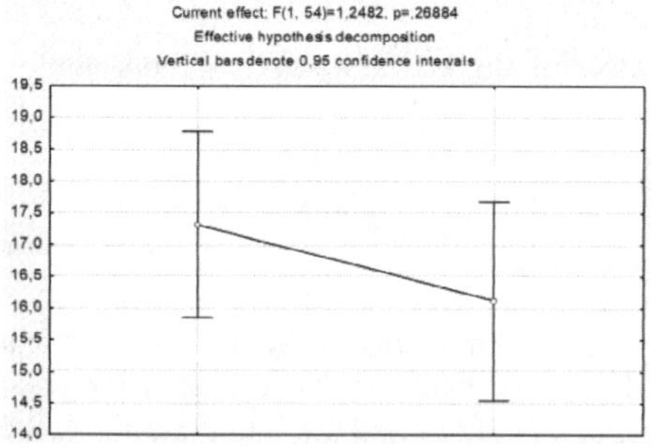

Figure 6. Diagram of the effects of the factor GENDER (test-indicators)

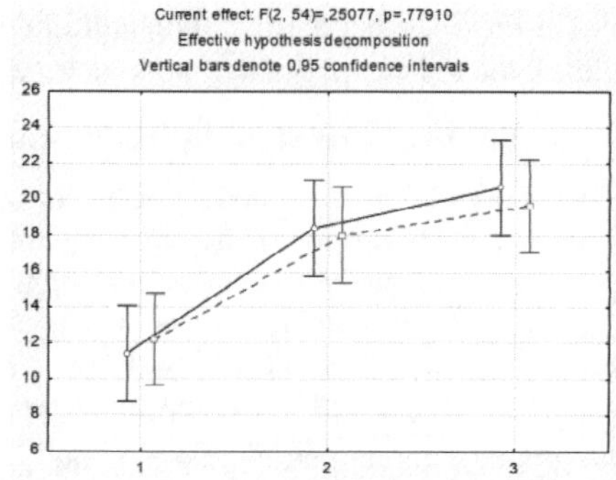

Figure 7. Diagram of the effects of the factor Methodology and FS (test-indicators)

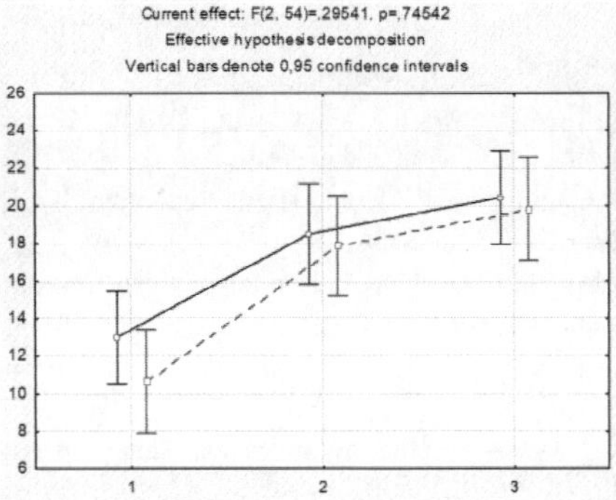

Figure 8. Diagram of the effects of the factors Methodology и GENDER (test-indicators)

It was noted a significant primary effect of the factor **Methodology**. Fischer's criterion for the factor Methodology is 22,1091, and the critical (table) value at a level of significance α<0,05 and degrees of freedom k1=2 and k2=54 is F0,05;2;54=3,13 [12], i.e. FA>Fα;k1; k2 , hence it follows that the significance of the factor Methodology [F(2,54)=22,1091; p=0,000000] is statistically substantiated and exerts a considerable effect on the outcomes of the instruction (Table 1). This is statistical proof of the positive result of the suggested new methodology. The influence of the other factor FS [F(1,54)=0,0315; p=0,859888] and GENDER [F(1,54)=1,2482; p=0,268836] has not been statistically proven.

There were not noted unexpected significances for the effects of interaction Methodology*FS [F(2,54)=0,2508; p=0,779101] and of the interaction Methodology*GENDER [F(2,54)=0,2954; p=0,745423].

Table 1. Significance of the effects on (test-indicators)

	Univariate Tests of Significance for Indicators Sigma-restricted parameterization Effective hypothesis decomposition				
Effect	SS	Degr. of Freedom	MS	F	p
Intercept	17467,33	1	17467,33	971,6061	0,000000
Methodology	794,95	2	397,47	22,1091	0,000000
FS	0,57	1	0,57	0,0315	0,859888
Gender	22,44	1	22,44	1,2482	0,268836
Methodology*FS	9,02	2	4,51	0,2508	0,779101
Methodology*Gen.	10,62	2	5,31	0,2954	0,745423
FS*Gender	1,10	1	1,10	0,0613	0,805352
Method.*FS*Gender	7,41	2	3,71	0,2062	0,814303
Error	970,80	54	17,98		

4. CONCLUSION

The use of serious games was suggested by many authors as an approach for overcome the difficulties in teaching computer programming. The games can be a fun and engaging way to learn programming concepts. However, it's important to note that they should be used as a supplement to a more structured learning approach, rather than as a replacement for formal instruction or practice.

This study compares two different approaches for using a serious game for teaching computer programming. The results showed that digital games enhanced students' understanding of the programming concepts and increased the students' confidence and motivation to learn.

ACKNOWLEDGEMENT

This publication is developed with the support of Project BG05M2OP001-1.001-0004 UNITe, funded by the Operational Programme "Science and Education for Smart Growth", co-funded by the European Union trough the European Structural and In-vestment Funds.

REFERENCES

Assaf, A. J. et al, (2022). Finding the Best Approach for Using Serious Games In Teaching Computer Programming. *Journal of Positive School Psychology*, *6*(7), pp. 4373-4401.

Bosse, Y. et al, (2017). Why is programming so difficult to learn? Patterns of Difficulties Related to Programming Learning Mid-Stage. *ACM SIGSOFT Software Engineering Notes*, *41*(6), pp. 1-6.

Cheah, C. S., (2020). Factors contributing to the difficulties in teaching and learning of computer programming: A literature review. *Contemporary Educational Technology*, *12*(2), ep272.

Doncheva, J., (2017). Principles of training in line with the new thinking and action, *SEA - Conf., 3 International Conference, Naval Academy, Constanta*, No 3, pp. 74, 2017.

Gee, J. P., (2003). What video games have to teach us about learning and literacy. *Computers in entertainment (CIE)*, *1*(1), pp. 20-20.

Kalogiannakis, M., Papadakis, S., & Zourmpakis, A., (2021). Gamification in science education. A systematic review of the literature. *Education Sciences*, *11*(1), 22.

Klopfer, E. et al, (2009). *Moving learning games forward.* Cambridge, MA: The Education Arcade.

Papadakis, S., Kalogiannakis, M., & Zaranis, N., (2016). Developing fundamental programming concepts and computational thinking with ScratchJr in preschool education: a case study. *International Journal of Mobile Learning and Organisation*, *10*(3), 187-202.

Prensky, M., (2001). Fun, play and games: What makes games engaging. *Digital game-based learning*, *5*(1), pp. 5-31.

Queirós, R. et al, (2020). Computer Programming Education in Portuguese Universities, *First International Computer Programming Education Conference (ICPEC 2020)*, Schloss Dagstuhl–Leibniz-Zentrum für Informatik, Dagstuhl, Germany. pp. 21:1–21:11.

Shoilekova, K., (2021a). Intelligent Data Analysis Using a Classification Method for Data Mining Knowledge Discovery Process. *Artificial Intelligence in Intelligent Systems: Proceedings of 10th Computer Science On-line Conference 2021*, Springer International Publishing, vol. 2, pp. 153-157.

Shoilekova, K., (2021b). Advantages of Data Mining for Digital Transformation of the Educational System, *Artificial Intelligence in Intelligent Systems. CSOC 2021. Lecture Notes in Networks and Systems*, vol 229. Springer, Cham, 2021, vol. 2, pp. 450-454.

Smith, S. P. et al, (2015). A meta-analysis of data collection in serious games research. *Serious games analytics: Methodologies for performance measurement, assessment, and improvement*, pp. 31-55.

Zhang, X. et al, (2013). Teaching introductory programming to is students: The impact of teaching approaches on learning performance. *Journal of Information Systems Education,* 24, pp. 147–155.

POSSIBLE SOLUTIONS FOR CAPITALIZING E-EDUCATION TOOLS IN MILITARY SCIENCE

Lucian Valeriu Scipanov[1], Daniel Roman[2], Petrişor Pătraşcu[3] and Dănuţa Mădălina Scipanov[4]
[1]*professor eng., the Naval Forces Department, the Command and Staff Faculty,*
[2]*associate professor, the Army Forces Department, the Command and Staff Faculty,*
[3]*associate professor, the Strategic Department, the Command and Staff Faculty,*
[4]*associate professor, the Logistics, finance, and accounting Department, the Command and Staff Faculty,*
"Carol I" National Defense University, Bucharest, Romania

ABSTRACT

This approach aims to examine the evolution of teaching-learning processes in higher military education following the pandemic crisis, based on the experience gained by teachers. It will consider the main changes and effects resulting from the increased use of mixed methods, including eLearning and e-Training tools, online platforms, and other e-Education resources. The authors will focus on the utilization of these tools in educational processes in general, with a particular emphasis on the field of military sciences, where their teaching and research activities are conducted. The technological advancements have also led to the recognition of the importance of incorporating artificial intelligence into decision-making and operational processes. Therefore, in the field of military sciences, it is acknowledged that the lessons learned from crisis situations should be integrated, enhancing the educational process through the appropriate use of eLearning and e-Training tools in future endeavors. Through this approach, the authors aim to enhance the quality and efficiency of the teaching-learning process by combining traditional and computer-assisted teaching tools. The desired outcome is the dissemination of best practices and experiences in e-Education specifically tailored to the field of military sciences. As military sciences primarily involve dynamic activities that require physical presence and active participation, the use of online platforms is considered relatively niche. Consequently, teaching and learning activities have been separated, and distinct stages have been identified, integrating e-Learning and e-Training tools, including artificial intelligence, alongside traditional methods, while maintaining participants' physical involvement in the process.

KEYWORDS

e-Education, e-Learning, e-Training, Military science

1. INTRODUCTION

This approach aims to examine the evolution of teaching-learning processes in higher military education during the pandemic crisis, drawing on the authors' experience in the field of military sciences. The observation and analysis of the field, as well as the assessment of resilience, serve as the foundation for this direction. The authors will track the evolution of eLearning tools in educational processes, with a particular focus on the field of military sciences where they conduct their teaching and research activities. Accordingly, the main changes and effects resulting from the increased utilization of eLearning tools, online platforms, and other teaching-learning tools specific to military sciences will be taken into consideration. The analysis aims to explore how the military educational process can adapt by applying principles of operational art in relation to the potential use of eLearning and e-Training tools.

The underlying premise of this approach is that technological advancements have necessitated the integration of artificial intelligence in decision-making and operational processes. Consequently, in the field of military sciences, it is acknowledged that the lessons learned from crisis situations must be leveraged to enhance the educational process, utilizing the most suitable e-Learning and e-Training tools for future military leader education.

Through this approach, the authors propose enhancing the quality and efficiency of the teaching-learning process by utilizing both traditional and computer-assisted teaching tools. The expected outcome is the dissemination of best practices and experiences in e-Education specific to the field of military sciences.

Given that military sciences primarily involve dynamic activities that require physical presence and active participation, the use of online platforms can be considered relatively niche. As a result, teaching and learning activities have been separated, and specific stages have been identified using e-Learning and e-Training tools, including artificial intelligence, in conjunction with traditional methods, while maintaining the participants' physical involvement in the process.

The authors assert that different teaching-learning methods can be categorized into classic tools and e-Education, which encompass methods involving physical presence, online methods, and mixed methods. Notably, all these approaches have observed varying degrees of contribution from modern tools in the development of educational processes.

2. LESSONS SPECIFIC TO THE MILITARY SCIENCE TEACHING-LEARNING PROCESS IDENTIFIED IN CRISIS SITUATIONS

This approach is based on the assumption that crisis situations have created special conditions for the achievement of didactic objectives, requiring a quick adaptation and the timely identification of some solutions to solve them. Within this direction of research, it is desired to create a documentary base of identified lessons that will be the basis for increasing the resilience of higher military education, as a result of the development and exploitation of eLearning platforms in the field of military sciences, which the authors know in their capacity as teachers.

The establishment of a documentary base regarding the management of evolving events impacting the way ordinary activities are carried out represents a valuable platform for capturing lessons learned and serves as a starting point for educational activities related to crisis situations. However, the novelty lies in the fact that crisis situations necessitate the identification of new solutions, thereby fostering system resilience, which, in the authors' opinion, is advantageous.

To address the urgent requirement for prompt access to crisis situation solutions and the training of decision-making commanders and specialists, it became necessary to disseminate the lessons learned. The current approach emphasizes the importance of complementing traditional teaching-learning methods with the utilization of computer-assisted education, specifically through the implementation of eLearning and e-Training tools, including artificial intelligence, to maximize quality and efficiency.

Working from home and online education represented a method to solve the situation of isolation of citizens during the pandemic, but also a solution to continue the educational process.

As a result of the research carried out by the authors, the analysis identified several educational realities. According to the authors, the impact of technologies led to a more realistic awareness of the role of computer networks, the development, and exploitation of e-Learning and e-Training fields, of artificial intelligence (AI), a fact that led to the development, implementation and timely exploitation of e-Education tools and in the field of military sciences. The main findings were:

- development and exploitation of online platforms

Existing online platforms have been developed and intensively utilized in educational processes. If online platforms were used for some courses, they became a valuable teaching-learning tool in crisis situations. The role of online platforms was realized not only for the established activities, for which they were designed but more, for their intensive use in the current educational process. Thus, the e-Learning component acquired a new form and value. Due to the need for documentation, in real-time, without substantial loss of time, access to virtual libraries and electronic databases has been carried out with greater intensity, their degree of efficiency being high. The e-Training component could not be leveraged effectively in this case, due to the fact that most training activities required physical presence. An important component in this field was computer-assisted training, the available online platforms being used predominantly during exercises to train the decision-making process and lead forces during simulated military actions.

- expansion of the volume of the documentary base

The need for digitization was re-emphasized by the expansion of the volume of the documentary base. Thus, they were developed and uploaded on the educational platforms managed by the military education institutions, the specialized manuals, of the courses as a support of the teaching-learning process, with the mention of the limited access to the classification level of the managed information. The aspect has directly generated an increase in knowledge in general and easy access to the documentary base in particular.

- increasing the speed of information dissemination

The rapid dissemination of data and information has influenced the educational process. Traditional formats such as round tables, conferences, and working groups have leveraged the advantages offered by online platforms. Scientific conferences initially took place online during crisis situations, followed by the acceptance of a mixed format involving both physical presence and online participation. Similarly, round tables facilitated debates in the online environment, leading to the unanimous acceptance of mixed debate methods that combine online and physical presence. The noteworthy development was the academic community's acceptance of the mixed format, allowing for a combination of online participation and physical presence in various events, highlighting the system's resilience in promptly embracing new communication possibilities.

- - identified, acquired, adapted, and developed teaching-learning methods

The training of teaching staff to work in the online environment was done on the fly, adaptively. New teaching methods were identified in the online environment, capitalizing on the experience of classical methods. These methods involved supporting students through active teaching-learning methods, short projects with fixed deadlines, online work in teams (Scipanov LV, 2021), assigning short tasks with concrete and immediate purposes, etc. In conclusion, the importance of e-Learning and e-Training tools in the teaching-learning process was realized through an effective capitalization, outlining the effectiveness of e-Education in military sciences (Scipanov LV, Scipanov DM, 2020) and the development of computer networks and the integration of educational laboratory experiments.

The authors emphasize the role of AI in the teaching-learning act. The development of computer networks and the realization of an education based on a laboratory experiment represents a modern method by which the operative art specific to the field of military sciences, the decoding of the contemporary operational environment is achieved and can provide answers to those hypothetical threat scenarios to which the social domains can be exposed to a given moment. A well-trained national safety and security specialist represents a chance for the survival of civil society in a crisis situation. In this context, the authors express their conviction that a Laboratory of decision-making experiments that capitalizes on the experience of lessons identified in crisis situations is a natural response to adapt to the realities of development and the opportunity to use e-Learning and e-Training in the field of military sciences.

As a result of these findings, the authors believe that certain teaching-learning methods can be differentiated using a mix of classic and e-Education tools, customized for the field of military sciences, as follows:

- the teaching-learning method through physical presence

Even if the method seems a classic one, thanks to technological development, it can also benefit from an AI component. Moreover, the teaching-learning method through physical presence also needs the use of e-Training methods with a greater weight than e-Learning, but without minimizing it.

- online teaching-learning method

Online teaching has been the focus of the academic community, especially for the education of the future. Crisis situations made the academic environment aware of the importance of this way of teaching-learning, so this method expanded. During the pandemic, one of the main measures taken to continue the educational process was the transition of teaching activities from traditional to online. In this sense, e-Learning has become the best option during the pandemic for students to respect one of the essential measures of the pandemic, namely social distance (Şuşnea E, et al., 2022). The disadvantage of this method is that it cannot make the most of e-Training methods, considering the fact that in military sciences, there are few situations where training is done in the virtual environment. Also, collective training in the digital environment does not cover all performance standards of the military (Pătraşcu P, 2021). Thus, the transition to the online environment of several military education programs may have future repercussions on the career of the military.

- the mixed/combined teaching-learning method

This method appeared as a result of the need to adapt to real situations, teaching under normal conditions but with the use of modern e-Learning and e-Training tools, and the use of AI. Thus, pursuing quantifiable learning outcomes (Scipanov LV, Bodescu A, 2023) in the process of optimizing teaching-learning in military sciences, the mixed method combines all the classical teaching-learning tools with those identified in crisis situations, particularizing the field of e-Education in education military superior.

In conclusion, the authors believe that the participants in the teaching-learning process have learned to manage online platforms much better, they have adaptively contributed to the development of specific teaching methods, and they manage the classroom much better in physical or online education. As a result, networks and databases were developed, and laboratories were designed and used specifically for military education based on e-Education tools. The authors re-emphasize that today, the field of military sciences uses the mixed teaching-learning method much more, compared to the situations before the pandemic crisis.

Next, having a useful platform of identified lessons that constitute the starting point in the educational activities, the authors propose a way to capitalize on experience in the field through a concrete contribution to the development of the combined teaching-learning method.

3. RESULTS AND PROPOSALS FOR CAPITALIZING ON THE COMBINED TEACHING-LEARNING METHOD

The educational activity in the field of military sciences is a specific one (Scipanov LV, Chirea N, 2020) precisely because of the existence of unpredictable situations that generate crises, such as the one generated by the COVID-19 pandemic. The nature of national safety and security issues requires solving in the area of sciences that address the problem of danger, implicitly those of a military nature, due to the particularities manifested in all non-specific situations. The danger, the loss of human lives, the existence of considerable destruction of critical infrastructures, the temporary or permanent decommissioning of institutional facilities, the impossibility of resuming daily activities, etc., are close to the lessons identified following military confrontations.

Exposure to the threat, aggression, and direct engagement of a destructive factor such as an invisible enemy, a physical enemy specific to war situations, or an unidentifiable enemy such as a military special operation situation, calls for a high level of training achievable through an effective educational act adapted non-specific situations. The level of preparation of the population and the realization of educational activity in the area of risk management makes military sciences not only the necessary support for the administration of a crisis situation but even more, through specific methods such as operative art, it can be the educational landmark for the other social sciences.

Identifying and promoting the best solutions, methods, systems, and services for e-Education in the field of military sciences, beyond the real benefits for the general public, is primarily an educational obligation in the field of national safety and security. In the past, due to the loss of human lives and the production of material damage, the resolution of a crisis situation was directed towards the application of solutions resulting from the issue of military conflicts. The crisis situation has taught us that there are other military response solutions to events in the non-military spectrum.

The lessons identified in crisis situations, such as the pandemic situation, cover a wide spectrum of social problems that are mainly directed to the area of decision-makers. The decision-making area or the institutional decision-maker who administers a crisis situation is the center of gravity in solving problems. Operative art as a method of analysis specific to military actions refers mainly to what we call the comprehensive approach and defines the dimensions of the risk-generating problem and can exploit the documentary base of lessons identified during previous military or non-military actions. However, operative art is also the direct beneficiary of the evolution of the educational act, the analysis methods are based on the previously presented teaching-learning methods, using combinations of classical tools and from the e-Education spectrum, which we have customized for the field of military sciences.

The benefits resulting from the understanding and decomposition of the few realities that have been presented, make e-Learning and e-Training genuine tools applicable to the educational processes in the field of military sciences.

The factors contributing to the resilience of the educational field, as observed through the identified lessons, include the development and utilization of online platforms, the expansion of the documentary base, the rapid dissemination of knowledge, the identification, acquisition, adaptation, and development of teaching-learning methods, and the establishment of computer networks and the implementation of laboratory-based education. These factors all share a common requirement: the utilization of e-Education learning-teaching networks.

The authors' approach, focused on implementing the lessons learned during the non-military crisis of the pandemic, has resulted in the conception, design, and development of a tool known as the Decision Experimentation Laboratory. This tool aims to facilitate working with and validating decisions in crisis or military conflict situations.

Particularly for learning based on the laboratory experiment, the authors find that it involves the existence of three dimensions: the hard dimension, the soft dimension, and a human dimension, which together interconnected lead to the elaboration of what can be called the professional network of social training management for crisis situations. The peculiarity of the laboratory of decision-making experiments consists in taking the operative art to a higher level of knowledge and understanding of the phenomena that take place in the sphere of societal domains (government, people, military - Figure 1).

Starting from this triptych, government – people - military, the authors amend and augment the Relationship of Political Aim, War-Narrative, and Time in a Limited War (Thomas Kopsch, 2010), augmenting the previously validated system through an artifice of process updating in relation to the lessons identified as a result of the implementation of e-Education tools in higher military education.

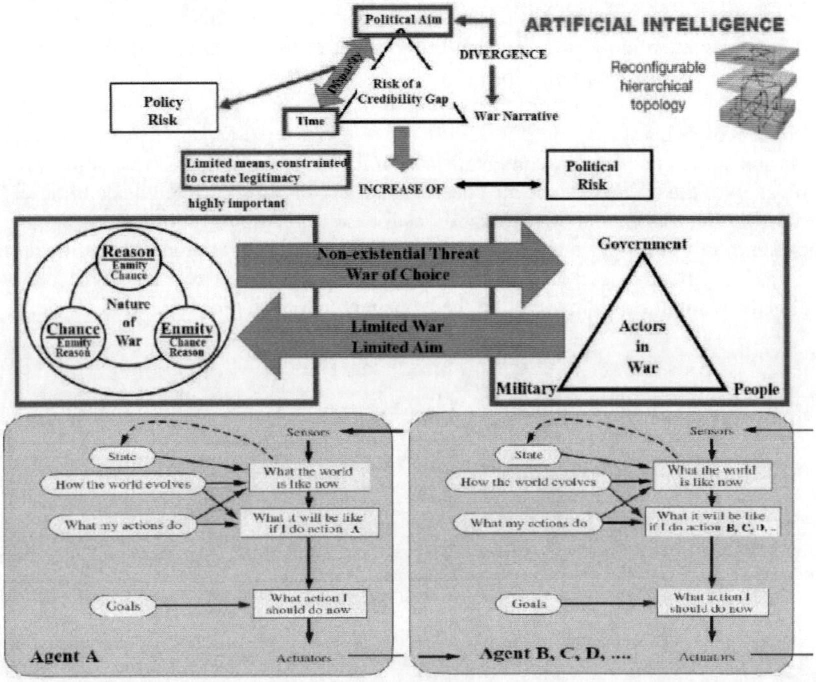

Figure 1. A new variant of the scheme for substantiating decisions by decomposing reality

Another consequence noted by the authors is the fact that the crisis situation generated social empathy. The decision-makers managed to maintain the support of the population and the military planners to be able to fulfill their functional duties. Capitalizing on this aspect in the projection of the laboratory of decision-making experiments, the resolution of crisis situations is currently done by keeping in contact with the decision-making factors - government - population (local administration) - the military component. Approaching society as a macro system of systems implies the transformation of knowledge about the nature of crisis or war situations, as well as a new understanding of the phenomena that take place in the sphere of societal domains. The algorithm of interactions between the Agents' development processes (A, B, C, D, E, ...) developed with the help of AI, is the basis for the development and foundation of new sets of decisions (Russell Stuart, et. al., 2010). Each Agent acts on the database and transposes himself in the analyzed situations through the prism of his own judgments and reasonings, respecting the same circuit of analysis. Later, the formulated solutions are taken into a mixer of hypothetical situations, a process supported with elements of artificial intelligence according to the model developed in Figure 1. Depending on the performance of the results, each Agent receives a new hierarchical position, following the classification of intervention priorities when making decisions. In this way, the consolidation of decisions is ensured based on the optimal resolution of the situations for which intervention was made in proportion to the nature of the evolution of the crisis situation or war/military conflict.

Decision-making experiments are based on threat scenarios that are generated based on historical events but also in the context of hypothetical situations generated by the evolution of technology and sciences, in general - of knowledge. But here is that today, knowledge adapted to technological development, generates an amendment of old theories, because the authors found that technology determined the development and use of e-Education tools in particular. For military sciences, e-Learning and e-Training components occupy an important place in the architecture of the laboratory of doctrinal experiments, along with artificial intelligence, which cannot be minimized in the analysis of risk factors.

An amendment to a theory promoted by the authors is brought to the inter-institutional information flow, which is supported by electronic platforms through which data and information are operated regarding the changes in the situations of the social fields involved directly or indirectly in assisting the solutions to the crisis. (The concept of inter-institutional workflow applied to education institutions might be a saving solution to the complex threats that modern society has to face nowadays.) (Roman Daniel, 2020)

The authors draw attention to the fact that the training of the military specialist for the decision-making component is based on an educational formative work algorithm, which is based on operative art, the main component of military art. Starting from the principles of military art, it can be considered that the educational algorithm consists of five stages:

- creation and activation of the documentary base of influencing factors (threat and aggression);
- identification of the single or multiple sources, respectively of the affected social field;
- identification and assessment of vulnerabilities for each entity/social domain;
- application of rules and measures/achievement of priorities, preparation of preliminary decisions;
- evaluating the effectiveness of decisions by tracking the values related to the performance of social domains, directly and indirectly, involved in the management of the crisis situation.

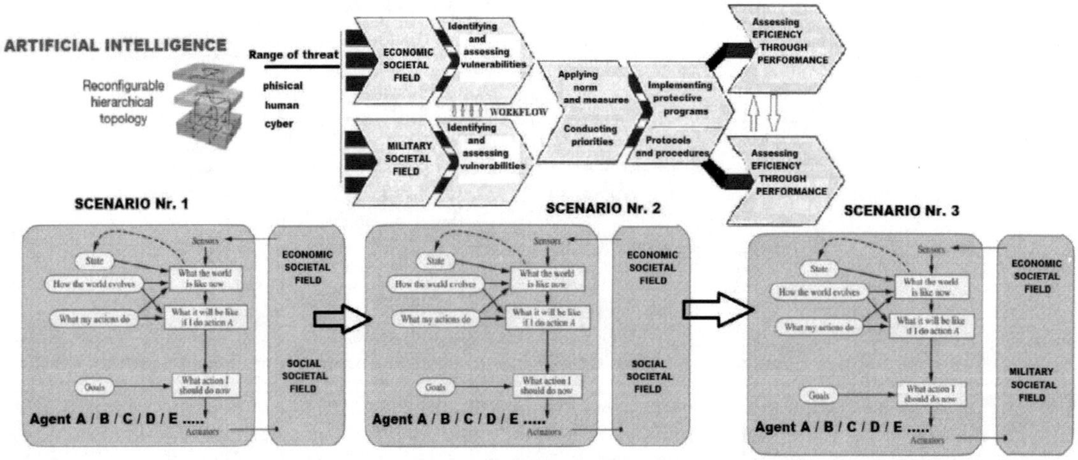

Figure 2. A new variant for the phased development of the workflow-scenario protection planning process of two societal fields

The resolution of a crisis situation or war/military conflict on the support of military art in an e-Learning format supported by elements of artificial intelligence consists of the implementation of at least three scenarios elaborated by the decision-making agents, (Figure 2). These three scenarios are linked by the same principle of collaborative work, referred to as the workflow-scenario process by the authors. (Roman Daniel, 2020). The authors mention that when developing each scenario, the same query matrix is kept with recording the values of the situation parameters. The transition from one situation S_i to another situation S_j in the collaborative work process represents the innovative element through which the redundant facts of the influence factors are filtered with the help of artificial intelligence. In this way, the elaboration of the decision is carried out by going through some hypothetical situations representing the future, or in other words, several possible futures are elaborated, from which one can not only choose but also build according to the model obtained in the Laboratory of decision-making experiments.

The design of the operation of the Laboratory of decision-making experiments according to the model of solving a crisis situation, elaborated in the workflow scenario, presents several identified advantages, as follows:

1. The work in the network of decision-making factors according to the workflow model (Roman Daniel, 2020) re-emphasizes several educational realities under the conditions of using eLearning and e-Training tools (common documentary base and high dissemination speed).
2. Another advantage is given by the digital translation of the values of the state parameters of the domains analyzed according to the dynamics of the crisis situation. Due to the real-time transfer of decisions between decision participants, a synergy of decisions can be achieved so that the decision-making team can avoid planning errors and a practical translation of formulated remedial solutions takes place. Another educational reality is re-emphasized, the use of AI in the decision-making process.
3. A technological advantage is provided by the speed of data processing. Another educational reality is re-emphasized, the need to develop and use online platforms and laboratory networks

For the balance of the evaluation and the transparency of the results regarding the operation of the laboratory, the authors also identified some weaknesses/disadvantages:

1. The conflicting situation of decisions; this disadvantage is mainly related to the cascading effects of the shock wave of the crisis. The conflictuality of decisions may arise due to causes such as non-compliance with the classification level of information, the existence of errors in the recording of reference data, technical errors, human errors, etc. Thus, the need to use artificial intelligence components in the decision-making process resulted. However, AI is one of the components of the decision-making experimentation laboratory that operates based on protocols that will be the subject of subsequent scientific articles. Only the educational process remains in the attention of this approach, mainly computer-assisted education. The authors believe that currently, decisions remain within the scope of human intelligence.
2. The uncertainty of the validation of the decisions resulting from the analysis through the use of e-Learning, e-Training, and AI tools. Although the answer is based on well-instrumented algorithms through technologies and AI, the decision remains to be applied through the intervention of the human factor. Moreover, from the authors' experience, a decisive response to solve a crisis situation does not present the certainty that it will lead to an execution exactly as planned.

The authors state that they are currently engaged in the development of technical support and the identification of optimal solutions for the practical implementation of the laboratory. Drawing from the model presented, it can be replicated, taking advantage of the principles derived from the lessons identified by the authors throughout their research.

4. CONCLUSION

As part of this effort, the authors created a documentary base of identified lessons, transposed from educational realities in the field of military sciences, as a result of the development and exploitation of e-Learning and e-Training tools, including the trend of using AI in the decision-making process:

- development and exploitation of online platforms,
- expansion of the volume of the documentary base,
- increasing the speed of information dissemination,
- identified, acquired, adapted, and developed teaching-learning methods,
- the development of computer networks and the realization of education based on a laboratory experiment.

Among the three identified teaching-learning methods—physical presence, online, and mixed/combined—the authors discovered that the mixed/combined method is increasingly employed in military sciences. The initial outcome of utilizing this method was the establishment of networks and databases. Consequently, there is a growing demand for the design and utilization of dedicated laboratories for military education utilizing e-Education tools. As a result, a potential future research direction could involve analyzing the various types of technological architectures suitable for equipping such laboratories.

The authors' conclusion is that education based on laboratory experiments encompasses three dimensions: the hard dimension, the soft dimension, and the human dimension. the human dimension, specifically in the analysis of crisis situations within the military sciences, requires a specialized tool known as operative art, which serves as a specific method of analysis for military actions. as such, the mixed teaching method effectively leverages the advantages of the five educational realities while mitigating certain drawbacks through the intervention of the human factor in the decision-making process. Without assigning a hierarchical value to the lessons derived from crisis situations in military sciences' teaching-learning process for training military specialists, the authors emphasize the significance of establishing documentary foundations by digitizing knowledge content. Within this context, the authors firmly believe that the decision experiment laboratory's multitude of advantages will contribute to the well-being and progress of modern education, which is centered around informatics, digitization, and artificial intelligence.

REFERENCES

Beck, K. and Ralph, J., 1994. Patterns Generates Architectures. *Proceedings of European Conference of Object-Oriented Programming.* Bologna, Italy, pp. 139-149.

Lehaci Niculai-Tudorel, Roman Daniel, Tuțuianu Diana, 2020, The covid 19 pandemic-a challenge for military education. The e-learning solution between necessity and efficiency, *EDULEARN20 Proceedings, IATED*, 2020, pp. 7361-7370.

Pătrașcu, Petrișor, 2021, Resilience and Critical Educational Valences during the Pandemic. In *International conference KNOWLEDGE-BASED ORGANIZATION*, Volume 27, No. 2, pp. 180-186.

Roman Daniel, 2020, The concept of workflow in the education system in the field of national security, *The 16th International Scientific Conference, eLearning and Software for Education,* Bucharest, April 23-24, 2020, Volume 1, DOI: 10.12753/2066-026X-20-032, p 253.

Russell Stuart, Norvig Peter, 2010, *Artificial intelligence a modern approach*, Pearson Publishing, Prentice Hall Press, One Lake Street Upper Saddle River, NJ, United States, ISBN-13: 978-0-13-604259-4, p. 54

Scipanov Lucian Valeria, Bodescu Alin, 2023, *Higher military education focused on quantifiable learning outcomes*, Bulletin of " CAROL I" National Defence University, 11 April, Bucharest, Romania, vol 12, no. 1, pp 132-142.

Scipanov Lucian Valeriu, Chirea Nicu, 2020, *Training, conceptual delimitations. adaptation of training to regional risks and threats*, Bulletin of " CAROL I" National Defence University, 20 October, Bucharest, Romania, vol 9, no. 3, pp 53-61.

Scipanov Lucian Valeriu, Scipanov Dănuța Mădălina, 2020, E-learning technologies, necessary but not sufficient in intelligence and military sciences, *Proceedings of the 15th International Conference on Virtual Learning*, October 31, University of Bucharest Models & Methodologies, Technologies, Software Solutions, e-Skills for the 21st Century, Bucharest, Romania, Volume 31, p.538, ISSN: 1844-8933-ISI Proceedings.

Scipanov Lucian Valeriu, Scipanov Dănuța Mădălina, 2020, Possibilities of choosing innovative E-learning technologies, *Proceedings of the 15th International Conference on Virtual Learning*, October 31, 2020, University of Bucharest Models & Methodologies, Technologies, Software Solutions, e-Skills for the 21st Century, p.531, ISSN: 1844-8933-ISI Proceedings.

Scipanov Lucian Valeriu, Stanciu Cristian-Octavian, Scipanov Dănuța Mădălina, 2021, Implications of the eEducation and eTraining trends on teaching-learning methods specific to military science, *Proceedings of the 17th International Scientific Conference "eLearning and Software for Education"*, Bucharest, April 22 - 23, 2021, Volume 1 | DOI: 10.12753/2066-026X-21-009 | Pages: 68-77.

Șușnea, E., Pătrașcu, P. and Buță, I.C., 2022, Students'learning challenges during the COVID-19 pandemic in Romania: analysis and improvement proposals. In *INTED2022 Proceedings*, IATED, pp. 6381-6390.

Thomas Kopsch, *Time and the Paradigm of Operational Art - Authority and Responsibility of the Operational Artist in the Political Military Discourse*, School of Advanced Military Studies United States Army Command and General Staff College Fort Leavenworth, Kansas 2017, p. 48.

ELECTRONIC VISUAL LEARNING AIDS FOR THE EFFECTIVE CLASS IN HIGHER EDUCATION TOWARDS METACOGNITIVE LEARNING

Dileta Jatautaite[1], Edita Butrime[2] and Vaiva Zuzevičiūtė[3]
[1]Vilnius University Business School, Saulėtekio av. 22, LT-10225, Vilnius, Lithuania
[2]Lithuanina University of Health Sciences, A. Mickevičiaus str. 9, 44307 Kaunas, Lithuania
[3]Mykolas Romeris University, Public Security Academy, Maironio st. 7, Kaunas; Ateities st. 20, Vilnius, Lithuania

ABSTRACT

Notwithstanding the discussions – valid! -about the necessity to preserve multilingualism and thus facilitate and cherish multifaceted cultural diversity, however, the English language has firmly established itself as a lingua franca globally. Consequently, employment arena requires employees to achieve high standards of English communication skills, knowledge and competences. The achievement of the level may be facilitated by an abundance of teaching and learning tools, aids at the disposal of a contemporary teacher in higher education. One of the tools in the tool kit of a teacher in higher education right after the pandemic is the wider spread usage of the electronic visual aids, which were incorporated into everyday teaching practices during COVID-19 pandemic out of necessity, however, may have proved themselves being useful further on. Visual aids, as pedagogy, due to the research and theories, offered, by esteemed thinkers in the field (Gardner, 1983) stated, were always considered as useful for the quality of teaching and learning of students in the classroom. However, once COVID-19 struck, even the teachers who had been reluctant previously to use the el. tools (including el. visual aids) previously, had been pushed to do exactly that. Many universities have managed to use the COVID-19 pandemic as an opportunity for modernization foreign language programs, curriculum content and helped modernize the way of work and study for the better. Implementing new technologies such as el. visual aids into the language learning and teaching environment, specifically for a new generation of students changed the landscape of pedagogy forever. According to the analysis of the findings from the empiric study, the results of which are presented in the paper, using el. visual aids may have certain positive effect on certain aspects of learning.

KEYWORDS

Electronic Visual Aids, Communicative Competencies, Foreign Language for Specific Purpose

1. INTRODUCTION

Contemporary globalized and technologized world is evolving fast, it takes requires energy and attention to stay attuned to the process. We are forced to adapt and keep with this pace otherwise we will be forced to become merely observers of the rich life in all walks of life: employment, education, even social services and leisure activities. Time and place of residence have taken on a completely different meaning due to new technologies, for example, it does not matter any longer where your workplace is or what nationality you are; you can work efficiently and effectively remotely using the latest electronic tools and means from any corner of the world and any time zone. Due to these circumstances, there is no denying the way the students study languages has changed, but in many ways the change was additionally prompted by the pandemic. Many universities, and that certainly stands true for the Lithuanian universities, have managed to use the COVID-19 pandemic as an opportunity for change and helped change the way of work and study. This shift in studying, including languages, languages has encouraged education to implement advance electronic tools usage practices in educational methodology. Interestingly, there had been colleagues before pandemic, who had been savvy in using the tools before 2020, however, also native and foreign language teachers were forced into a crash course on modern technology. As a result: more than ever students and teachers have the skills and knowledge to work more effectively and successfully with new technologies.

Thus, the **aim** of the article is to analyze the implementation of electronic visual aids facilitating the foreign language teaching and learning.

The **object** of the research – usage of electronic visual aids for teaching and learning for foreign language in a contemporary higher education (further on - HE) class.

The **objectives** of the research are based on three questions such as: What is the possible educational value of using electronic visual aids for teaching and learning a foreign language, based on the theoretical and empiric analysis? What are the possible educational gains/added value of using electronic visual aids for teaching and learning a foreign language, based on the theoretical and empiric analysis?

Research methods: critical analysis of scientific literature, empiric study, based on the quantitative research methodology in social sciences (questionnaire).

2. CONTEMPORARY CLASS IN HIGHER EDUCATION AND THE EL.VISUAL AIDS: OVERVIEW OF SOURCES

Many language teachers and students would never have had the opportunity or reason to become so clued up on modern technologies if it had not been for pandemic. Due to it, during at least a year and in some instances, longer, many language courses were provided entirely online by universities. It turned out to be popular among tech-savvy generation students. It means that if after the pandemic modern universities return to using only traditional teaching strategies with the same set of teaching materials, they would definitely side track the contemporary students' expectations and general ambience. In other words, today language teaching and learning paradigm has shifted, probably, forever. During this unprecedented pandemic time, students and teachers have embraced new skills and developed new behaviors which have improved the way they operate and process in the classroom, which, we think, is the most significant issue in this context. However, there is still a problem of attaining good foreign language skills, knowledge, and competencies that needs to be addressed. In addition, there is a demand to learn the language as quickly as possible with a good language command otherwise you will turn to an unlucky wretch: no job in any multinational business company, or work as a lawyer or an officer. The world has shrunk so much that it is no longer enough to speak one language. English has become a lingua franca in the Global world (the term lingua franca is defined as a language systematically used to make communication possible between people without the use of their first language). Consequently, employers require from their employees high standards of English communication skills, knowledge and competencies. In terms of career prospects speaking another language has become an undeniable plus, o rather a 'must' on the resume. Several studies show a direct correlation between the importance of the first salary and the ability to speak a foreign language.

However there are other voices, for example, the Equal Employment Opportunity Commission (EEOC) "views rules requiring employees to speak English at all times at work as a burdensome condition of employment" (shrm.org). We totally agree with the standing, by the organization: the EEOC asserts the importance of restricting employees' ability to speak their native language though it can cause the "atmosphere of inferiority, isolation, and intimidation based on national origin" (shrm.org). It means that in an age of borderless communication and travel, it seems we are almost pushed to use the aforementioned lingua franca, which may have – and does! -have a detrimental effect of individual's right to communicate in their national language, foster their national identity. Though, besides the English there are many other languages that belong to a group of foreign languages for a specific purpose, for example, German, French, Spanish and many others that are not the native language of a learner. Notwithstanding our general concern for linguistic and cultural diversity, as teachers in higher education we have our professional duties. The duties include both to teach the foreign language and to do that in a professional way, attuned to students' expectations and general ambience. Sometimes (actually, quite often) we therefore need an assistance from our colleagues in IT departments or the colleagues who have the IT competence due to other reasons, such as general interest in technologies for education. Generally, cooperation between teachers and professionals in other fields is essential for qualitative educational services in a contemporary higher education (Zuzeviciute and Butrime, 2016).

3. TEACHING AND LEARNING FOREIGN LANGUAGE (ENGLISH) FOR SPECIFIC PURPOSES: HISTORIC OVERVIEW AND THE CONTEXT

The second foreign language term was introduced several decades ago; Klein, for example, indicates: "foreign language' is used to denote a language acquired in a milieu where it is normally not in use" (1986: 19). However, English is the most studied foreign language in the world, with 1.5 billion foreign language learners.

Using new technologies such as el. visual aids into the language learning and teaching environment, in particular for Generation Z and Millennials is relatively easy, as they are technology savvy (Jatautaite and Kazimianec, 2019). As the Pew Research Center put it: cutoff point for the end of the Millennial generation is behind us. Millennials (according to the organization) are those who are born between 1981 and 1996. Millennials are described as the ones who are essentially used to taking the advantage of social media and consistently using connection to the internet. Hence, the people who are born in the late 90s cannot imagine their lives without the use of electronic tools. Pew states that the generation that comes after Millennials is named as the post-Millennial generation or Generation Z (Loria, 2018).

The importance of usage of visual aids was emphasised since the beginning of the professional pedagogy era (e.g., see: ideas of 'pictures' by J.A. Comenius in 17th century, here from Gedvilienė, Zuzevičiūtė, 2007). However, the most influential theory of recent times was suggested by Gardner (1983), where the idea of multiple intelligences was presented with the accompanying idea of the need to address learning by emphasis on verbal, audio, visual, musical, interpersonal, intrapersonal and other aspects of human thinking, and thus learning. The el.visual aids (almost always incorporating also verbal, audio, other aspects) expand further the available assortment for teachers (and learners). There are many studies on the effect of using visual aids based on using electronic devices (Dalali and Mwila, 2022; Lopez, 2022; Martiniello et. al., 2022). Lopez explains visual aids as items "of illustrative matter, including, but not limited to a film, slide, graph, chart, or model, designed to supplement written or spoken information so that it can be understood more easily" (2022). Waite (2006) stated that electronic devices could facilitate the students' participation in different social and cultural environments and to improve communicative competencies by using authentic, contextual, and cultural materials. According to Jatautaite and Kazimianec (2019) technology assisted visual aids in learning and teaching curriculum may better engage learners in communicative activities in a much more effective way. Generally, visual aids help to better focus attention on the content to be learned and remember the concepts they have just learned, providing associative learning by linking the concept to the visual representation (Stancampiano, 2013). Moreover, visuals aids benefit the learners better store verbal information in the brain than merely verbal communication (Antonova et. al., 2019). Besides, while using visual aids teachers have additional opportunities to engage learners and help to clarify difficult concepts (Nitu et. al., 2017). Quecan (2021) states that using visual aids may also enhance the motivation of language learners. Ahmed (2018) adds: visual aids increase students' involvement and interaction, students become more involved in language learning process and they are more attentive to process, which leads to increased understanding. Today visual tools comprise multimedia pedagogical tools, which range from simple static characters that provide information to complex and dynamic animated three-dimensional characters that narrate information while gesturing (Castro-Alonso et.al., 2021). Visual (electronic) aids are available in many forms: such as pictures, photographs, drawings, flashcards, wall charts, pictures stories, maps, and movies – even readily accessible objects brought to the classroom ranging from "a bottle opener to small toys" (Allen, 1983; quoted in Atas, 2019). What concerns the research of this article, it deals mainly with visual aids, which may be incorporated into teaching and learning either in a form of or: "electronic textbooks and materials, smartphones, tablets and internet extensions, computers, smartboards, TVs and projectors" (Ünal, 2022). This explanation of visual aids best suits the research problem due to the advancements in technology since the number, assortment, inventory, format, also media of visual aids for education have increased and diversified.

Dudley-Evans and St. John (1998) provide a comprehensive description of the foreign language for specific purpose as language teaching designed to meet the specific needs of the learners through employing effective teaching methodologies and teaching activities. This concept is based on the focusing of personal language needs on specific goals. In this respect, the teaching of a specific purpose language, unlike the teaching of a general language, focuses on a specific area, the boundaries of which are determined by the needs of the learner. Moreover, this concept is based on the importance of being aware of what and how to teach it. In this regard, objectives, content and teaching context are determined by focusing on the specific goals of the learner, as it was identified several decades ago in studies, which provided grounds for a generation of teachers and learners by now (Hutchinson and Waters, 1987: 18-19). For example, teaching language for specific purpose to business management students is mainly focused on developing writing skills, to the military is centered on speaking and listening skills or basically communicative approach, teaching language for future law enforcement officers encompasses both formal register (for academic, professional settings) and in-formal register (for communication with citizens and especially residents whose English may be either rudimentary or street-wise based). Moreover, in terms of teaching law enforcement students, it depends upon the purpose of the course whether they need written and verbal communication skills. That is, some of the students later in their professional life will work in court, prosecutor's office and may be will even participate in legislature (even internationally), or at least organize and implement operations internationally; they will use formal register, surely, written communication will comprise a huge part of their work. Some other students will directly work 'in the field' of law enforcement, which will inevitable include using the informal register, primarily, verbal communication. On the written side, law enforcement officers must be able to communicate the details, for example, describing findings in a crime scene or report clearly and thoroughly in participating in events, activities of international and local organizations. On the verbal side, they must communicate positively and with sensitivity to diversity (nationality and culture of people and populations they work with). Thus, the foreign language for specific purpose focuses mainly on the use of language (Farmati et. al., 2022). However, we, the authors of this paper will argue that also teaching grammar and language structures are very important if a possible career in legislature (especially, internationally) is kept in mind. Certain (advanced) level of command of professional foreign language is an absolute necessity in those cases. This is especially valid in a contemporary situation for an increased need for international cooperation against the direct threats to national and international security due to the war against the Ukraine.

Current state of methodological educational approaches in synergy with rapid developments in business, military, and law enforcement poses certain expectations (advanced!) for a second foreign language competencies (in our case: English for Specific Purpose).

In summary, language teaching for specific purpose is defined as an approach that aims to meet the academic and vocational/professional needs of learners and is student-centered, multi-disciplinary and an effective combination of theory and practice (Anthony, 2018).

When the definition is so ambitious, teachers and learners must have wide opportunities to use specifically technology supported visual aids in communicative language learning and teaching process. Traditional approaches focus on learning the language without too much specific focus. According to the communicative approach, the use of language as a means of communication is more important than its rules. Therefore, communicative competence is one of the main aims in communicative approach. The communicative competence of teaching and learning the foreign language for specific purpose is alignment to reality. Grammar is important but in communicative approach the learner should also use communication strategies and be aware of the features of discourse and sociolinguistic rules of interactions (Savignon, 2017), Mart (2017) claims that rules of both language use and grammar have an important role in gaining communicative competence. Communicative competence has four main components: linguistic competence, sociolinguistic competence, discourse competence, strategic competence. Thus, while teaching and learning the foreign language for specific purpose "there should be a close link between the tasks performed by learners in the language classroom and in the outside world", which is authentic learning (Van den Branden, 2006; 6). Therefore, while, in teaching foreign language for specific purpose teachers incorporate technological (electronic) equipment or any digital device in designing curriculum, instruction (Garofalo and Farenga, 2021), they both facilitate (at least potentially) learning and at the same time attune the classroom experiences to the everyday life practices and routines of contemporary young people, the students.

The critical analyses of sources highlight the importance of using visual aids (of any kind and format), because of the multifaceted nature of human learning (as it was suggested several decades ago by H. Gardner. The IT assisted teaching, including using el. visual aids, which stormed into classes in HE during the last few years of the pandemic may have a positive role in enriching teaching and learning experiences. This is especially true, while learning foreign language for specific purposes, because of the possibilities for diverse assortment of 'windows' into realities of professions: e.g., analysis of news broadcasts on the current situation in markets or a particularly intriguing criminal case. These 'windows' allow teachers to start discussions on realities out there, to attune class and profession and thus to ensure relevance between the curriculum and the everyday routines of young people.

It seems logical here to share our position as professionals in contemporary HE, specifically, regarding the need to use el. visual aids in a responsible way. That is – even if ideas of esteemed and distinguished theoreticians in psychology and education are focused on a variety of channels, formats and media for presenting information, caution is advised. While learning is individual, and thus a variety of channels, formats and media for presenting of information is necessary in order to accommodate different individuals in class, however, learning also requires time for processing and reflection. That potentially may lead – and should- to building metacognitive learning practices among higher education students (Masoodi, 2020; Zuzeviciute, 2011).

We want our students to be reflective about their own learning, in order to equip them to be reflective in their professions, and to participate in life long learning in meaningful and responsible, also rewarding way. Therefore it is very important both to use el. visual aids and also to refrain from using them too much, too often. There should be a balance between input and reflection, and the balance between time to consume information, and the time to reflect on it; the time to build structures and the time to fill in those structures with additional materials.

Thus, teachers in contemporary HE class should not overplay they hand in using el.visual aids, even if the benefits are evident: aids help to link materials in textbooks to everyday realities, to motivate students, to break up routines in class. And yet, for the in-depth learning, also some routine is necessary, as it is necessary to have some time to reflect on materials, to discuss materials externally and internally. Metacognitive learning needs to be fostered together with communicative competencies.

Our professional practice and experience enables us to share that in order to build an engaging yet metacognitive learning orientated class, there should be limits in using el. visual aids, our suggestion being to allocate up to 5-7% of time for actual input with el. visual aid; provided it is logical to use aids in general logics of a given class. Surely, the analysis, reflections follow, thus the total amount of time may reach up to 10-12% of time in a given class.

Moreover, not every class and not every theme needs to be illustrated in that particular format. Variety, flexibility and routine need to be balanced in order to foster meaningful learning, which encompasses retaining information and then, preferably, building metacognitive learning practices.

4. SOME FINDINGS OF AN EMPIRIC STUDY

Once it was established that el. visual aids seem to have support of researchers, esteemed scholars, two issues become evident.

Firstly, language teachers, notwithstanding the 'fight or perish' experiences during pandemic, still in many cases need support of IT department colleagues or other colleagues who have gained competencies in the field because of other reasons (for example, they had participated in projects in the field, or just had been interested in IT assisted teaching) (Zuzeviciute, 2021). The support teachers provided for each other was invaluable, however, the discussion along these lines exceeds the scope of this paper.

Secondly, while there seems to be a unanimous agreement among scholars regarding the effectiveness and important role of el. visual aids in educational settings, it was decided to analyse whether Lithuanian students share the same or similar enthusiasm regarding potential benefits. Therefore, three phased pilot empiric study was designed and implemented; here we are presenting data from the second phase, while the results of the first phase are presented elsewhere: Jatautaite, Zuzeviciute, 2022, paper accepted, pending publication (Proceedings of ALTA'22. Advanced learning technologies and applications. Digital Competencies). This phase of the study aimed at examining students' perspective on foreign language learning for specific purpose using el. visual aids with a focus whether aids may facilitate learning the language for specific purpose, whether students think that using aids motivates them to learn faster and in a more effective way. The students were invited to share their perspective in an anonymous questionnaire, with several questions dedicated to the profile of respondents and 16 questions dedicated to the usage of el. visual aids in class. Students from two Lithuanian universities were invited to share their perspective. In the sample (N= 91, N from U1=51, N from U2=40), men and women were represented almost equally (43 men, 48 women) in the sample, they were undergraduate students (18-23 years old, thus, members of Generation Z). The study was carried out during the last quarter of 2022 and the first quarter of 2023; all of the respondents studied English as a foreign language for specific purposes: Business English and English for law enforcement. Due to the limitation for the paper, certain findings are presented rather than all of them.

Findings indicate that students seem to be reflective, they notice that the el. visual aids (which in a majority of cases include audio, other aspects, further enriching learning) are being used in the HE classrooms, 95 percent of students indicate that to be the case. Regarding the benefits, the perspective of students seems to indicate as a main benefit the positive effect on developing competence of comprehension via listening; professional vocabulary was ranked second. Positive impact on developing speaking skills was ranked third, and positive impact on readings skills was ranked fourth; the positive impact on writing skills was ranked as the last one (Fig. 1).

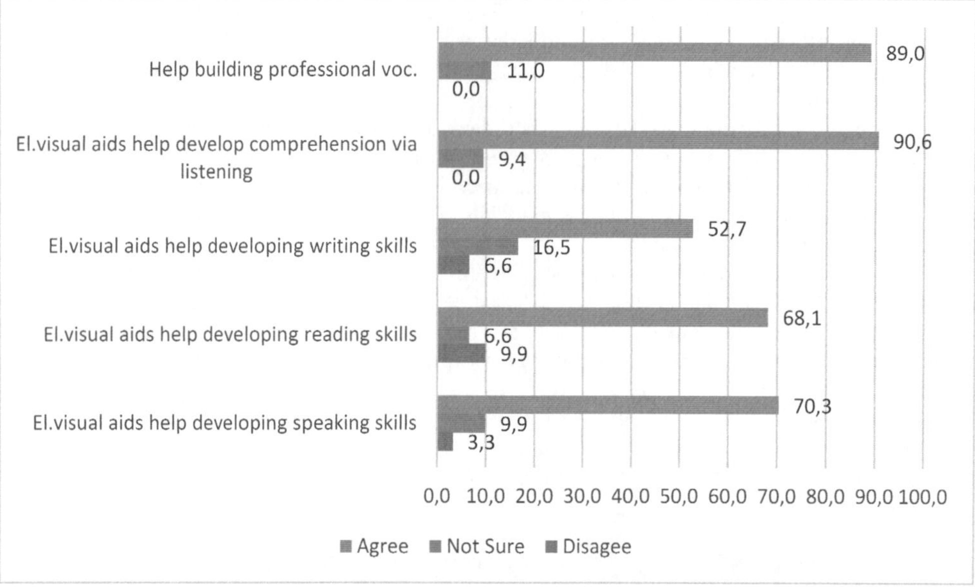

Figure 1. The benefits of using el. visual aids in contemporary HE classroom

In general, almost all respondents indicate the el. visual, which in most cases (e.g., clip from News, such as BBC, CNN or similar) will encompass verbal communication, musical inserts, illustrations (e.g., footage of the event of a crime scene), the explanatory text; that is, contemporary el. visual aids provide a multitude of learning channels for our students. Thus, not surprisingly, almost everyone (97%) indicates a general consent that el. visual aids help developing communication skills.

Regarding other aspects, emphasized by scholars, such as motivation, joy of learning, the latter was noted by more students than the former; but only just above the half of respondents thought el. visual aids being helpful for retaining information, especially, the terminology (Fig.2).

The findings about students' relative hesitance regarding the el. visual aids being useful for retaining information, and for making learning easier are aligned to our own observations regarding the limits of educational value of using el. visual aids. It seems that even our students note that the studies in HE inevitable include discipline and hard work, though they indicate observing that rich assortment of tools, modes, forms and for enriching pedagogical practices is used. While visual aids help break up routine in class and to link textbook materials to realities, but learning in always linked to discipline, internal and external reflection and effort. While further discussing the findings in the light of theoretical considerations and in the context of our own professional experience, also the findings of our own previous studies we may posit with a considerable degree of confidence that even if students do not indicate certain tool or method being useful, it does not negate the usefulness per se. Certain reluctance to indicate benefits may be explained by the cat that students are just concentrated on other issues: succeeding in a particular course-unit, covering curriculum. It is only after a specifically organized reflection that they start indicating certain benefits (or weaknesses).

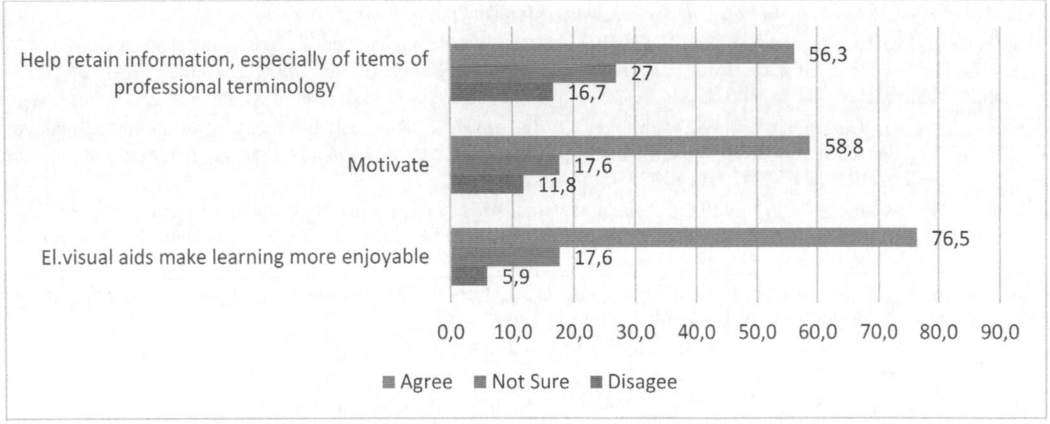

Figure 2. Other benefits of using el.visual aids in HE class

While our main goal is to support students' academic growth, we, teachers, do not always have time or duty to organize such in-depth discussions; moreover, not all students in all cases need that. They have their own goals specific to their curriculum, and, if they are not covering pedagogy course, the discussion is not even needed; if they observe benefits and indicate them, we should be content, and for more subtle analysis turn to other methods of research.

5. CONCLUSION

As a response to globalization and Covid-19 pandemic, language teachers were forced to reconsider the ways teaching and learning of the foreign language for specific purpose. They had to organize teaching in order to promote appropriate pedagogical flexibility, creativity in classrooms through communicative approach with the distance education as a looming reality. Using visual aids (of any kind and format), because of the multifaceted nature of human learning is important (as it was suggested several decades ago by H. Gardner). The IT assisted teaching, including using el. visual aids may have a positive role in enriching teaching and learning experiences. This is especially true, while learning foreign language for specific purposes, because of the possibilities for diverse assortment of 'windows' into realities of professions. These 'windows' allow teachers to start discussions on realities out there in professional context, to attune class and profession and thus to ensure relevance between the curriculum and the everyday routines of young people. Teachers in contemporary HE class should be reflective in using el.visual aids, even if the benefits are evident. While el.visual aids help to link materials in textbooks to everyday realities, to motivate students, to break up routines in class, yet, for the in-depth learning, also some routine is necessary. Students need to have some time to reflect on materials, to discuss materials externally and internally. Metacognitive learning needs to be fostered together with communicative competencies. Not every class and not every theme needs to be illustrated by el.visual aids. Variety, flexibility and routine need to be balanced in order to foster meaningful learning, which encompasses retaining information and then, preferably, building metacognitive learning practices. Empiric study revealed that contemporary students note that HE class is enriched by using el. visual aids, they note benefits of using aids for building professional vocabulary, development of communication competencies, they agree that using aids made the class more enjoyable. The findings of our study show that some of aspects outlined by scholars resonate among Lithuanian students, but some other - less so, thus further study in the field may be warranted.

REFERENCES

Ahmed, A. O. A. (2018). EFL Teachers' and Students' Approaches in Using Teaching Aids: A case Study. *Arab World English Journal (AWEJ)* Special Issue on CALL 4, pp.109 - 124.

Allen, V. F. (1983). *Techniques in vocabulary teaching.* Oxford: Oxford University Press.

Antonova, I., Snezhko, Z. and Anohina, Y. (2019). The Way We Teach Medical French Using Authentic Audio-Visual Aids for Exchange of Experience. *Journal for the Education of Gifted Young Scientists*, 7(4), pp. 953-965. https://doi.org/10.17478/jegys.621922.

Castro-Alonso, J. C., Wong, R. M., Adesope, O. O. et al. (2021). Effectiveness of multimedia pedagogical agents predicted by diverse theories: a meta-analysis. *Educational Psychology Review.* https://doi-org.am.enformation.ro/10.1007/s10648-020-09587-1.

Dalali, B. G., and Mwila, D. P. M. (2022). Effects of Visual Aids in Enhancing Teaching and Learning Process in Public Secondary Schools in Ilemela Municipality, Tanzania. *Journal of Research Innovation and Implications in Education*, 6(1), pp. 379–390.

Dudley-Evans, T., St John, M. J., and Saint John, M. J. (1998). *Developments in English for specific purposes: A multi-disciplinary approach.* Cambridge university press.

Farmati, C., Mohamed, Y. E. O. U. and Benzehaf, B. (2022). English for Specific Purposes: A Holistic Review of Empirical Research. *International Arab Journal of English for Specific Purposes*, 5(1), pp. 116-131.

Gardner, H. 1983. *Frames of Mind: A Theory of Multiple Intelligences.* New York: Basic Books.

Garofalo, S. G., and Farenga, S. F. (2021). Cognition and spatial concept formation: Comparing non-digital and digital instruction using three-dimensional models in science. *Technology, Knowledge, and Learning*, 26(1), 231-241.

Gedviliene, G., and Zuzeviciute, V. (2007). *Edukologija.* Kaunas, VDU.

Hutchinson, T., and Waters, A. (1987). *English for specific purpose.* Cambridge University Press.

Jatautaite, D., and Kazimianec, J. (2021). The Importance of Developing Foreign Language Competences for Military Cadets. *Siuolaikines Visuomenes Ugdymo Veiksniai*, 6, pp. 249-273.

Klein, W. (1986). *Second Language Acquisition.* Cambridge: Cambridge University Press.

López, Y. (2022). *The Influence of Visual Aids in High School ELL and General Education Social Studies Classrooms* (Doctoral dissertation). Kean University.

Loria K. MAR 19, (2018), 20:30 IST https://www.businessinsider.in/.

Martiniello, N., Eisenbarth, W., Lehane, C., Johnson, A. and Wittich, W. 2022. Exploring the Use of Smartphones and Tablets Among People with Visual Impairments: Are Mainstream Devices Replacing the Use of Traditional Visual Aids? *Assistive Technology*, 34(1), 34-45, DOI: 10.1080/10400435.2019.1682084.

Masoodi. M. (2020). Metacognitive Awareness: A Qualitative Comparative Case. *Contemporary Research on*

Nitu, K. S. and Dahiya, H. (2017). Audiovisual aids: An essential tool for teaching. *International Journal of Nursing Education*, 9(1), 64-66. https://doi.org/10.5958/0974-9357.2017.00012.

Organization Management and Administration, 8 (2), pp. 58-72, DOI: org/10.33605/croma-022020-004

Quecan, L. (2021). *Visual Aids Make a Big Impact on ESL Students: A Guidebook for ESL Teachers.* Master's Projects and Capstones. 1157. https://repository.usfca.edu/capstone/1157.

Savignon, S. J. (2017). Communicative competence. In John I. Liontas (ed.), *The TESOL Encyclopedia of English Language Teaching*, 1–7. Hoboken, NJ: Wiley.10.1002/9781118784235.eelt0047.

Stancampiano, J. (2013, March 1). Benefits of Visual Facilitation in the classroom. Retrieved from https://www.trainingindustry.com/contentdevelopment/articles/the-benefits-of-visual-facilitation-in-the-classroom.as.

Ünal, K. (2022). A study on the effect of visual and auditory tools in foreign language teaching. *Journal of Research in Social Sciences and Language*, 2(2),108-117. https://doi.org/10.20375/0000-000f-7453-1.

Van den Branden, K. (2006). Introduction: Task-based language teaching in a nutshell. In K. V. d. Branden (Ed.), *Task-based language education from theory to practice* (pp. 1-16). Cambridge, UK: Cambridge University Press.

Zuzevičiūtė, V. (2011). Learning at university: challenges, strategies, perspectives for lifelong learning. Saarbrücken: Lambert academic publishing, ISBN 9783846515501

Zuzeviciute, V. (2021). Memoirs of a higher education teacher, who survived two years of teaching during the Covid pandemic // *ALTA'21: Advanced Learning Technologies and Applications = Pazangios mokymosi technologijos ir aplikacijos. From Distance to Hybrid Learning: learning practices on/for time: conference proceedings*, 1st of December, 2021 / edited by D. Rutkauskiene. Kaunas: Kaunas University of Technology. ISSN2335-2140. 2021, pp. 145-153.

Zuzeviciute, V., and Butrime, E. (2016). Contemporary university and virtual learning environments - pathway towards effectiveness or alienation? // *ICERI 2016 Proceedings: 9th annual international conference of education, research and innovation*: Seville, Spain. 14-16 November, 2016. Valencia: IATED, 2016. ISBN 9788461758951. ISSN 2340-1095. p. 8839-8846. DOI:10.21125/iceri.2016.0990.

TYPES OF AFFORDANCES AND THEIR ROLE IN QUANTITATIVE UX/UI RESEARCH. A SCOPING REVIEW OF AFFORDANCES-FOCUSED LITERATURE

Ionut-Dorin Stanciu[1], Oana Teodora Cramariuc[2] and Irina Georgiana Mocanu[3]
[1]The Technical University of Cluj-Napoca, Department of Psychology and Pedagogy
15 Daicoviciu Street, Cluj-Napoca, Cluj, Romania.
[2]The IT Center for Information and Technology
Av. Radu Beller, Bucharest, Romania.
[3]Politehnica University, Department of Computer Sciences
313 Splaiul Independentei, Bucharest, Romania

ABSTRACT

Human-computer interaction is one of the most prolific fields of the last decades, due to both the rapid technological developments and the increased mass availability of these developments. HCI is also a highly interdisciplinary field, which brings together researchers from engineering, computer sciences, psychology, sociology, and even fields like medicine, education, and arts and humanities, and makes use of a huge variety of quantitative, qualitative, and mixed research methods. One of the key constructs in technological design and in HCI is that of "affordances", introduced by the American psychologist James Gibson in the late '70s - early '80s. This construct underpins the design and development of almost all technological devices intended for human use and is central to what is nowadays known as user-centered design. However, it is exactly this complexity and interdisciplinary nature of the field that allows for a vast and sometimes worrying inconsistency in the guiding theories and research methods. Against this backdrop, and in alignment with the increased calls for reproducibility and clear research reporting, we performed a scoping review of quantitative primary research studies in UX/UI that focused on 'affordances' use and/or analysis. Our findings show a scarcity of studies that are thorough in reporting important information, such as clearly mentioning the research design, analysis methods used, measurements, procedure, and sample. Furthermore, another worrying aspect, acknowledged in previous studies, is the significant overlapping of the concept of "affordances" with that of (system/device) functionality. We hope that our review will be helpful for researchers new to the field or for those interested in searching for papers that place 'affordances' at their core.

KEYWORDS

Affordances, User Experience, User Interaction, Quantitative Research, Research Designs and Methods

1. INTRODUCTION

Our daily lives are being increasingly impacted by the tremendous development of technologies. The pervasiveness of technological devices makes it that most of us interact on a daily basis with a huge variety of electric and electronic devices, and rely significantly on them to function and interact in society, at work, or in the most personal and intimate areas of our lives. Not surprisingly, the design of these technologies influences how the user interacts with them, which has significant and substantial consequences, both for the individual and for the technology developer/manufacturer.

2. THEORETICAL BACKGROUND AND RATIONALE

One of the key concepts in technological design is that of 'affordances'. The term was introduced by the American psychologist James J. Gibson, to explain how organisms perceive the surrounding environment and the objects that populate it, how the associated meaning is built, and how action is taken based upon this perception and meaning. In Gibson's original words, "The affordances of the environment are what it offers the animal, what it provides or furnishes, either for good or ill" (Gibson, 1979, p. 127).

Attesting to its importance is that, albeit originating from psychology, the concept of affordances has found uses in other areas of science or professional practices. Beyond its origin grounds in ecological psychology, it can be encountered in more related sciences, like pedagogy (Chow et al., 2015; Meinel & Krohn, 2022), arts and humanities (Gieseking et al., 2014; Scarlett & Zeilinger, 2019), as well as in more distal fields, such as music (Clarke & Clarke, 2011), or political sciences (J. L. Davis, 2020). However, regardless of the field or discipline in which it is used, it almost always signals an interdisciplinary approach, like is the case, for instance, for neurophysiology's applications in, and implications for art and design (A. Chatterjee & Cardillo, 2022).

Concomitantly with its widespread and transdisciplinary adoption, we also witness a need for increased precision, substantial and consistent empirical support, and direct and unambiguous application (i.e., operationalization), especially in fields such as engineering and computer sciences. While visionary, Gibson's concept of 'affordances', as it was introduced, was nevertheless in its infancy and required more refinement, at least with respect to people's interaction with technology (see, for instance, Oliver, 2005, for an open call for clarity and appropriateness). Gibson, himself, most likely, had no way to foresee how impactful his 'affordances' hypothesis would become. Moreover, Gibson's perspective allowed for ambiguity in the meaning of affordances; "an affordance is neither an objective property nor a subjective property; or it is both if you like" (Gibson, 1979, p. 129).

Therefore, unsurprisingly, technological design was not only among the fields that most readily embraced the concept of affordances but it was also the field in which it found the most clearly defined applications (Khosrow-Pour D.B.A. & Bhattacharyya, 2021; Ling, 2004; Nardi, 1996; Sears & Jacko, 2007). With specific regard to human-computer interaction and the subsequent technological design of information processing devices, Donald (Don) Arthur Norman, is perhaps the most prominent researcher and leading figure. His work furthered the meaning of affordances and opened new avenues for research and practice, in UX and UI in general, and especially in what is known as 'user-centered design'. Specifically, while Gibson's conceptualization of affordances disregarded the human agent's role in determining if something is or is not an affordance (or, more precisely, limited it to identifying the affordances, i.e., the Gibsonian affordance is independent of the user, it either exists or not and the user either acknowledges and access it or not), Norman emphasizes the central and crucial role of the user's own interpretation in deciding if something is an affordance or not (Donald A. Norman, 1988; Donald A. Norman, 2013).

In technological design, the above-mentioned ambiguity in the Gibsonian acceptance of the meaning of affordances needed to be dealt with, since, arguably, it is realistically and objectively impossible for a technology developer to faithfully observe the 'objective-subjective' dualism of Gibsonian affordances straightforwardly. Don Norman's conceptualization elegantly and effectively addressed this issue, by including the human agent and agency in the building of the meaning of affordances. In Norman's conceptualization, the user's mental processing and models (which, in certain cases, can be pre-existing and/or culturally embedded/loaded) play a role in identifying the 'potential' affordances and in determining their status. Furthermore, the user also evaluates the ease of use of said affordances which ultimately decide the propensity for action (or 'actionability', for short). Finally, Norman's view allows for the user to even conclude (more or less as a 'stretch of imagination') that an affordance exists, even though, objectively, as a physical object, it does not.

The concept of affordances was significantly developed and refined after its introduction by Gibson. However, it is beyond the purpose of this paper to present a thorough and comprehensive review of these developments or of their applications in specific domains. In this respect, we warmly direct the reader to several excellent and thorough reviews, such as Chong and Proctor's (2020) On the Evolution of a Radical Concept: Affordances According to Gibson and Their Subsequent Use and Development, or Tabuenca et al.'s (2021) Affordances and Core Functions of Smart Learning Environments: A Systematic Literature Review on the characteristics of desirable affordances for smart learning environments, Ronzhyn et al.'s (2022) Defining affordances in social media research: A literature review, Xue and Churchill's (2019) A review of empirical studies of affordances and development of a framework for educational adoption of mobile social media, or Bahari's (2022) Affordances and challenges of technology-assisted language learning for motivation: A systematic review. These and other valuable reviews and meta-analyses did excellent scientific work in explaining the evolution of the conceptualization of *affordances* and their consequent use in technological design.

Nevertheless, while the theoretical bases of 'affordances' are relatively well covered in the literature, we could not find thorough and comprehensive reviews regarding the research methods and the methodologies used in affordance research. This, in itself, is not surprising, considering the highly interdisciplinary nature of the field and its rapid evolution, but it is a shortcoming when considering the calls for reproducible research. Therefore, against the above backdrop, we performed a literature review with a specific focus on quantitative research that was explicitly interested in affordances, and operationalized and modelled them as part of their analyses.

3. LITERATURE REVIEW

The databases covered by our search were Elsevier's Science Direct and Scopus, Springer Link, Emerald Journals, Wiley Journals, ProQuest, Taylor and Francis, Sage, Nature Journals, and Google Scholar* (with the notable mention that the latter was rather a starting point since it isn't a fully-fledged academic database and academic articles indicated by it were, in virtually all instances, indexed in one or several of the other, more established academic databases). This initial search yielded 3821 results, ranging from 1985 to 2023 (we did not impose a year limit), which, after filtering out the duplicated files and keeping only journal articles and conference papers, went down to 3465.

The second filtering implied searching for publications' texts for "quantitative" research and for "user experience (UX)" or "user interface (UI)". Specifically, in order to keep only studies reporting quantitative research, we used the Boolean phrase ("scale" OR "questionnaire" OR "inventory") to search the articles' main text, which resulted in 79 papers, that were manually processed for the information of interest. The aggregated Boolean query was "("scale" OR "questionnaire" OR "inventory") AND ("affordance") AND ("user experience OR UX" OR "user interface OR UI")". This second step kept a number of 1819 publications.

Our 3rd and final selection kept only papers that mentioned specifically the keyword "affordances" in their research questions (we filtered the text using regular expressions) and in which at least one of the constructs used in the model referred directly to affordances. Ultimately, a selection of 79 was kept to be individually processed for information extraction (including manually).

We excluded papers that used machine learning approaches (while promising in terms of delivering insights on users' behavior and predicting decisions, they do not derive meaning directly from the participants' responses, i.e., prediction is central, whereas the underlying factors/causes are not explicitly analyzed), and studies that did not measure/operationalized affordance clearly and/or did not provide quantifiable and verifiable associated validity and reliability indices.

Refining the search and filtering the initial selection or publications based on criteria like the type of publication (e.g., journal article, conference paper, book chapter, book, etc.), type of research (e.g., primary vs secondary), and by using the meta-data provided by the academic database (*e.g.*, abstracts, keywords, subject/discipline) is not only common practice in such reviews (Ronzhyn et al., 2022, and others; see, for example, Xue & Churchill, 2019), but it is rather necessary to identify relevant publications.

4. RESULTS

Out of the 79 studies that entered the manual extraction of information, only 26 studies meet all criteria of completeness with respect [1] being focused on affordances (*i.e.*, the research questions mention 'affordances' and the model includes an operationalization of affordances), [2] be or include quantitative analysis, and [3] to clearly specify the analysis method and the measurements used.

Tables 1 and 2, below, present the results of our literature review. Namely, Table 1 presents the research design, the analysis, the types of affordances, and the role that affordances had in the respective studies. Table 2 is a continuation of Table 1, in that it presents the roles of the variables in the models developed in the reviewed studies (*e.g.*, predictors, mediators, moderators, or outcomes).

Table 1. Literature review table

Study	Research design	Analyses	Affordances types	Affordances roles
Belitski et al. (2023)	associative	seemingly unrelated regression equations	cultural; human capital; technology;	predictors
Chan et al. (2019)	associative	hierarchical regression	SNS affordances (accessibility, information retrieval, editability, association)	predictors
Chatterjee et al. (2020)	associative	structural equation modeling (CB-SEM)	process management A; organizational memory A; collaborative A;	predictors
C. Chen et al. (2022)	associative	hierarchical regression	affordance-based gratifications (being-there; agency enhancement; community-building; responsiveness; browsing; novelty; activity; play;)	outcomes
X. Chen et al. (2021)	associative	structural equation modeling (VB/PLS-SEM)	utilitarian; hedonic; connective;	predictors
Chris Zhao & Zhu (2014)	associative	structural equation modeling (CB-SEM)	motivational affordances (autonomy; competence; relatedness; leadership; affect)	moderators
D'Ambra et al. (2022)	mixed methods	coding & thematic analysis; VB/PLS-SEM	e-textbook affordance actualization (portability, accessibility, searchability, highlighting, copying, browsing, hedonic value, utilitarian value)	predictors; outcomes
Dubé & McEwen (2017)	quasi-experimental	ANOVA	computer affordances	outcomes
Fengliang & Jianhong (2021)	associative	structural equation modeling (CB-SEM)	live streaming commerce affordances (visibility; metavoicing; guidance shopping)	predictors
Huang & Zhou (2020)	associative	structural equation modeling (VB/PLS-SEM)	social gamification affordances (interactivity; cooperation; competition)	predictors
Irshad (2022)	quasi-experimental	MANOVA	spatial affordances in VR	predictors/moderators
Kim et al. (2022)	associative	path analysis	group play	predictors
Litster et al. (2021)	associative	path analysis	facilitating affordances (helping vs hindering)	mediators
Mao et al. (2023)	associative	multiple regression	social media affordances (presence; persistence)	predictors
Pibernik et al. (2019)	quasi-experimental	ANOVA	time affordances	predictors
Shin & Hwang (2020)	associative	path analysis	technological (traceability; security); emotional (privacy); affective (transparency; trust; satisfaction; reliability)	predictors; outcomes
Stoeckli et al. (2020)	mixed methods	multiple procedures (quant: Q-methodology)	lower- vs. higher-level affordances	outcomes
Suh & Wagner (2017)	associative	structural equation modeling (VB/PLS-SEM)	gamification affordances (rewardability; competition; visbility of achievement; quality of knowledge contribution; quantity of knowledge contribution)	predictors; outcomes
Sundar et al. (2015)	quasi-experimental	multiple procedures (t tests; path analysis)	technological affordances (customization; chatting)	predictors
Sun, Wang, et al. (2020)	associative	structural equation modeling (VB/PLS-SEM)	association; visibility; persistence; editability	predictors
Sun, Liu, et al. (2020)	associative	structural equation modeling (VB/PLS-SEM)	communication visibility (message transparency; network translucence)	predictors
Tsai & Ho (2013)	associative	structural equation modeling (CB-SEM)	affordance-based beliefs (diversity; intuition)	predictors

Study	Research design	Analyses	Affordances types	Affordances roles
Wang & Li (2021)	mixed methods	t tests and coding & thematic analysis	technology affordances (externally scripted vs non-scripted)	predictors
Wong et al. (2020)	mixed methods	t tests and coding & thematic analysis	perceived affordance	predictors
Yu et al. (2023)	associative	structural equation modeling (VB/PLS-SEM)	persuasion-related (cognitive (synchronicity; communication visibility; anonymity; triggered attending); motivational presence (telepresence; social presence; parasocial presence), co-experience (participation; cognitive communion; reasonant contagion)))	predictors
Zhou & Xu (2022)	associative	multi-level SEM, path analysis	annonimity; bandwidth; social presence	predictors

Notes to table: We used the terms 'predictors' and 'outcomes' as substitutes for independent and, respectively, dependent variables (even though it may be argued that the latter is the more natural terminology for the quasi-experimental research), because there were no purely experimental, controlled studies, and to keep a maximum consistency and ease of identification.

Table 2. Literature review table (continuation)

Study	Predictors	Moderators	Mediators	Outcomes
Belitski et al. (2023)	cultural; human capital; technology;			net entry; surival rate (firms); high growth;
Chan et al. (2019)	SNS affordances		criminogenic ooportunities; offending likelihood (inclination to bullying)	SNS bullying
Chatterjee et al. (2020)	harmonious IT affordance (HITA); organizational courage		actualized HITA; exploitative innovation	exploratory innovation
C. Chen et al. (2022)				Instagram use (problematic; habitual;)
X. Chen et al. (2021)	utiltarian A; hedonic A; connective A;		social identity; platform identity;	social commerce engagement
Chris Zhao & Zhu (2014)	motivation (external; introjected; identified; integrated; instrinsic)	affordances; task granularity		participation effort
D'Ambra et al. (2022)	e-textbook affordance actualization	device preferences	e-textbook engagement	e-textbook affordance effect
Dubé & McEwen (2017)	cognitive abilities; computer congiguration			ability to identify computer affordances
Fengliang & Jianhong (2021)			flow experience	impulse purchase intention
Huang & Zhou (2020)	social gamification affordances		psychological outcomes (recognition; social overload)	behavioral outcomes (green IT services use)
Irshad (2022)	spatial VR affordances; time			heart rate
Kim et al. (2022)			reciprocity	bridging social capital; bonding social capital; social status
Litster et al. (2021)	pre-gameplay variables (attitude; knowledge; game use)		helping affordances; hindering affordances	post-gameplay academic (mathematics) performance
Mao et al. (2023)	perceived social media presence; perceived social media persistence;	social tie strength		intention to communicate
Pibernik et al. (2019)	time affordances			perceived usability

Study	Predictors	Moderators	Mediators	Outcomes
Shin & Hwang (2020)	traceability, security, privacy		confirmation; trust; transparency; reliability	satisfaction
Stoeckli et al. (2020)	chatbot themes and constrains			affordances
Suh & Wagner (2017)	rewardability; competition; visibility of achievement		perceived hedonic value	quality of knowledge contribution; quantity of knowledge contribution
Sundar et al. (2015)			sense of control; perceived usability	attitudes toward website
Sun, Wang, et al. (2020)	ESMP affordances		knowledge acquisition; knowledge provision	creative performance
Sun, Liu, et al. (2020)	communication visibility		perceived social value; perceived information value; perceived hedonic value	excessive ESM use at work
Tsai & Ho (2013)	affordance-based beliefs; IDT-based beliefs		TAM-based beliefs (perceived usefulness; perceived ease of use;	attitudes towards smartphone usage)
Wang & Li (2021)	technology affordances (externally scripted vs non-scripted); others			role emergence
Wong et al. (2020)				level of expectation
Yu et al. (2023)			political mindfulness	political engagement intention
Zhou & Xu (2022)				dialogic communication measures (mutuality, pro-pinquity, empathy, risk, commitment)

5. DISCUSSION AND CONCLUSION

5.1 Summary of the Findings and Contribution to the State-of-the-Art

In summary, we identified that most of the studies focusing on affordances had an associative design (N = 18, *i.e.*, 69%), followed by an equal number of quasi-experimental and mixed methods studies (N = 4, 15%), with a note that two of the mixed methods studies also included associative designs. With respect to the roles that the affordances played in the studies' models, these were predominantly predictors (N = 17, *i.e.*, 65%). Studies that used affordances as outcomes or as outcomes and predictors, concomitantly, followed in equal proportion (12%, N = 3). Finally, there was only one of each study that used affordances as mediators or moderators.

We caution against interpreting these results as evidence that there are only very few studies that used quasi-experimental designs or performed various experimental manipulations. Quite to the contrary, that is not true at all, as a huge number of studies (not included here because they didn't meet the criteria for thorough reporting of the measurements or for not using scales for operationalizing the constructs) did, indeed, use quasi-experimental designs. However, as uncovered during our selection process, the problem was that most studies did not have a thorough reporting procedure and/or a rigorous description of their measurements (*e.g.*, lacking important information regarding the validity and reliability).

If anything, supported by our data, we want to drive home the idea that the field could use more experimental or quasi-experimental designs that go beyond case studies and in-house designed measurements. To that effect, the first significant conclusion that we derived from our literature review is that the number of complete reports, compared to the number of studies published on the subject is overwhelmingly low. Second, we noticed that many studies entertain a tendency to use 'affordances' as an umbrella term, covering or overlapping with, what are, in fact, 'system' or 'device' functionalities. This overlap, in itself, is understandable and perhaps unavoidable, but it becomes worrisome when is doubled by shortcomings of the theoretical underpinnings that should guide the operationalization of those affordances. Third, the number of quantitative primary studies, regardless of the thoroughness of their reporting, is significantly smaller than that of qualitative studies. This aspect, again, is not unusual, as many studies report on niche projects and many researchers are under pressure to publish quickly.

Nevertheless, qualitative research and case studies are notoriously difficult to reproduce and, often, it is not even desirable or relevant to do so. Rather differently, quantitative research not only requires thorough reporting of the methodology (and the measurements used) but actually invites replications. Furthermore, it is very often the case that many quantitative studies use the same measurements, exactly because they showed good psychometric properties in previous studies. In that regard, we must acknowledge that the literature does include notable instances of excellent qualitative studies, and the number of thorough qualitative research appears to be increasing (see, for example, the qualitative studies of McCarthy et al., 2022; or Lainema et al., 2023 among others).

5.2 Limitations and Future Directions

Our study must acknowledge a series of clear limitations. First, no literature review can assert absolute confidence in identifying all relevant publications, except, perhaps, when the initial pool of studies is very small, to begin with. A large number of starting publications leads directly to increased risks of unintentionally omitting relevant studies. Second, the writing styles and manuscript structures vary greatly and directly reflect the diversity of contributing fields. Specifically, researchers from various fields observe different styles of writing and journals in various disciplines often have quite different requirements for structuring the manuscript and standards pertaining to reporting. This makes it very difficult to pinpoint and extract exact information using regex-based separation of text sections. Moreover, regex-based information extraction, as natural as it may be, is also prone to errors of omission, when synonyms or paraphrasing is used instead of 'classical' or traditionally coined expressions, headings, and phrases.

In future refinements, we plan to use machine learning to train a model capable of identifying the sections of text of interest, based on semantic similarity. Furthermore, we plan to continue this analysis by performing topic modelling, such as, for example, based on Latent Dirichlet Allocation (LDA), to get more insight into the similarities and dissimilarities of the studies of affordances. As can be seen from the reporting of the selection process, the majority of studies identified during the initial stages of our literature search used a qualitative design and/or were case studies. Manually going through each of these would be a daunting and ultimately, unrealistic task. However, a machine learning approach would address a second avenue of interest to us, i.e., uncovering the conceptualisation of affordances and their roles in qualitative studies, as well. Lastly, we must note that it seems, at times, that there is a 'real and present danger' that pretty much everything under the sun can potentially be construed in terms of affordances, and we wholeheartedly align with the existing calls for a rigorous approach to operationalizing affordances (J. Davis & Chouinard, 2017).

ACKNOWLEDGEMENT

This work was supported by a grant of the Ministry of Research, Innovation and Digitization, CCCDI - UEFISCDI, project number PN-III-P2-2.1-PTE-2021-0255, within PNCDI III.

REFERENCES

Bahari, A. (2022). Affordances and challenges of technology-assisted language learning for motivation: A systematic review. *Interactive Learning Environments*, 1–21. https://doi.org/10.1080/10494820.2021.2021246

Belitski, M., Korosteleva, J., & Piscitello, L. (2023). Digital affordances and entrepreneurial dynamics: New evidence from European regions. *Technovation*, *119*, 102442. https://doi.org/10.1016/j.technovation.2021.102442

Chan, T. K. H., Cheung, C. M. K., & Wong, R. Y. M. (2019). Cyberbullying on Social Networking Sites: The Crime Opportunity and Affordance Perspectives. *Journal of Management Information Systems*, *36*(2), 574–609. https://doi.org/10.1080/07421222.2019.1599500

Chatterjee, A., & Cardillo, E. R. (2022). *Brain, Beauty, and Art: Essays Bringing Neuroaesthetics Into Focus*. Oxford University Press.

Chatterjee, S., Moody, G., Lowry, P. B., Chakraborty, S., & Hardin, A. (2020). Information Technology and organizational innovation: Harmonious information technology affordance and courage-based actualization. *The Journal of Strategic Information Systems*, *29*(1), 101596. https://doi.org/10.1016/j.jsis.2020.101596

Chen, C., Cohen, O., & Sundar, S. S. (2022). Differentiating Problematic from Habitual Instagram Use: A Uses and Grats 2.0 Perspective. *Social Media + Society*, *8*(3), 205630512211163. https://doi.org/10.1177/20563051221116339

Chen, X., Liu, Z., Wei, S., & Liu, Y. (2021). Understanding the Role of Affordances in Promoting Social Commerce Engagement. *International Journal of Electronic Commerce*, *25*(3), 287–312. https://doi.org/10.1080/10864415.2021.1943170

Chong, I., & Proctor, R. W. (2020). On the Evolution of a Radical Concept: Affordances According to Gibson and Their Subsequent Use and Development. *Perspectives on Psychological Science*, *15*(1), 117–132. https://doi.org/10.1177/1745691619868207

Chow, J. Y., Davids, K., Button, C., & Renshaw, I. (2015). *Nonlinear Pedagogy in Skill Acquisition: An Introduction*. Taylor & Francis.

Chris Zhao, Y., & Zhu, Q. (2014). Effects of extrinsic and intrinsic motivation on participation in crowdsourcing contest: A perspective of self-determination theory. *Online Information Review*, *38*(7), 896–917. https://doi.org/10.1108/OIR-08-2014-0188

Clarke, D., & Clarke, E. (2011). *Music and Consciousness: Philosophical, Psychological, and Cultural Perspectives*. OUP Oxford.

D'Ambra, J., Akter, S., & Mariani, M. (2022). Digital transformation of higher education in Australia: Understanding affordance dynamics in E-Textbook engagement and use. *Journal of Business Research*, *149*, 283–295. https://doi.org/10.1016/j.jbusres.2022.05.048

Davis, J. L. (2020). *How Artifacts Afford: The Power and Politics of Everyday Things*. MIT Press.

Davis, J., & Chouinard, J. (2017). Theorizing Affordances: From Request to Refuse. *Bulletin of Science, Technology & Society*, *36*, 027046761771494. https://doi.org/10.1177/0270467617714944

Dubé, A. K., & McEwen, R. N. (2017). Abilities and affordances: Factors influencing successful childtablet communication. *Educational Technology Research and Development*, *65*(4), 889–908. https://doi.org/10.1007/s11423-016-9493-y

Fengliang, X., & Jianhong, L. (2021). Research on the influence of live streaming commerce affordances on consumers' impulse purchase intention based on SPSS25.0 and Amos23.0. *2021 2nd International Conference on E-Commerce and Internet Technology (ECIT)*, 95–99. https://doi.org/10.1109/ECIT52743.2021.00028

Gibson, J. J. (1979). *The Ecological Approach to Visual Perception*. Psychology Press.

Gieseking, J. J., Mangold, W., Katz, C., Low, S., & Saegert, S. (2014). *The People, Place, and Space Reader*. Taylor & Francis.

Huang, J., & Zhou, L. (2020). Social gamification affordances in the green IT services: Perspectives from recognition and social overload. *Internet Research*, *31*(2), 737–761. https://doi.org/10.1108/INTR-03-2020-0121

Irshad, S. (2022). Investigating the user experience of IDN based virtual reality environments for solving complex issues. *New Review of Hypermedia and Multimedia*, 1–23. https://doi.org/10.1080/13614568.2022.2150325

Khosrow-Pour D.B.A., M., & Bhattacharyya, S. (Eds.). (2021). *Encyclopedia of Information Science and Technology, Fifth Edition:* IGI Global. https://doi.org/10.4018/978-1-7998-3479-3

Kim, S. S. Y., Huang-Isherwood, K. M., Zheng, W., & Williams, D. (2022). The art of being together: How group play can increase reciprocity, social capital, and social status in a multiplayer online game. *Computers in Human Behavior*, *133*, 107291. https://doi.org/10.1016/j.chb.2022.107291

Lainema, K., Syynimaa, K., Lainema, T., & Hämäläinen, R. (2023). Organizing for collaboration in simulation-based environments: An affordance perspective. *Journal of Research on Technology in Education*, *55*(2), 307–323. https://doi.org/10.1080/15391523.2021.1962451

Ling, R. (2004). *The Mobile Connection: The Cell Phone's Impact on Society*. Elsevier Science.

Litster, K., Lommatsch, C. W., Novak, J. R., Moyer-Packenham, P. S., Harmon, M. J., Roxburgh, A. L., & Bullock, E. P. (2021). The Role of Gender on the Associations among Children's Attitudes, Mathematics Knowledge, Digital Game Use, Perceptions of Affordances, and Achievement. *International Journal of Science and Mathematics Education*, *19*(7), 1463–1483. https://doi.org/10.1007/s10763-020-10111-8

Mao, C. (Molly)., Xu, S., & Kee, K. F. (2023). What influence virtual workers' communication?: Exploring network inertia and media affordance theories. *The Social Science Journal*, 1–16. https://doi.org/10.1080/03623319.2023.2189394

McCarthy, S., Rowan, W., Kahma, N., Lynch, L., & Ertiö, T. P. (2022). Open e-learning platforms and the designreality gap: An affordance theory perspective. *Information Technology & People*, *35*(8), 74–98. https://doi.org/10.1108/ITP-06-2021-0501

Meinel, C., & Krohn, T. (2022). *Design Thinking in Education: Innovation Can Be Learned*. Springer International Publishing.

Nardi, B. A. (1996). *Context and Consciousness: Activity Theory and Human-computer Interaction*. MIT Press.

Norman, Donald A. (1988). *The psychology of everyday things*. Basic books.

Norman, Donald A. (2013). *The design of everyday things* (Revised and expanded edition). Basic Books.

Oliver, M. (2005). The Problem with Affordance. *E-Learning and Digital Media*, *2*(4), 402–413. https://doi.org/10.2304/elea.2005.2.4.402

Pibernik, J., Dolic, J., Vistid, A., & Mandic, L. (2019). Time Affordance Components impact on User's Experience. *2019 International Symposium ELMAR*, 151–153. https://doi.org/10.1109/ELMAR.2019.8918661

Ronzhyn, A., Cardenal, A. S., & Batlle Rubio, A. (2022). Defining affordances in social media research: A literature review. *New Media & Society*, 146144482211351. https://doi.org/10.1177/14614448221135187

Scarlett, A., & Zeilinger, M. (2019). Rethinking Affordance. *Media Theory*, *3*(1), 01–48.

Sears, A., & Jacko, J. A. (2007). *The Human-Computer Interaction Handbook: Fundamentals, Evolving Technologies and Emerging Applications, Second Edition*. CRC Press.

Shin, D., & Hwang, Y. (2020). The effects of security and traceability of blockchain on digital affordance. *Online Information Review*, *44*(4), 913–932. https://doi.org/10.1108/OIR-01-2019-0013

Stoeckli, E., Dremel, C., Uebernickel, F., & Brenner, W. (2020). How affordances of chatbots cross the chasm between social and traditional enterprise systems. *Electronic Markets*, *30*(2), 369–403. https://doi.org/10.1007/s12525-019-00359-6

Suh, A., & Wagner, C. (2017). How gamification of an enterprise collaboration system increases knowledge contribution: An affordance approach. *Journal of Knowledge Management*, *21*(2), 416–431. https://doi.org/10.1108/JKM-10-2016-0429

Sun, Y., Liu, Y., & Zhang, J. (2020). Excessive Enterprise Social Media Use Behavior at Work: Role of Communication Visibility and Perspective of Uses and Gratifications Theory. *IEEE Access*, *8*, 190989–191004. https://doi.org/10.1109/ACCESS.2020.3032035

Sun, Y., Wang, C., & Jeyaraj, A. (2020). Enterprise social media affordances as enablers of knowledge transfer and creative performance: An empirical study. *Telematics and Informatics*, *51*, 101402. https://doi.org/10.1016/j.tele.2020.101402

Sundar, S. S., Go, E., Kim, H.-S., & Zhang, B. (2015). Communicating Art, Virtually! Psychological Effects of Technological Affordances in a Virtual Museum. *International Journal of Human-Computer Interaction*, *31*(6), 385–401. https://doi.org/10.1080/10447318.2015.1033912

Tabuenca, B., Serrano-Iglesias, S., Martin, A. C., Villa-Torrano, C., Dimitriadis, Y., I. Asensio-Perez, J., Alario-Hoyos, C., Gomez-Sanchez, E., L. Bote-Lorenzo, M., Martinez-Mones, A., & Kloos, C. D. (2021). Affordances and Core Functions of Smart Learning Environments: A Systematic Literature Review. *IEEE Transactions on Learning Technologies*, *14*(2), 129–145. https://doi.org/10.1109/TLT.2021.3067946

Tsai, J.-P., & Ho, C.-F. (2013). Does design matter? Affordance perspective on smartphone usage. *Industrial Management & Data Systems*, *113*(9), 1248–1269. https://doi.org/10.1108/IMDS-04-2013-0168

Wang, C., & Li, S. (2021). The trade-off between individuals and groups: Role interactions under different technology affordance conditions. *International Journal of Computer-Supported Collaborative Learning*, *16*(4), 525–557. https://doi.org/10.1007/s11412-021-09355-5

Wong, C. Y., Ibrahim, R., Hamid, T. A., & Mansor, E. I. (2020). Measuring Expectation for an Affordance Gap on a Smartphone User Interface and its Usage Among Older Adults. *Human Technology*, *16*(1), 6–34. https://doi.org/10.17011/ht/urn.202002242161

Xue, S., & Churchill, D. (2019). A review of empirical studies of affordances and development of a framework for educational adoption of mobile social media. *Educational Technology Research and Development*, *67*(5), 1231–1257. https://doi.org/10.1007/s11423-019-09679-y

Yu, T., Chen, Y., & Luo, X. (Robert). (2023). How do live-streaming platforms facilitate persuasion in political campaigns? Theory and empirical evidence from the perspective of affordance actualization. *Information & Management*, *60*(3), 103775. https://doi.org/10.1016/j.im.2023.103775

Zhou, A., & Xu, S. (2022). Computer mediation vs. Dialogic communication: How media affordances affect organization-public relationship building. *Public Relations Review*, *48*(2), 102176. https://doi.org/10.1016/j.pubrev.2022.102176

ENSURING HIGHER LEARNING OUTCOMES THROUGH E-LEARNING: AN EXAMPLE OF KATHMANDU UNIVERSITY SCHOOL OF EDUCATION, NEPAL

Niroj Dahal, Bal Chandra Luitel and Binod Prasad Pant
Kathmandu University School of Education, Department of STEAM Education, Nepal

ABSTRACT

In this reflective paper, the authors discuss their experiences of developing e-learning courses to teach and learn to create an engaged pedagogy in various semesters at Kathmandu University School of Education, Nepal, that involves their (PGD, Master, MPhil, and Ph.D.) students of the 2020-2022 batches for ensuring higher learning outcomes. This paper describes the ongoing learning from research done in the context of the authors' planning, delivering, and receiving feedback from the students during the pedagogical practices through e-learning. The authors give examples of some of their PGD, Master, MPhil, and Ph.D. e-learning practices that they delivered to students in the 2020–2022 batches. We attempt to share e-learning practices that we were practicing for on-campus, online, and distance learning to ensure higher learning outcomes in higher education in Nepal and worldwide.

KEYWORDS

E-Learning, Reflective Learning Outcomes, On-Campus, Online, Distance Teaching, Learning

1. E-LEARNING AT KUSOED: A BRIEF HISTORY

The educational sector has significantly advanced through the application of technology and cutting-edge tools (Katsaris & Vidakis, 2021). In order to improve its methods of teaching and learning, Kathmandu University School of Education (KUSOED) in Nepal has a history of embracing e-learning and educational technology. The first e-learning center was established by KUSOED in 2005 to advance the use of technology in education. The center began training professors and students on how to use digital tools for education, such as learning management systems, multimedia tools, and web-based resources. The initiative between KUSOED and the Norwegian University of Science and Technology (NTNU) in 2006, which aimed to develop capacity in the use of technology for education and to promote e-learning on a small scale, was a milestone in e-learning. In addition to acting as a resource hub for other nations in the South Asian region, KUSOED was crucial to the success of the initiative's implementation in Nepal. The KUSOED MOODLE learning platform was introduced in 2012 by KUSOED with the help of NORHED-QUANTICT for KUSOED students and faculty members; the platform offers access to online courses, resources, and learning materials. Additionally, it provides teachers and other education professionals in Nepal with professional development courses. In recent years, KUSOED has increased its e-learning options and integrated cutting-edge tools like virtual and augmented reality into its teaching and learning methods. The school has also taken part in several studies on the application of technology in education and has written articles and reports on the subject (See more Dahal & Pangeni, 2019; Dahal et al., 2022a; Dahal, 2022).

Near the end of the section, during the pandemic, however, schools and universities were closed all over the country, so we had to focus our educational efforts online (Dahal et al., 2022b; Karakose et al., 2022). Still, students and teachers everywhere kept learning at home using synchronous and asynchronous digital platforms like Zoom, Google Meet, and others (Dahal et al., 2022b; Karakose et al., 2021). Even during the pandemic, KUSOED students were busy taking online classes using synchronous online learning in MOODLE.

2. E-LEARNING AT KUSOED

At KUSOED, e-learning practices was already the standard operating procedure that has been in practice since 2012 AD. However, for other institutions in Nepal, e-learning practices gained momentum after COVID-19. This momentum presents many challenges and opportunities for university faculties. E-learning practices have become increasingly popular in educational systems worldwide because they provide students with more engaging and effective instruction customized to their specific needs (Czepula et al., 2018). In general, e-learning is a learning system that combines formalized instruction with electronic resources. E-learning, also called online learning or distance education, is the use of technology to teach and educate people who are not in a traditional classroom setting (Ally, 2004). This type of learning involves using computers, mobile devices, and the internet to get to course materials, talk to teachers and other students, and do assignments and tests (Huang et al., 2021). Clark and Mayer (2016) say that e-learning has several advantages over traditional classroom-based learning, such as more flexibility and accessibility, personalized learning experiences, and cost savings for both learners and educational institutions. E-learning has also been shown to be just as effective, and in some cases even more effective, than traditional classroom-based learning (Dahal, 2022; Means et al., 2013). But e-learning also has some problems, such as the need for learners to be self-directed and motivated, the possibility of social isolation, and the need for effective instructional design and delivery to keep learners interested and interested in what they are learning (Dennen & Bonk, 2019). Because of this, e-learning is becoming a more popular way to teach and learn, and its pros and cons are driving research and development in the field.

Since integrating MOODLE as a learning management system for the first time in Nepal in 2012 AD, the Kathmandu University School of Education has been using the NORHED-Quantic project to develop the infrastructure for online teaching and learning. This necessary technological support gives us the confidence to implement the curriculum in a hybrid or fully online format. Beginning in 2013, students enrolled in the academic programs were required to create an account on the e-learning portal and enroll in the relevant courses. The course instructors were not accepted any assignments or projects sent via e-mail. Along with mandatory training on the Moodle platform, this requirement lays the foreground for online learning, which is essential in today's dynamic educational landscape. E-learning is facilitated at KUSOED by a course facilitator with the learning management system (LMS) that allows instructors for all programs to create their self-study materials. Because of the faculty-created self-learning resources for independent study, e-learning strategies can now be implemented (see Dahal, 2016). These resources for the independent study were crafted according to modules that each includes a module summary, key learning materials, learning outcomes, learning activities (pre-class activity, during class activities and after-class activities), assessments (assignments, quizzes, forum discussions), learning supports, and recordings of classes (if any). Figure 1 for the interface of e-learning at KUSOED.

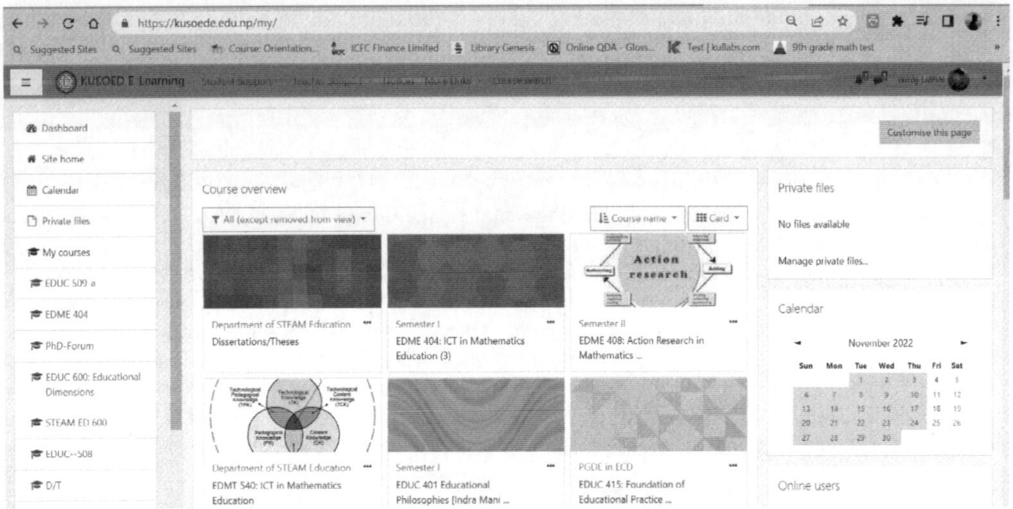

Figure 1. Interface of e-learning of KUSOED

This is just one example of how e-learning practices are evolving due to recent innovations at KUSOED. The needs of students are always evolving, and we adapt to meet those needs if we are to create best practices in higher education. This paper argues that e-learning is one of the most effective pedagogical processes because it maximizes the use of active learning strategies and increases the prospect of positive learning outcomes for students (Pizzi, 2014). This paper is based on how e-learning practices are being implemented at Kathmandu University School of Education, Nepal, to ensure learners' higher learning outcomes?

3. E-LEARNING PRACTICES: A CONTEXT OF KUSOED

Kathmandu University School of Education was the pioneer in initiating e-learning in Nepal. For the establishment of an e-learning platform, the first step was taken in 2012 for the students who were enrolled in a One Year M Ed in Leadership and Management. The combined approach to education and training was maintained with the above involvement. This initiation helped us run all of our regular day-to-day teaching and learning activities without interruption, even as a global pandemic gripped the world. Given the wide range of students we teach in Nepal and beyond, it would seem impossible to personalize instruction to meet the needs of each student. Because of its hybrid nature, e-learning can be tailored to individual needs in terms of time, learning style, and other factors; however, it necessitates some adjustments to be put into practice. As a result of these shifts, we can investigate hybrid approaches to education. These included interacting with learners, switching learners' partners, using games as learning tools, conducting experiments online, blending and driving one's mixtures online, and learning entirely online. See Figure 2 of the Online Module's Contents.

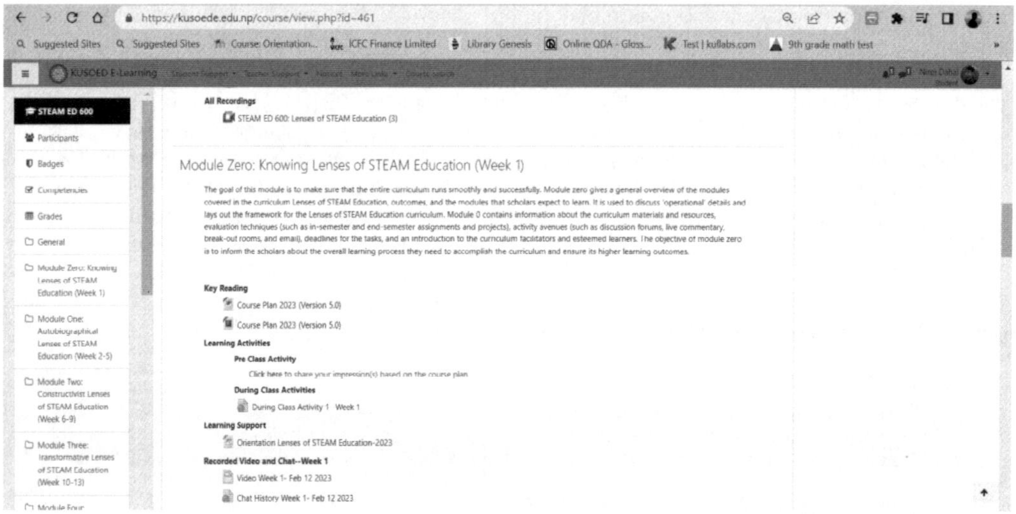

Figure 2. Online Module's Contents

First and foremost, face-to-face learning entails traditional instructor-led sessions enhanced with technological means to give students more agency over their own learning pace (Dahal et al., 2022b). Role-playing, guidance, practice, and feedback on actual tasks are all useful outcomes. The students move in a circle from one learning activity to another, either in a teacher-led class or independently in online courses such as ZOOM or Google Meet breakout rooms. One example is the flipped classroom, where students complete pre-classwork online before coming to class in person. Likewise, gamification is one of the best ways to keep students engaged in learning. Gamification techniques such as points and levels encourage students to study on their own time by creating an atmosphere of friendly competition. See Figure 3 in the next page for students' engagement during online classes as during class activity 1.

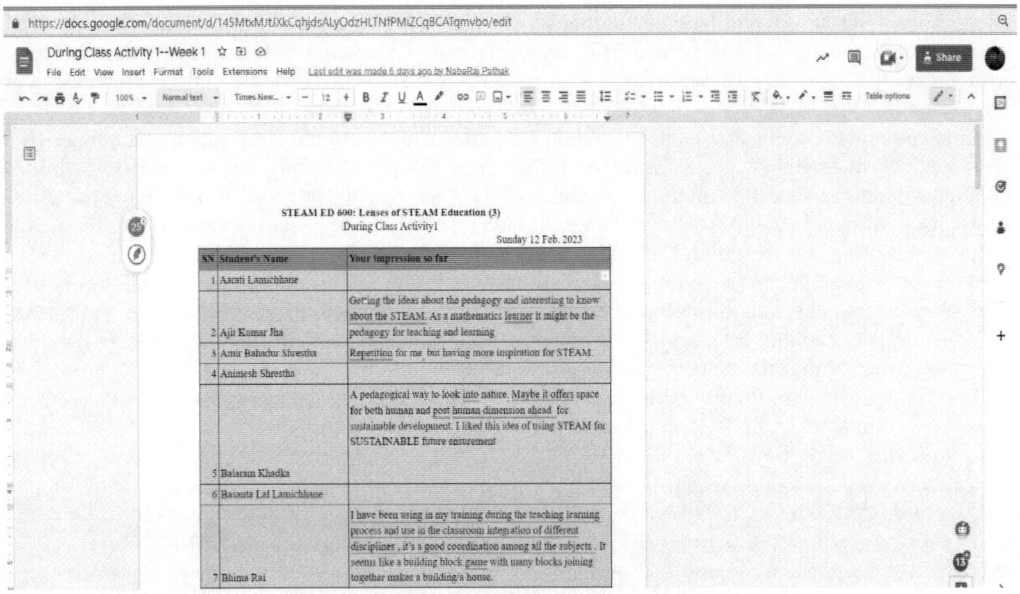

STEAM ED 600: Lenses of STEAM Education (3)
During Class Activity1
Sunday 12 Feb. 2023

SN	Student's Name	Your impression so far
1	Aarati Lamichhane	
2	Ajit Kumar Jha	Getting the ideas about the pedagogy and interesting to know about the STEAM. As a mathematics learner it might be the pedagogy for teaching and learning
3	Amir Bahadur Shrestha	Repetition for me but having more inspiration for STEAM.
4	Animesh Shrestha	
5	Balaram Khadka	A pedagogical way to look into nature. Maybe it offers space for both human and post human dimension ahead for sustainable development. I liked this idea of using STEAM for SUSTAINABLE future ensurement
6	Basanta Lal Lamichhane	
7	Bhima Rai	I have been using in my training during the teaching learning process and use in the classroom integration of different disciplines , it's a good coordination among all the subjects . It seems like a building block game with many blocks joining together makes a building/a house.

Figure 3. Students' Engagement During Class Activity

Google docs activities offer students to be active throughout the online and/or blended classes. In the blended learning model, professional training for teachers may conduct online before, during, or after the main event. Most of the students at our institution are well-known for m-learning. The term "m-learning" refers to using mobile devices to deliver educational materials to students. Learning is more active and lasting with engagement in the m-learning approach. Webinars, seminars, blog writing, and video tutorials are all examples of self-blended learning materials that self-directed students can use to learn more about a topic. We have been using LMS, wherein LMS can aggregate content from multiple sources into one convenient location by attracting students' interest and inspiring them to expand their knowledge. Self-paced learning is at the heart of e-learning, which occurs online but is not limited. In this learning process of LMS, students can communicate with their teacher in real time via chat, e-mail, or a message board. Although it offers schedule flexibility and individualized instruction, this form of e-learning does not include any live instructor for student interaction. To facilitate self-directed learning using media and, ultimately, classroom discussion, LMS is ideal. We have been using these platforms to ensure students' higher learning outcomes as a custom learning system.

4. OUR BEST PRACTICES OF E-LEARNING TO ENSURE HIGHER LEARNING OUTCOMES

E-learning is not a novel approach to improving educational outcomes at KUSOED. E-learning effectively covers multiple courses with self-motivated and engaged students on-campus and online learning. The key, however, is for students to take ownership of their own academic achievement. The question of best integrating at-home and classroom-based learning is at the heart of e-learning. To disseminate our most effective strategies for blended education, we have summarized the various approaches, a wide range of activities, and the incorporation of ICTs. See Figure 4, which shows how we welcome the students on the e-learning portal of the courses.

Based on our experiences, we would have already failed if the students had not grasped the value of a e-learning approach that incorporates both online and face-to-face instruction from the start. Instead, we should spend some time justifying our decision to use a blended approach to teaching. One possible reason is a desire to make better use of class time by allowing students to work at their own pace and discuss what they have learned. Likewise, one more benefit of e-learning is the variety of activities that can be incorporated into the curriculum and pedagogy. Even though some students learn best from books, others thrive when given real-world tasks; it is common to hear the proverb, "Try to be everything to everyone." While there is no single best approach to learning, our experiences on e-learning show that increasing user engagement through various activities (such as group discussions, online quizzes, games, and even role-playing) ensures students' higher learning outcomes across the curriculum and pedagogy.

Students in the twenty-first century have lots of information and communication technology (ICT) and other media at their disposal and they need to use them. To put it simply, the Internet is vital. Instead of asking a monotonous question on the whiteboard, why not show an interesting video? Take it one step further by giving students a video to watch on their own with discussion questions for the next day in class. Instead of just telling students something, let them find proof for themselves. We could also mix a short in-class presentation with an interactive online module. Students are more likely to remember what they have learned when they have a chance to use it. In e-learning, students listen to lectures or watch videos and participate in the process. It is suitable for everyone if students use what they have learned as soon as they leave the classroom, whether that classroom is on-campus or virtual.

Further. see Figure 4, Figure 5, Figure 6 and Figure 7 for students' engagements and evaluations by peers and course facilitators in the e-learning portal of KUSOED.

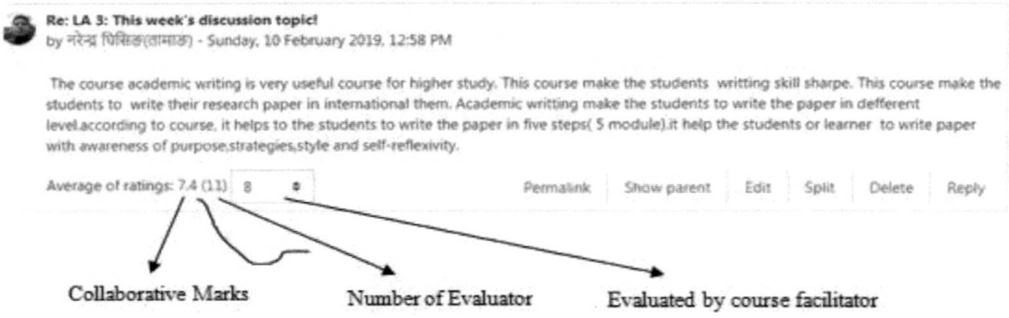

Figure 4. Responses to Forum

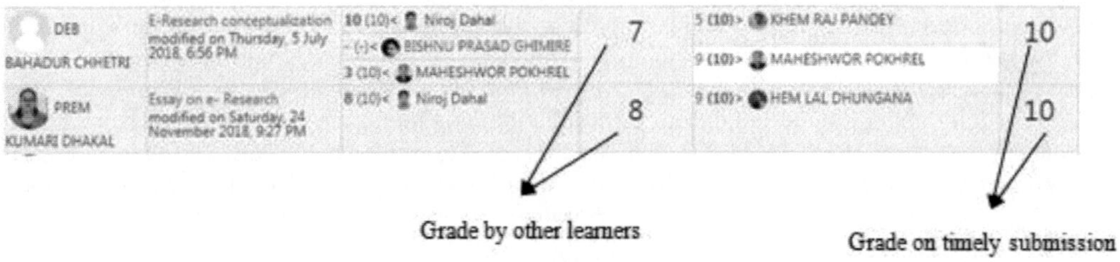

Figure 5. Evaluation of peer work

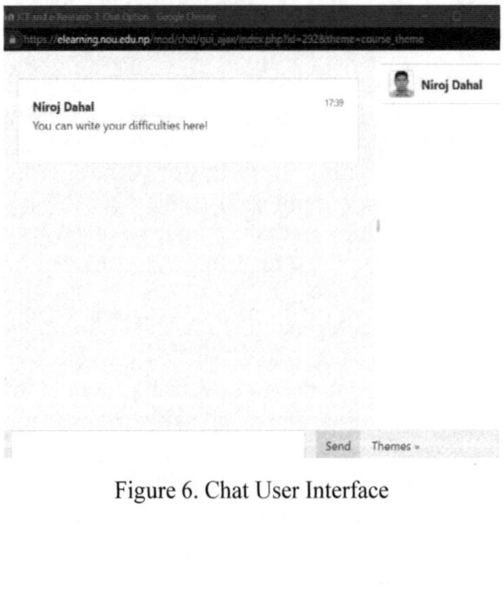

Figure 6. Chat User Interface

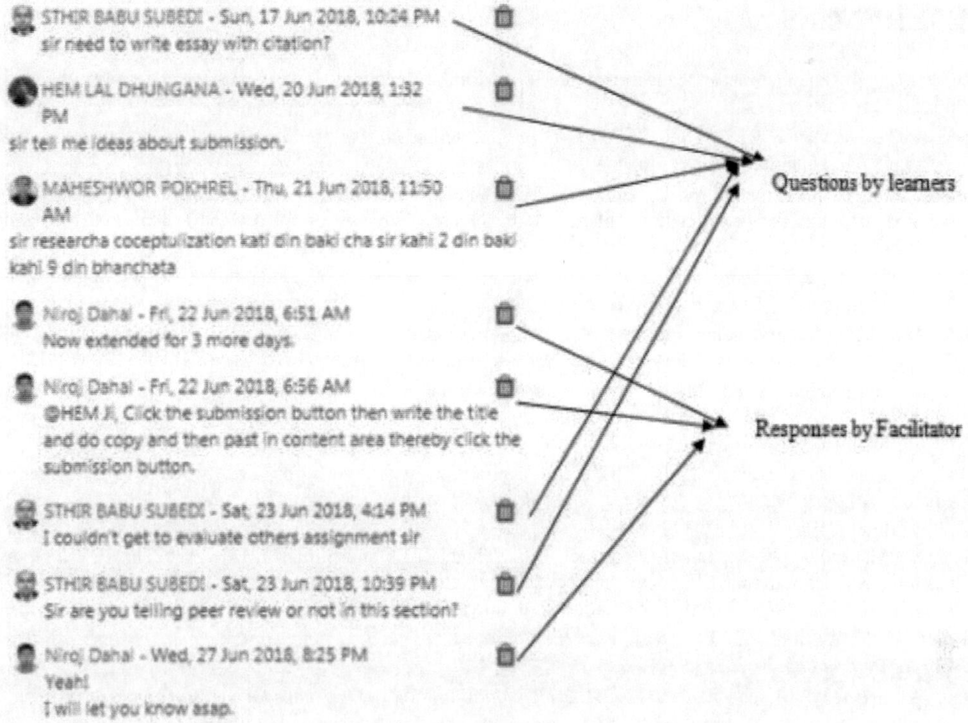

Figure 7. Participation of Students in the Discussion on Comments

5. FUTURE OF E-LEARNING AT KUSOED

This paper attempted to examine the e-learning practices at KUSOED as a combination of teaching methods and materials. Unlike direct or face-to-face (f2f) learning in the classroom, online methodologies (via e-learning with the help of MOODEL as LMS) serve as a formal education method at KUSOED and other institutions in Nepal. The effectiveness of mixed learning in learning processes is anticipated to increase through the reviewed characteristics, components, issues, and challenges. Next, e-learning via MOODLE has a substantial impact on all aspects of education. It bridges the various modes of teaching and learning by encouraging university faculties and teachers to promote students' quality engagement in all subject areas. This paper explores possible e learning strategies to facilitate instruction by ensuring higher learning outcomes. In the same way, in recent years, everything has been about switching from traditional teaching and learning to online and/or other digital ways. This has made getting better results easier for some university faculties and teachers and incredibly hard for others. This paper attempts offers options for university faculties to deliver content in an e-learning format. For this reason, we discussed various e-learning modes for achieving higher learning outcomes. However, university faculty must be allowed to select the e-learning mode that best suits their teaching and learning styles. University faculties must also modify their perspectives on e-learning instruction to effectively use currently available yet-to-be-released techniques for promoting pedagogy, learning, and learners' empowerment.

Near the end, given its growing popularity over the past few years and the possibility of further expansion, e-learning has a promising future, with trends and developments shaping the future of e-learning at KUSOED. Personalized learning, gamification, mobile learning, virtual and augmented, collaborative learning, data-driven learning, adaptive learning, artificial intelligence, remote learning, and continuous learning are some of the trends and developments in e-learning for the future at KUSOED. As e-learning evolves, it will provide learners with new opportunities to develop their skills and achieve their career goals.

REFERENCES

Ally, M. (2004). Foundations of educational theory for online learning. *Theory and Practice of Online Learning, 2*, pp. 15-44.

Clark, R. C., and Mayer, R. E. (2016). *E-learning and the science of instruction: Proven guidelines for consumers and designers of multimedia learning*. John Wiley & Sons.

Czepula, A. I. D. S., Bottacin, W. E., Júnior, E. H., Pontarolo, R., and Correr, C. J. 2018. Active methodology and blended learning: An experience in pharmaceutical care. *Currents in Pharmacy Teaching & Learning*, *10*(1), pp. 106-111.

Dahal, N. (2016). *Self-learning module of ICT in mathematics education*. Kathmandu University, School of Education, Nepal. https://doi.org/10.13140/RG.2.2.12766.10562

Dahal, N. (2022). Understanding and uses of collaborative tools for online courses in higher education. *Advances in Mobile Learning Educational Research*, *2*(2), pp. 435-442. https://doi.org/10.25082/AMLER.2022.02.012

Dahal, N., and Pangeni, S. K. (2019). Workshopping in online courses: Insights for learning and assessment in higher education. *International Journal of Multidisciplinary Perspectives in Higher Education*, *4*(1), pp. 89–110. https://doi.org/10.32674/jimphe.v4i1.1275

Dahal, N., Luitel, B. C., and Pant, B. P. (2022a). Exploration of the workshop activity for peer assessment in online courses of mathematics. *Advances in Mobile Learning Educational Research*, *2*(2), pp. 475-482. https://doi.org/10.25082/AMLER.2022.02.016

Dahal, N., Manandhar, N. K., Luitel, L., Luitel, B. C., Pant, B. P., and Shrestha, I. M. (2022b). ICT tools for remote teaching and learning mathematics: A proposal for autonomy and engagements. *Advances in Mobile Learning Educational Research, 2*(1), pp. 289-296. https://doi.org/10.25082/AMLER.2022.01.013

Dennen, V. P., & Bonk, C. J. (2019). *The Handbook of Online Learning: Innovations in Higher Education and Corporate Training*. Routledge.

Huang, R., Spector, J. M., and Yang, J. (2021). *Educational technology: A primer for the 21st century*. Springer.

Karakose, T., Ozdemir, T. Y., Papadakis, S., Yirci, R., Ozkayran, S. E., and Polat, H. (2022). Investigating the Relationships between COVID-19 Quality of Life, Loneliness, Happiness, and Internet Addiction among K-12 Teachers and School Administrators—A Structural Equation Modeling Approach. *International Journal of Environmental Research and Public Health*, *19*(3), p. 1052. http://dx.doi.org/10.3390/ijerph19031052

Karakose, T., Polat, H., and Papadakis, S. (2021). Examining teachers' perspectives on school principals' digital leadership roles and technology capabilities during the COVID-19 pandemic. *Sustainability*, *13*(23), pp. 134-148. http://dx.doi.org/10.3390/su132313448

Katsaris, I., and Vidakis, N. (2021). Adaptive e-learning systems through learning styles: A review of the literature. *Advances in Mobile Learning Educational Research*, *1*(2), 124-145. https://doi.org/10.25082/AMLER.2021.02.007

Means, B., Toyama, Y., Murphy, R., Bakia, M., and Jones, K. (2013). *Evaluation of evidence-based practices in online learning: A meta-analysis and review of online learning studies*. US Department of Education.

DEVELOPMENT OF MODULAR E-LEARNING MATERIALS AS OER FOR A MOOC

Fabian Dillenhöfer[1], Bernd Künne[1], Silke Frye[1], Tobias Haertel[1], Laura Altland[1], Frederike Kossack[2], Beate Bender[2], Alina Sersch[3], Peter Gust[3], Yannick Lattner[4], Matthias Müller[4], Dominik Hinse[5], Klaus Pantke[5], Jens Bechthold[6], Julia Frank[7] and Tim Richard[7]

[1]*TU Dortmund University, D-44227 Dortmund, Germany*
[2]*Ruhr-University Bochum, D-44780 Bochum, Germany*
[3]*University of Wuppertal,* D-42119 Wuppertal, Germany
[4]*Fachhochschule Dortmund – University of Applied Sciences and Arts, D-44139 Dortmund, Germany*
[5]*Hamm-Lippstadt University of Applied Sciences, D-59063 Hamm, Germany*
[6]*South Westphalia University of Applied Sciences, D-59494 Soest, Germany*
[7]*Bochum University of Applied Sciences, D-44801 Bochum, Germany*

ABSTRACT

This paper is (I) a report about the development of modular E-Learning materials as OER for a MOOC and the changes regarding the creation surrounding Covid-19, as well as (II) an evaluation of the changes on the initial situation of higher education met in 2019 in contrast to the current situation for teachers and learners that is given through the availability of the developed modular MOOC. Advantages and limitations for the creation of E-Learning materials within a syndicate are discussed and a survey conducted at six German universities is evaluated in order to indicate the encountered changes precisely. Moreover, the framework including the modularity of the learning materials in combination with the principle of Constructive Alignment is described and analyzed. The empirical data regarding the students and teachers experience with the framework and learning materials collected through the survey are discussed elaborately. Last but not least, a conclusion on the consolidation and continuous use of the learning materials is drawn, which shows that transparent learning objectives help to raise the students' motivation.

KEYWORDS

E-Learning, Higher Education, Creating E-Learning Courses, Engineering Education, MOOC

1. INTRODUCTION

Technical drawings are a fundamental tool for communicating technical product specifications (TPS). In times of advancing technological development, digitization and globalization, clear communication continues to gain in importance. Not only engineers, but also designers, production employees or employees in quality assurance have to deal with them. The field of technical drawings is usually a basic subject for engineering students at undergraduate university level.

Since the teaching topic of technical drawings is basic and mostly standardized, it is well suited for Open Educational Resources (OER) and an overall Massive Open Online Course (MOOC). Open Educational Resources (OER) are educational materials under an open license that allows free access as well as free use, editing, and redistribution by third parties without or with minor restrictions (Unesco, 2019). For this reason, six German universities united to create and advocate an MOOC consisting of modular OER.

The aforementioned project was planned before the Covid-19 pandemic to promote OER, while the project itself started during the pandemic. This led to various changes relating different groups and factors that are described within this paper. At the same time the number of MOOC across the world doubled, according to *Class Central*, from around 10,000 in 2019 to over 20,000 in 2022 (Shah, 2021).

The work described in this paper is (I) a report about the development of modular E-Learning materials as OER for a MOOC and the changes regarding the creation surrounding Covid-19, as well as (II) an evaluation of the changes on the initial situation of higher education met in 2019 in contrast to the current situation for teachers and learners that is given through the availability of the modular MOOC consisting of OER. For this purpose, students, teachers and didactic experts have been included in the pretest and the final evaluation to get a complete overview. Moreover, it is analyzed in what way the OER are compatible with a MOOC and a modular framework that fulfill the needs of numerous professions covering some topics at various levels in different depth. The aim is to give an overall view of the extent to which the above concepts can be combined and what conditions must be in place in order to fulfill certain framework conditions based on empirical results.

2. STATE OF THE ART

The teaching of technical drawing in the academic context is influenced in many ways and "a shared technical drawing assessment grid" was proposed in 2011 (Metraglia, 2011). This proposal plans to design a "grid for the evaluation of Technical Drawing learning levels" for the European Qualifications Framework (EQF) (Metraglia, 2011). Didactic concepts could base on these learning levels. To design competency-orientated learning, the principle Constructive Alignment is used to align learning objectives, assessment tasks and teaching/learning activities (Biggs, J. & Tang, C., 2011). Teachers have generally planned their courses from the perspective of subject content - students, based on their learning on the concrete requirements for the examination. These different perspectives can lead to a situation where learning and teaching are not aligned and this is perceived as unsatisfactory by both sides. The concept of Constructive Alignment helps not only to coordinate learning objectives and learning methods, but also to include the examination in the planning of a course. These learning objectives are the basis for the design of all learning and teaching activities and learning outcome assessments (Biggs and & Tang, 2011). The different cognitive levels (learning objectives taxonomy) are defined according to Bloom's origin and range from pure knowledge through understanding to evaluation and judgment (Bloom, 1974; Anderson et al., 1994).

Another framework for planning phases and work steps for designing virtual learning modules is provided by Arnold et al. to create E-Learning material (Arnold et al., 2004). Here, a chronological procedure with the different aspects of the concept like professional needs, the didactic structure, formal structure including the learning space and the operational structure are taken into account. Also, some important requirements that need to be addressed are mentioned: accessibility, necessity for quality management, adaptability, metadata, evaluation, didactic concept, and a concept for consolidation (ibid.).

The combination of e-learning and learning on-site is a central didactic task in the design of modern, education. "Blended Learning is the strategic combination of online and in-person learning" (Picciano et al., 2021). It is a successful learning and teaching concept if it works to coordinate those formats both in terms of content and organization. This was also established by the 2016 Horizon Report, which highlighted blended learning as a leading key trend in higher education (Johnson et al., 2016, p. 18). Blended Learning combines positive effects of both learning strategies showing "improved outcomes, though blended conditions tended to involve more instructional time, instructional resources, and elements encouraging peer interaction among learners" (Picciano et al., 2021). In contrast, the hybrid learning concept uses presence and digital learning taking place simultaneously (Liat and Einat, 2021).

Blended learning concepts can allow students more flexibility and individuality. They have the opportunity to work through the e-learning content on their own schedule and at their own pace. And teachers can also benefit from more flexibility in the execution of their teaching, which cannot be created in this way by pure face-to-face teaching. An analysis of national and international lecturer and learner surveys conducted by the University of Marburg revealed that lecturers want to provide more digital teaching opportunities for students in the future (Breitenbach 2021).

Nevertheless, a transformation phase of digital teaching is going on and further measures need to be taken to promote the integration of digital teaching into university teaching and to ensure the quality of the developed concepts, especially with regard to increasing the digital competence of teachers as well as addressing clarity, didactics, infrastructure, and legal issues (Krempkow et al. 2022; Marchwacka et al. 2023; Breitenbach, 2021). The provision of digital teaching and learning content with orientation aids represents a first step toward structured digital teaching. Some work to create sticky MOOC (Oakley et al., 2016) and discussing the challenges of creating a MOOC (Sánchez-Vera et al., 2015) were established among others.

3. INITIAL SITUATION, PRE-EVALUATION AND SHIFT

Considering these challenges, the learning and teaching content for technical drawing was to be digitized within the framework of a project of various German universities, described in this paper. In addition, making the content available as OER on the state-wide, cross-university online portal for study and teaching (ORCA.nrw) is a goal of the consortium. The target group consists of both teachers and learners in higher education at undergraduate level from engineering faculties.

Two surveys were conducted to assess the status quo according to the subject of technical drawing and to derive a framework for content development. First, 259 students from six universities of the consortium were surveyed about the difficulty and scope of the content as well as media use. On the other hand, semi-structured interviews were conducted with members of the consortium as well as the members of the project monitoring committee to obtain an overview of the didactic concepts currently implemented.

In the following, the most important results of 2020 (see Dillenhöfer, 2022), based on the components of Kerres' framework model of didactics (Kerres, 2018), are explained in more detail.

Conditions: The involved consortium states that there is a wide range of courses with different technical depth in the field of technical drawing. The focus is on the undergraduate level, namely 1st and 2nd semester, and medium to large numbers of participants (from 20 to 500 per university module, over 10,000 students will be addressed state-wide) are expected.

Teaching/learning content: The amount of content covered is perceived by students as average. When evaluating the degree of difficulty of the content, the answers tend to scatter on the possible scale from 1 (low) to 5 (too high) with a normal distribution around the mean value 3. This can be attributed to the heterogeneous prior knowledge due to different educational biographies (e.g. training, basic internship, no prior knowledge).

Media used: Opinions also differ in terms of the media used. There is a basic satisfaction, but also a need for improvement. So far, video recordings, slides, scripts, and textbooks are among the materials used.

Organization of learning: In the surveys, traditional formats like lectures and other concepts of front-of-class teaching predominate over digital formats. Digital formats are already widespread in online semesters due to the switched initial situation as a result of the Covid-19 pandemic. However, they are geared to face-to-face teaching and are generally designed from the perspective of the lecturer.

These findings result in the following requirements for the teaching and learning materials to be developed:

- No presupposition of previous knowledge (because of the early stages of study)
- Introduction of categories for grading the technical depth of the content (because of the different courses as target groups)
- Reduction of support effort and planning for scalability (because of the large number of participants)
- Planning and designing self-learning materials from the students' perspective (to ensure student-centered teaching)
- Enabling diversity of materials and media (to enable individual learning paths)

Furthermore, the analysis of the current state of digital media revealed that, despite the university-specific differences in the content offered, most subject areas are similar. For example, the topics of freehand sketches, dimensioning, sectional views or fits are represented at each of the universities. Therefore, a modularization of the contents in the later procedure seems to make sense. Also, very few universities have formulated concrete, competence-oriented learning objectives. However, these are essential for the Constructive Alignment.

However, the initial situation changed through the Covid-19 pandemic. Not only consortia meetings could only be held digitally, but also requirements for content creation switched. For instance, non-digital teachable skills like the creation of technical drawings by hand have become attentive. On the other hand, digital learning and the digital skills have become more common on both sides (teachers and students). Another driving factor was the variety of digital learning software that grew fast after the pandemic started. In conclusion, there were positive impacts for the creation of E-Learning materials.

There are some requirements restricted by the OER-portal ORCA.nrw, which overlap mostly with the E-learning requirements (see chapter 2). The most important requirements regarding the online course are listed as follows:

- (digital) accessibility,

• necessity for quality management,
• public license (CC BY-SA 4.0, CC BY or CC 0),
• compatibility with Blended Learning (see chapter 2),
• Constructive Alignment, compatibility with Blended Learning (see chapter 2),
• curricular integration,
• and a concept for consolidation (ORCA, 2022).

The consortium declared some hypotheses for the future work on the change on the situation regarding the technical drawing topic in higher education:

Hypothesis 1: By working in a large consortium with different teaching approaches, a versatility of learning materials is available. This simplifies the creation of modular teaching/learning content and ensures quality assurance for continuity. However, it is difficult to defy modular boundaries that are overall valid globally.

Hypothesis 2: Since metadata have become more and more important in times of fast-growing availability of OER-materials and MOOCS, it is crucial both that learners get the material they are searching for and that is suitable for them. Thornhill and Hyman mentioned user-focused metadata as making "the process of storing, sharing and tracking digital assets much easier" (Thornhill and Hyman, 2023).

Hypothesis 3: Through making teaching and learning goals more transparent, learners' acceptance on these goals are raising, which also elevates motivation.

It is expected that the misconceptions between teachers and students decrease, and the formulated competence targets coincide (Dillenhöfer, 2022). Metadata are important to find the well-suited material. Because the material was not fully created for the last winter semester, it was only an add-on to the teaching material at the universities and was not implemented inside the learning material. This will change in the future, where the material should be directly integrated into the topics the different universities deal with.

4. CREATION OF E-LEARNING MATERIALS

According to the structure plan, the data of both the participating project partners and some other statewide universities are collected, analyzed and processed. Afterwards, the main content topics will be developed in a structured manner, the lecture and examination framework conditions as well as common requirements and special areas will be defined. Content creation is assigned to all editors according to their expertise. Subsequently, the digital teaching/learning offerings are first conceptually created and then finalized before quality assurance measures are implemented. The project partners exchanged information and coordinated with each other. The requirements for the different stakeholders were then processed. Finally, work groups for each important topic were formed, so that the creation of the material could be done by means of their particular expertise. These groups then formulated learning goals with the help of didactic experts, which checked the compatibility with Constructive Alignment. While the topics are almost the same or similar at many universities across the board, there are some different focuses on certain topics, so it is difficult to define modular boundaries. On the other hand, many topics overlap and need to be linked together. Another challenge is that students have very different experiences when they enter university.

The aforementioned situation necessitates several levels of difficulty to meet the requirements. In order to do so, three competency levels namely easy (remember & understand), medium (apply & analyze) and difficult (evaluate & create) have been formulated according to blooms taxonomy (see chapter 2, see table 1). With these subcategories it is possible to adapt the learning materials to the different levels of experience and to the extent of dealing with a certain topic in different levels of complexity, according to the student's needs. Thereby, teacher can build their own course with the elementary topical components without losing actual eference to the main topic and learning goals.

The whole course focuses on a high level of interaction. Students should be able to evaluate themselves. To achieve this goal, tests are used to test whether the student reached a complexity level regarding a special topic or not. Another part where these tests are used is a learning map. This has been created so that (I) the students are able to get an overview of the learning landscape, as well as (II) the progress they made overall and to see clearly what is coming next or what are the future goals to be achieved. The progress is being tracked via different h5p-elements fading in and out, where the students see both a progress bar and a student icon going down a progressing learning path. Additionally, the tests for the taxonomy level reached fade in and out if one has completed the previous test with over 90 % of correct answers successfully. This gives the whole course a touch of a gaming approach, which should lift the students' motivation. Also, new (partially) virtual learning scenarios, namely Blended Learning, Inverted Classroom (Flipped Classroom) and Game Based Learning, considered were, so teachers are able to combine the material that they take from the course offer with the scenarios they use.

Table 1. Taxonomy levels according to Bloom divided into three subcategories in terms of modularity

Complexity level ↓

Remember	Copying, defining, finding, locating, quoting, listening, repeating, retrieving, …	easy
Understand	Annotating, associating, tagging, summarizing, relating, categorizing, …	
Apply	Articulate, reenact, loading, choosing, determing, displaying, judging, executing, …	medium
Analyze	Calculating, deconstructing, linking, organizing, attributing, explaining, …	
Evaluate	Arguing, validating, testing, assessing, criticising, debating, grading, predicting, …	difficult
Create	Building, adapting, composing, directing, simulating, solving, managing, negotiating, …	

5. EVALUATION

The evaluation is categorized into two layers: macro-evaluation (general MOOC/OER) and material-evaluation. The aim is to get reliable empirical results on the acceptance of learning goals and the perception of the whole course. Other factors besides acceptance that are being evaluated are: learning goals, didactics, and media related. In order to determine the degree of agreement regarding the questions, the Likert scale, developed by Rensis Likert (1931), from one to five (strongly agree-strongly disagree) is implemented (e.g. Croasmun and Ostrom, 2011; Gay, L. et. al., 2009).

The survey was conducted online and is structured into five parts:

1. General information (e.g. number of semesters)
2. Topic OER
3. Importance of the topics and personal assessment of the abilities with regard to the topics
4. Use and rating special topics
5. Personal comments

51 students participated in the final evaluation survey. Most of them are inside the first semester and the majority are studying mechanical engineering. Most of the students have no experience or previous knowledge regarding the topics. In general, the students think OER are positive and should be used more frequently in higher education. This is where teachers strongly agree. But in contrast, only 20 % of the students worked through the given topics, although the topics were important for the exam. Therefore, there are some limitations of part 4 of the survey structure (use and rating special topics) regarding the scalability of these results because less students answered these questions.

The evaluation regarding the learning material shows that every student agreed or strongly agreed, because these materials are coherent and a good support for the exam, well interactively designed, raised the knowledge, presented good in terms of media and well-suited for the lecture, except for one topic out of ten. Likewise, the results were positive for the acceptance of the learning objectives and the quality of the learning materials (see Fig.1). Few students (25 %) mentioned that the topic of creating technical drawings by hand is rather unimportant. This changed compared to the initial survey, where the majority (> 80 %) of students stated that drawings by hand are unimportant. The scope was judged as appropriate in each case. Exactly the same results as at the initial survey are reached regarding the scope for the topic CAD (Computer aided design), where the students wish a higher availability of learning material. Students also want more learning material for other topics, even though most didn't use the offered OER materials. The students also estimated their abilities and knowledge regarding all topics mostly as quite good or very good, so they estimate their ability as good.

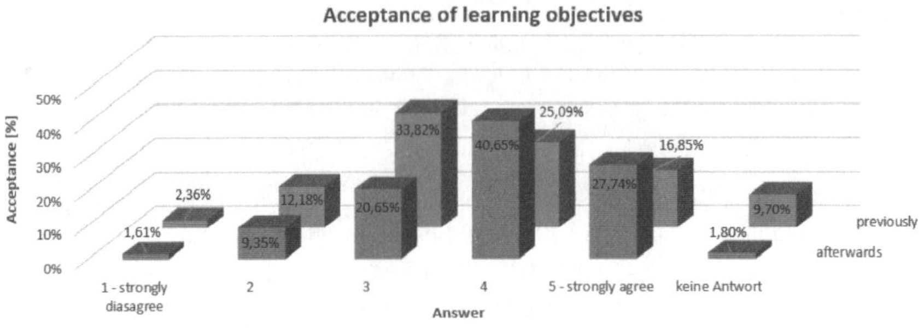

Figure 2. Acceptance of learning objectives by the students previously in 2020 and afterwards in 2023

Teachers stated in interviews that the modular learning material kit is transparent and easy to understand. They are satisfied with the extend of the offer, the structure and the division, how the material is split into subtopics of different taxonomy levels. However, the students found it more difficult to find the right taxonomy level that is needed for their study.

6. DISCUSSION

According to the initial survey, many factors and circumstances have changed. This led to digital skills that students and teachers have developed over the past few years. With respect to the media both teachers and students are more open to new possibilities and technologies, as well as better in adopting new skills for new software solutions (Raju, 2021). This also reflects in the final evaluation survey, where almost everybody agreed that the media used are good to work with and helped to reach teaching/learning goals. Furthermore, the acceptance of some topics have been grown because of transparent learning objectives. That means that the implementation of the Constructive Alignment to tackle the problems found in the initial survey worked quite well. Now students seem to understand, why certain skill sets are not only crucial for future topics, but also for professional skills needed in the workforce later on. However, students and teachers review OER material good in general. The positive status could be explained by the expansion of available learning materials as well as the possibility of free use and public discussion.

On the other hand, using OER materials voluntary is not working well, because according to the final evaluation, only few students used these learning materials. Course administrators stated that over 300 students are registered in the course and many other students use the experimental course without authentication. The tracking of activities shows that many students executed the test for checking the taxonomy level. It seems that the students who used the material did not participate in the survey and vice versa. At this point, it can be recognized that voluntary surveys do not work fine as they should be. Reinforcing factors are needed to raise the participation. Moreover, when the material will be integrated into a special lecture, it is assumed that more students attend in the survey, as well as the overall OER learning material.

Another result described in chapter five is that students wish more learning materials in terms of some topics, but they do use the offered OER materials. These must be put inside the official teaching material; otherwise, Students will disregard these topics since it may seem unimportant for the exam. Therefore, the students focus more on preparing for the exam instead of trying to achieve skills for future profession. In this course, this topic is only discussed inside the motivation/introduction, which many students skip. A possible solution would be making the introduction mandatory, so that the following topics/material only fades in, when the introduction is finished.

Metadata are an important tool to help finding the material one is searching for. These Metadata were prepared but have not been entirely applied yet. These will be included fully when the course will be uploaded to the ORCA.nrw platform where also OER-searching indexes are implemented. Then it should be also easier for students to find the material that fits their needs.

In general, it was an advantage to create the MOOC/OER learning materials with many partners, because of the impact of different perspectives and experiences. All topics are covered, and the work can be split according to each expertise. Splitting the work helps to cover the whole topical landscape, so that the offer is complete, and every teacher is able to serve themselves. Building work groups helps to effectively editing and produce the learning materials. The partners can then start a review process on the learning material so quality is being ensured.

Besides the assurance of quality, consolidation is important for future use. This is mainly complied through the complete topical landscape that is covered, the consideration of various requirements of different groups involved, the quality provisions and measures such as evaluation, offer materials with public license so one can edit and adapt it, and publishing results. Moreover, all universities need to be informed that these offers are now available and open to use freely.

The creation of an E-learning MOOC a lot of work at the beginning, but if it is used frequently and the tests are evaluated automatically, development is worth it. Then, there is less work for teachers to assess the students. Using learning analytics can also help both teachers and students to obtain interesting information about the learning progress, so they are able to supervise precisely and quickly if problems occur.

Moreover, it is difficult to create modular learning materials that are coherent. When someone picks out the parts they need for their degree, there is a risk that the materials lose the connection to the other materials or the course. This has to be taken into account when creating modular learning materials.

7. CONCLUSION

This paper presented a framework on how to integrate OER into MOOC with regards to modularity. At the beginning, the initial situation including the requirements of the various groups was demonstrated and how this situation shifted through the Covid-19 pandemic. Furthermore, it is shown how the modular aspect can be realized with the didactic model of Constructive Alignment. Creating E-Learning materials or respectively a MOOC with many partners is a big advantage and should be supported, so that hypothesis 1 is correct.

Hypothesis 2 deals with the importance of metadata. It was not possible to analyze this hypothesis fully, because the MOOC was not implemented into the learning portal where many education offers are present. Thus, the searching and finding have not been evaluated yet. But since teachers found these metadata helpful in pretests, and students had difficulties finding the right material, it is estimated that hypothesis 2 can be confirmed. However, further work needs to be done when the learning material is actually implemented inside the lectures. Students hesitate to use it according to interviews, so it should not be available as a coexisting offer.

Last but not least, according to hypothesis 3, misconceptions, such as the acceptance of learning goals, between teachers and students decreases and the formulated competence targets coincide, if the goals are made more transparent. This can be confirmed, because the acceptance regarding some not accepted learning objectives in 2019 has been grown (see chapter 5).

Another factor is the omnipresent license information and logos. These ever-appearing distractions could impede the learning process. It should be researched how these influences on the learners when they access the material and see these information everywhere at the beginning or at the end.

Finally, the course will be published on ORCA.nrw soon and should be available during the summer term 2023.

ACKNOWLEDGEMENT

We gratefully acknowledge financial support through the DH.NRW with funds provided by the Ministry for Culture and Science of the state of North Rhine-Westphalia (MKW) under Grant No. 214-5.01.03.02 - 145504.

REFERENCES

Anderson, L. W.; Sosniak, L. A.; Bloom, B., 1994. Bloom's taxonomy: a forty-year retrospective, Chicago (Ill.): University of Chicago press. Online verfügbar unter http://lib.ugent.be/catalog/rug01:000925762.

Arnold, P., Kilian, L., Thillosen, A., Zimmer, G., 2004. E-Learning. Handbuch für Hochschulen und Bildungszentren; Didaktik, Organisation, Qualität. 1. Aufl. Nürnberg: BW Bildung und Wiss. Verl.

Biggs J. & Tang, C., 2011.Teaching for Quality Learning at University: What the Student Does (4th ed.): Maidenhead: Open University Press.

Bloom, B. S., Engelhart, M. D., Furst, E. J., Holl, W. H. & Krathwohl, D. R., 1974. Taxonomie von Lernzielen im kognitiven Bereich. Beltz.

Breitenbach, A., 2021. Digitale Lehre in Zeiten von Covid-19: Risiken und Chancen. Marburg, 18 S. - URN: urn:nbn:de:0111-pedocs-212740 - DOI: 10.25656/01:21274

Croasmun, J. T.; Ostrom, L., 2011. Using Likert-type scales in the social sciences. In: J Adult Educ 40, S. 19–22.

Dillenhöfer, F., Künne, B., 2022. Analysis of learning objects for optimization and digital transfer report if intermediate results on learning path ideas: first suvey. In M. B. Nunes & P. Isaias (Hrsg.), Proceedings of the International Conference E-Learning 2022 (Postprint, S. 169–172). IADIS.

Gay, L. R., Mills, G. E., & Airasian, P., 2009. Educational research: Competencies for analysis and applications. Columbus, OH: Merrill; World Journal of Social Sciences and Humanities.

Johnson, L., Adams Becker, S., Cummins, M., Estrada, V., Freeman, A. & Hall, C., 2016. NMC Horizon Report 2016 – Higher Education Edition. Austin/Texas: The New Media Consortium.

Kerres, M., 2018. Bildung in der digitalen Welt: Wir haben die Wahl. In: denk-doch-mal.de Online-Magazin für Arbeit-Bildung-Gesellschaft (02-18). Online verfügbar unter http://denk-doch-mal.de/wp/michael-kerres-bildung-in-der-digitalen-welt-wir-haben-die-wahl/.

Krempkow, R., Wilhelm,E., Zawacki-Richter, O. (Hg.), 2022. Beiträge zur Hochschulentwicklung. 1. Auflage. Norderstedt: BoD - Books on Demand (Zeitschrift für Hochschulentwicklung Jg. 17, 2).

Liat, E., Einat. G.: Hybrid learning spaces - a three-fold evolving perspective. 2021. DOI 10.13140/RG.2.2.30059.26407.

Marchwacka, . A., Kugler, J., Schaal, T., Tolks, D., 2023. Digitale Hochschullehre im ersten COVID-19-Semester. Ergebnisse einer Befragung von Lehrenden in Public Health, Medizin und Pflege. In: Präv Gesundheitsf 18 (1), S. 22–29. DOI: 10.1007/s11553-022-00937-1.

Metraglia, R., Baronio, G., Villa, V., 2011. Learning Levels in Technical Drawing Education: Proposal for an assiignment grid based on the European qualifications framework (EQF), Interantional Conference on Engineering Design, ICED11.

Oakley, B., Poole, D., Nestor, M. A., 2016. Creating a sticky MOOC. In: Online Learning 20. DOI: 10.24059/olj.v20i1.731.

ORCA.nrw, 2022. For Teachers, call for tenders; https://www.orca.nrw/lehrende/ausschreibungen; last accessed on 03/19/2023.

Picciano, A.G., Dziuban, C.D., Graham, C.R, Moskal, P.D. (Eds.), 2021. Blended Learning: Research Perspectives (1st ed.), Routledge. https://doi.org/10.4324/9781003037736.

Raju, R., Md Noh, N. H., Ishak, S. N. H., Eri, Z. D, 2012. Digital Tools Acceptance in Open Distance Learning (ODL) among Computer Science Students during COVID-19 Pandemic: A Comparative Study; Asian Journal of University Education, v17 n4 p408-417.

Sánchez Vera, M., Leon, M., Davis, H., 2015. Challenges in the Creation, Development and Implementation of MOOCs: Web Science Course at the University of Southampton. In: Comunicar 22, S. 37–44. DOI: 10.3916/C44-2015-04.

Shah, D., 2021. By The Numbers: MOOCs in 2021, A decade has gone by since MOOCs' popularization. They've now reached 220M learners. Here are the stats. https://www.classcentral.com/report/mooc-stats-2021; last accessed on 03/19/2023.

Thornhill, T., Hyman, A., 2023. The importance of good-quality, user-focused metadata: How the Transport for London corporate archives benefit from voluntary standards and meaningful metadata. Journal of Digital Media Management, Henry Stewart Publications.

Unesco, 2019. Open Educational Resources; https://www.unesco.org/en/open-educational-resources; last accessed on 03/19/2023.

TOWARDS AN EFFECTIVE BASELINE OF IT GOVERNANCE MECHANISMS IN HIGHER EDUCATION INSTITUTION

Abdelilah Chahid, Souad Ahriz, Kamal El Guemmat and Khalifa Mansouri
Department of Mathematics and Computer Science, Higher Normal School of Technical Education
Mohammedia, Morocco

ABSTRACT

In recent years, higher education institutions have set up technological infrastructures made up of a variety of applications, different platforms, academic systems and cloud applications. This situation leads to a management, cost and performance problem that requires the adoption of effective governance mechanisms consistent at the implementation level. On the other hand, the adoption of these information technology governance (ITG) mechanisms in higher education is not treated in the same way in terms of efficiency and implementation in the literature. In order to determine what the main different implementations of these mechanisms and their effectiveness, we propose a literature review (LR). This review is based on a comparative analysis of the different approaches resulting from multiple case studies targeting hundreds of universities in different countries. Although these mechanisms have long been used in the private sector, this article shows that they are increasingly adopted in the literature and practice of the academic sector. The qualitative and quantitative results of this LR revealed that most of the good practices that have shown effectiveness are implemented within universities. At the end, our article presents the recommendations of the implementation of these mechanisms in order to set up a more efficient ITG.

KEYWORDS

IT governance, Universities, Mechanisms, ITG, Higher education institutions, Practices

1. INTRODUCTION

The evolution of information technologies is becoming more and more vital to guarantee a good performance and efficiency to the organization (Ribeiro & Gomes, 2009). To cope with innovation and change and adapt to new technologies, universities frequently need to improve their IT governance mechanisms. An IT governance framework can be deployed with a set of mechanisms such as structure, process and relational mechanism (De haes & Grembergen, 2004; Grembergen & al., 2011; Peterson, 2004; Weil & Ross, 2005).

Higher education institutions are a type of organization where the technological infrastructure consists of different platforms and heterogeneous technologies used in the academic process to support the activity of research, education and learning. All these technologies require effective IT governance (Noor Azizi, 2008; Reschiwati & al., 2021) which can only be implemented with a good definition of the different implementations and the efficiency of the most important mechanisms.

The treated problem aims to show the effectiveness of these IT governance mechanisms and their different implementations in the literature. Several authors have argued that Higher Education Institutions (HEI) should use ITG mechanisms (Jairak & al., 2015; Khouja & al., 2018; Meçe & al., 2020) but little research attempts to provide the efficacy of these ITG mechanisms. Our research is an exploratory study, adopting a strategy based on surveys of the state of the art in hundreds of universities and case studies in various countries in order to draw the most effective and implemented good practices from these mechanisms. The qualitative and quantitative results of this literature review revealed that the majority of good practices, which have shown effectiveness, are implemented within universities. However, other studies have found that universities are still between the first and third level of maturity regarding ITG (Krisanthi & al., 2014; Pawanel & al., 2019; Sadikin & al., 2014).

This article will be structured as follows: Section 2 will introduce the study methodology; Section 3 will present a literature review of the most implemented and efficient mechanisms. Section 4 will discuss recommendations of the results found. Section 5 will summarize the study and discuss the perspectives and proposals for future research.

2. METHODOLOGY

In order to carry out our literature review, we consulted specific online database, such as Direct Science, Web of Science, IEEE Explore, and AIS and Library and Google scholar. We drew on the work of (De Haes & al., 2009; Almeida & al., 2013) in order to identify a list of the most used ITG good practices. Subsequently, this list was used as a reference to carry out a research in the literature in order to identify the most implemented and effective good practices in HEI. We read 60 articles on ITG mechanisms within universities among which 18 articles were identified as relevant by respecting the following selection criteria. They must:

- Be related to ITG,
- Explicitly mention ITG mechanisms,
- Contain a clear and unambiguous definition of the ITG mechanisms mentioned in the studies (De Haes & al., 2009; Almeida & al., 2013).

The steps of the research are mentioned in Figure 1.

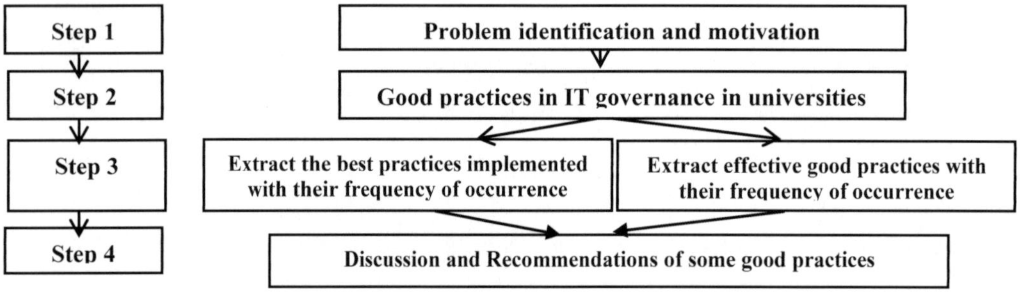

Figure 1.Research steps

3. GOVERNANCE MECHANISMS

Effective IT governance is a set of best-practice mechanisms classified into three categories: structure, processes and relational mechanisms. (Lunardi & al., 2016) has shown that this efficiency is crucial for an organization to achieve its performance objectives. Previous study has shown that effective IT governance relies heavily on the ITG practices of these mechanisms, ranging from decision making at the board level to the implementation of IT control (Wu & al., 2015).

We summarize, in Table 1, the approaches used by universities in different countries. In Figures 2 and 3 we present the percentage by approach of effective and implemented good practices found in LR. While tables 2 and 3 summarize the good practices according to their frequency of occurrence and present their respective references.

3.1 Governance mechanisms implemented and effective

Reading the selected articles allowed us to identify the most relevant ITG mechanisms within universities in various countries, as well as their purpose (see Table 1)

Table 1.Summary of the different approaches

Author	Item	Country	Treated mechanism
Bianchi & al., (2021)	1	Brazil, Spain, Portugal, Netherlands	Structure, Process, Relational
Dlamini (2013)	2	United States	Structure, Process, Relational
Fattah & al. (2021)	3	Indonesia.	Process
Jack mc credie (2008)	4	Canada, USA	Structure, Process, Relational
Bhattacharjya & al. (2006a)	5	Australia	Structure, Process, Relational
Bhattacharjya al. (2006b)	6	Australia	Structure, Process, Relational
Bianchi & al. (2016)	7	Many countries	Structure, Process, Relational
Noor Azizi (2008)	8	Malaysia	Structure, Process, Relational
Bianchi Isaías Scalabrin & al. (2018)	9	Malaysia	Process
Ko & Fink (2010)	10	Australia	Structure, Process, Relational
OM Adjei1 & al. (2020)	11	Ghana	Structure, Process, Relational
Binyamin & Husnayati (2016)	12	Malaysia	Structure, Process, Relational
Hicks & al. (2010)	13	Australia	Structure, Process, Relational
Montenegro & Flores (2015)	14	Ecuador	Structure, Process, Relational
Fattah & Setyadi (2021)	15	Indonesia,	Structure, Process, Relational
Putz Rosane Zaneti & al. (2017)	16	Brazil	Structure, Process, Relational
Bianch (2017)	17	Brazil, Dutch and Portugal	Structure
Dodi Setiyawan (2019)	18	China	Structure, Process

3.2 Frequency of good practices implemented and effective

From the analysis of the different approaches (Table 1), we noticed that there are qualitative and quantitative studies which have dealt with the implementation of good practices without showing their effectiveness. This allowed us to have a vision of the best most used practices most used in universities. While others have shown efficiency at the level of structure, process, relational. On the other hand, rare studies have proposed conceptual models for a possible evaluation. Figures 2 and 3 show the percentage of good practice studies by approach.

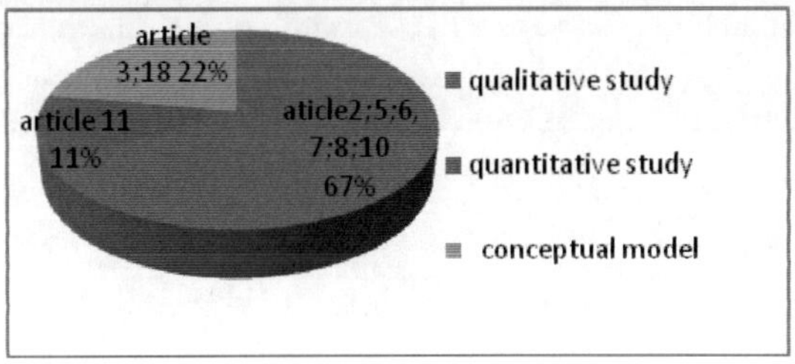

Figure 2. Percentage by approach of good practices implemented

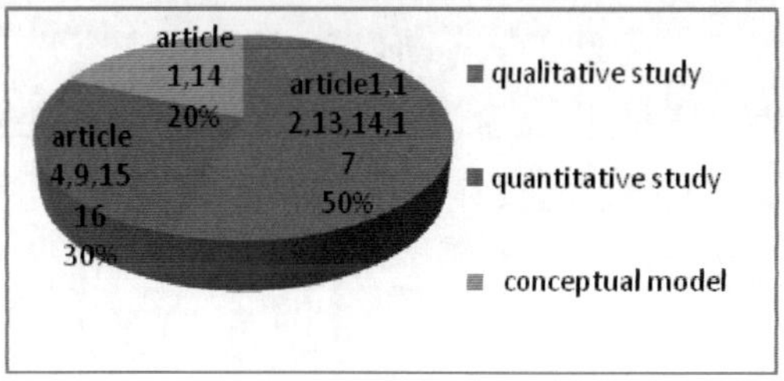

Figure 3.Percentage by approach of Good Effective Practices

Table 2. Frequency of good practices implemented (GPI)

Mechanism	Good practice		Item 2	3	5	6	7	8	10	11	18	Frequency
Structure	S1: Roles and responsibilities		x		x	x			x	x		5
	S2 : IT Strategy Committee		x		x	x		x	x	x		6
	S3 : IT steering committee				x		x	x	x	x	x	6
	S4 : Structure of the IT organization		x		x	x	x	x	x	x		7
	S5 : CIO to the Executive Committee		x			x		x	x	x		5
Process	P1: Information system planning strategy		x		x	x		x				4
	P2: Portfolio Management		x			x	x	x	x			5
	P3: IT budget control and reports		x					x				2
	P5:Service level agreements									x		1
	P4:ITG framework	Cobit			x	x	x		x	x	x	6
		Pmbok,			x	x						2
		Itil,			x	x	x		x	x	x	6
		ISO 17799			x	x			x			3
		Bsc		x	x		x			x		4
		p-cmm				x						1
		Cobra			x							1
		ISO / IEC 27002								x		1
		ISO IEC 38500.									x	1
Relational	R1: IT leadership		x		x	x			x	x		5
	R2: Shared understanding of business / IT objectives,		x						x			2
	R3: Internal communication		x		x	x	x	x	x	x		7
	R4: Active participation and collaboration between the main stakeholders,				x			x			x	3
	R5: Sharing knowledge on IT governance					x					x	2
	R6: IT staff training		x			x		x				3

Table 3.Frequency of Good Effective Practices (GEP)

Mechanism	Good practice		Item 1	16	15	14	13	17	4	9	12	Frequency
Structure	S1: Roles and responsibilities			x	x	x	x		x			5
	S2: IT strategy committee,		x	x	x	x	x				x	6
	S3: IT steering committee,		x			x			x		x	4
	S6: Project management office		x	x								2
	S7: Business process management office		x									1
	S4: IT organization structure		x	x		x	x	x			x	6
	S5: CIO on the Executive Committee		x	x		x	x				x	5
Process	P1: Information system planning strategy,		x	x					x		x	4
	P2: Portfolio Management,		x						x		x	3
	P3: IT budget control and reports		x						x		x	3
	P4: ITG framework	Cobit		x		x			x		x	4
		Pmbok,								x		1
		Itil,	x			x			x	x		4
		ISO 27001							x	x		2
		Bsc		x	x		x		x		x	5
		Prince2								x		1
Relational	R1: IT leadership		x						x		x	3
	R2: Shared understanding of business / IT objectives,					x						1
	R3: Internal communication		x		x	x					x	4
	R4: Active participation and collaboration between the main stakeholders,			x		x						2
	R5: knowledge sharing on IT governance		x			x					x	3
	R8: Job rotation										x	1
	R6: IT staff training		x								x	2
	R7: A partnership with the software industry		x									1

4. RESULTS AND DISCUSSION

In this section, we examine the quantitative and qualitative results of ITG mechanisms, comparing their effectiveness in implementing the structure, process, and relational mechanism. We also identify efficiency limitations mentioned in Table 4 and provide recommendations to overcome them. Figures 4, 5, 6, and 7 confirm our research by showing that the most effective best practices are being implemented in universities and that their proportional values are similar.

4.1 Structure

Figure 4 demonstrates that most effective best practices for organizational structure are implemented in universities, except for S6 and S7. This is due to the highly complex existing structure remaining unchanged and the size of these universities.

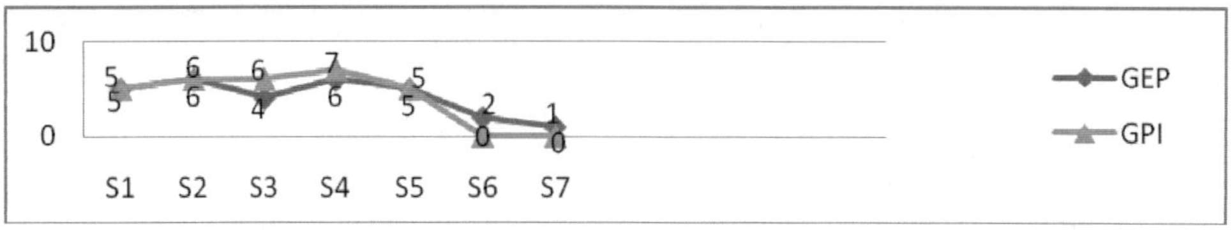

Figure 4. Comparison between GEP and GPI structure

4.2 Process

Figure 5 shows that most of the implemented processes showed efficiency in universities with the exception of P5. However, the proportional values are very close. This comes down to using common norms and standards to avoid risk in process execution. Figure 6 shows that COBIT, ITIL, BSC and ISO frameworks are the most used practices.

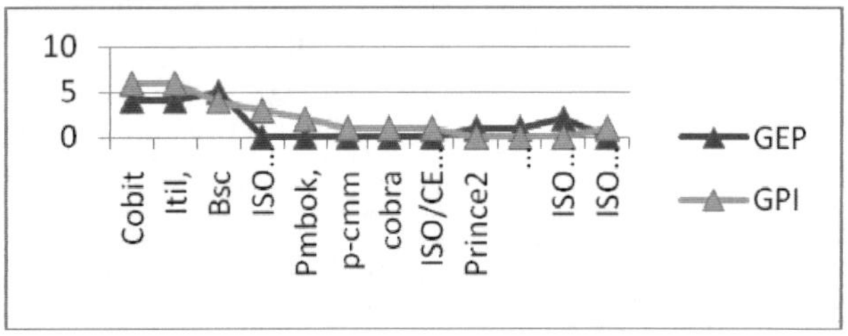

Figure 5.Comparison between GEP and GPI of processes

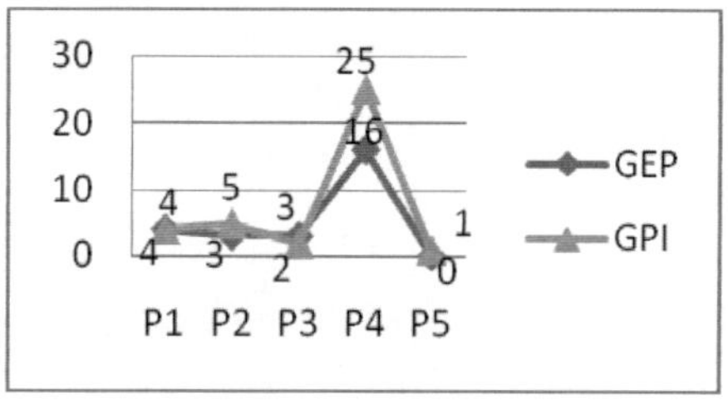

Figure 6. Comparison between GEP and GPI of the ITG framework process (P4)

4.3 Relational

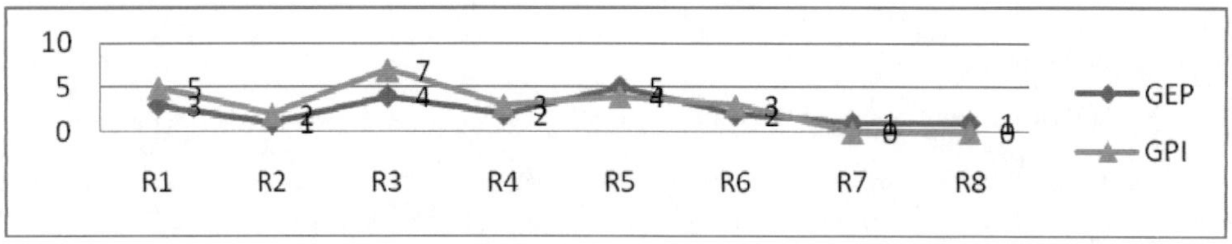

Figure 6.Comparison between the GEP and GPI relationship

According to Figure 7, the good practices R1 and R3 have shown efficiency and are the most implemented in universities. This is due to the important role of these good practices as a decisive factor in the success of governance. On the other hand, R7 and R8 are considered to be effective for a possible implementation.

4.4 The recommendations for an effective adoption of the baseline of IT governance

By examining the best governance practices presented in this review, as well as the qualitative and quantitative results, the researchers' personal conclusions and opinions, we have developed a baseline to adopt the effective recommended practices for higher education institutions, as illustrated in Figure 8. For an effective implementation of this baseline, a set of guidelines and recommendations are presented in Table 4.

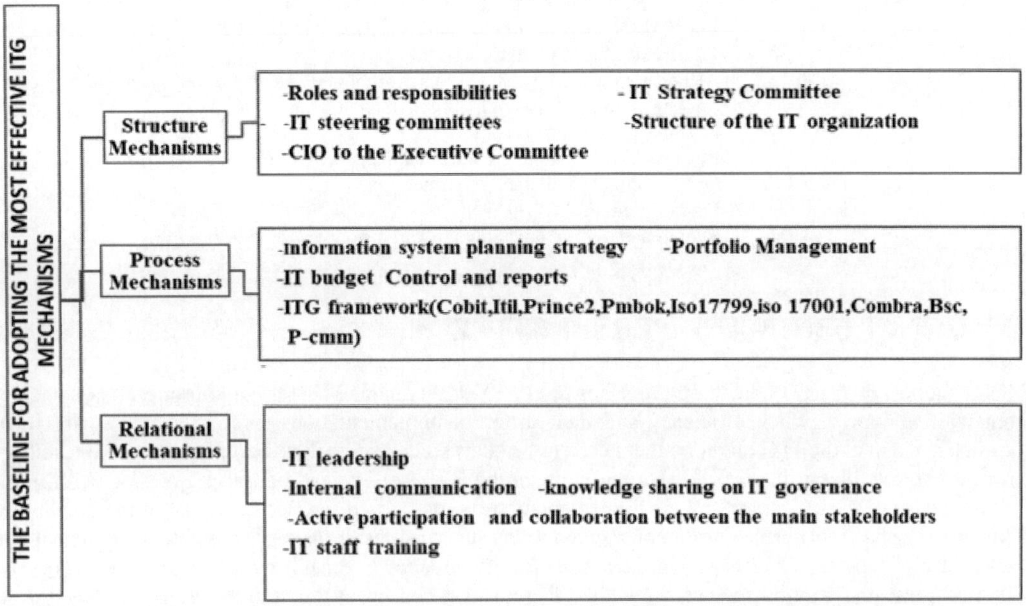

Figure 7. The baseline for adoption of the most effective IT governance mechanisms

Table 4.The recommendations of the most effective and implemented good practices

Mechanism	Good practice	Recommendations
Structure	S1	-The criteria for the distribution of governance tasks -Clearly define roles and responsibilities (Ko & Fink, 2010). -The decision concerning information and computer science is very complex (Yanosky & al. 2008)
	S2	- Identification of the needs of the different levels of IT actors (Attaranel & al., 2017)
	S3	-The criteria for defining the roles and responsibilities of the Committees (Albrecht & al., 2004) -Overlaps in committee decisions (Ko & Fink, 2010).
	S4	Centralizes: - Data storage management -Change in the size of the University Federal: - Data storage management -Management of coordination and communication between strategies and centralized roles and decentralized IT operations (Bianchi & al., 2017)
	S5	-DSI prerogative within the committee -Requirement of the choice of CIO within the committee (Dlamini, 2015)
Process	P1	-Criterion for the choice of priorities and investments in the IT domain -Resilience and agility to environmental change (Lemoine, 2020) -Quality of strategic plan
	P2	-Difficulty of defining the priorities of the projects which really impact the university (Ahriz & al., 2018) -Political requirement that affects the choice of projects

	P3	-Define a coherent financing strategy - Estimate of capital budget and IT project priorities
	P4	-Define the success factors of the framework - The challenges of implementation -The need for human resources -The choice of the ideal setting
Relational	R1	-Skill level and experience (Dlamini, 2015) -Mature relationship between IT and business
	R2	-Share common IT goals -Understand the purpose of IT in universities
	R3	-Informal channels of communication (Bianchi & al. ,2021) -Multiple means of communication
	R4	-A cooperation strategy as a means ofbringing together the various interests and realizing vision (Noor Azizi, 2008; Albrecht & al. ,2004)
	R5	-Requires good communication -Resilience to change in the information system
	R6	-Orient training to the field of strategy -Cost of training -Requirement of material and software resources for training

5. CONCLUSION AND PERSPECTIVE

Some authors of the ITG literature (De Haes & al., 2009; Almeida, 2013) use and propose ITG mechanisms but without showing their efficiency and their different implementations. We argue that such a fact is a knowledge gap in the ITG literature that needs to be addressed. The qualitative and quantitative results of this review of the literature not only solved this problem of efficiency but also revealed a proportional convergence of the efficiency and the implementation of good practices. Because most of the good practices, which have shown efficiency, are implemented within universities, it makes it possible to achieve a positive relationship between effective IT governance and governance mechanisms. However, the different case studies presented in this review show that there is no consensus on a framework or standard for IT governance to be used in higher education institutions. The multiplicity of information system frameworks leads to a myriad of organizational and technical problems (El Yamami & al., 2018). Additionally, good practices S6, S7, R7, and R8, which have shown effectiveness, have not been implemented in universities, while P4 has been well implemented without demonstrating effectiveness. To overcome implementation challenges, we have proposed recommendations that constitute a guideline for effective ITG in (HEI).

The main limitation of this research is that it only focuses on certain universities, which prevents the generalization of results. As future work, we need to focus on how to evaluate IT governance mechanisms in EES in Morocco using design science research (DSR). The implementation of effective mechanisms from different case studies could lead to this evaluation to reveal the strengths and weaknesses of the mechanisms discussed in this article.

REFERENCES

Adjei, PO, & Yaokumah, W. Information Technology Governance Structures, Processes and Mechanisms in Ghanaian Universities.

Ahriz, Souad, Yamami, A. E., Mansouri, Khalifa, & al. (2018). Cobit 5-based approach for IT project portfolio management: Application to a Moroccan university. *International Journal of Advanced Computer Science and Applications (IJACSA),* 9 (4), 88-95.

Ajayi Binyamin, Hussin Husnayati, & al. (2016). IT governance from practitioners' perspective: sharing the experience of a malaysian university. *journal of Theoretical & Applied Information Technology*, 88 (2).

Albrecht B.; Pirani, J.A. Using an IT Governance Structure to Achieve Alignment at the University of Cincinnati. Available *online: https://library.educause.edu/resources/2004/5/using-an-it-governance-structure-to-achieve-alignment-at-theuniversity-of-cincinnati* (accessed on 10 May 2004).

Almeida Rafael, Pereira Rúben, & Da silva, Miguel Mira (2013). IT governance mechanisms: A literature review. In*: International Conference on Exploring Services Science*. Springer, Berlin, Heidelberg, 186-199.

Attaran, Mohsen, Sharmin, & Celik, Bilge Gokhan (2017). Promises and challenges of cloud computing in higher education: a practical guide for implementation. *Journal of Higher Education Theory and Practice*, 17(6), 20-38.

Bhattacharjya, Jyotirmoyee & Chang, Vanessa (2006). An exploration of the implementation and effectiveness of IT governance processes in institutions of higher education in Australia. *IT Audit: Strategic Measures for Performance, Value & Quality*, 153-163.

Bhattacharjya, Jyotirmoyee & CHANG, Vanessa (2006). Evolving IT governance practices for IT and business alignment: A case study in an Australian institution. *Facing the Information Society*.

Bianchi, Isaias Scalabrin, Sousa, Rui Dinis, Pereira, Rúben, & al. (2017). IT governance structures in Brazilian, Dutch and Portuguese universities. *Procedia Computer Science,* vol. 121, 927-933

Bianchi, Isaías Scalabrin & Sousa, Rui Dinis (2016). Frameworks used for IT governance at universities: An exploratory study.

Bianchi, Isaias Scalabrin & Sousa, Rui Dinis (2016). IT Governance mechanisms in higher education. *Procedia Computer Science*, vol. 100, 941-946.

De Haes, Steven & VanGrembergen, Wim (2004). IT governance and its mechanisms. *Information systems control journal*, vol. 1, 27-33.

De Haes Steven & VAN GREMBERGEN Wim (2009). An exploratory study into IT governance implementations and its impact on business/IT alignment. *Information Systems Management*, 26 (2), 123-137.

Dlamini Reuben S. (2015). The role of the strategic and adaptive Chief Information Officer in higher education. *Education and Information Technologies,* 20 (1), 113-140.

EL Yamami Abir, Mansouri Khalifa, Qbadou Mohammed, & al. (2018). A New Pattern for the Deployment of IT Governance Frameworks in Organizations. *International Journal of Engineering &Technology*, 7 (4), 3459-3465.

Fattah, Anwar & Setyadi, Resad. Determinants Effectiveness Information Technology Governance in Higher Education Institution (HEI) using partial least squares structural equation modeling approach (PLS-SEM). In: *Journal of Physics: ConferenceSeries*. IOP Publishing, 2021. p. 012007.

Fattah, Anwar, Saragih, Hoga, et Setyadi, Resad. Determinants Effectiveness of Information Technology Governance and IT Performance in Higher Education Institution (HEI): A Conceptual Framework. *International Journal of Science, Technology& Management,* 2021, vol. 2, no 1, p. 36-47.

Hicks, Michael, Pervan, Graham, et Perrin, Brian. A case study of improving information technology governance in a university context. In: *IFIP Working Conference on Human Benefit through the Diffusion of Information Systems Design Science Research.* Springer, Berlin, Heidelberg, 2010. p. 89-107.

Jairak K.; Praneetpolgrang P.; Subsermsri, P. (2015). Information technology governance practices based on sufficiency economy philosophy in the Thai university sector. Inf. Technol. People 28, 195-223.

Khouja, M.; Rodriguez, I.B.; Ben Halima, Y.; Moalla, S. (2018). IT Governance in Higher Education Institutions. Int. J. Hum. Cap. *Inf.Technol. Prof*., 9, 52-67.

KO, Denise & FINK, Dieter (2010). Information technology governance: an evaluation of the theory-practice gap. Corporate Governance: *The international journal of business in society*.

Krisanthi, Gusti Ayu Theresia, Sukarsa, I. Made, & BAYUPATI, I. Putu Agung (2014). Governance audit of application procurement using COBIT framework. *Journal of Theoretical and Applied Information Technology,* 59(2), 342-351.

Lemoine, Pamela A. & RICHARDSON, Michael D (2020). Planning for higher education institutions: Chaos and the covid-19 pandemic. Educational Planning, 27 (3), 43-57.

Lunardi, G. L., Maçada, A. C. G., Becker, J. L., & Van Grembergen, W. (2016). Antecedents of IT Governance Effectiveness: An Empirical Examination in Brazilian Firms. *Journal of Information Systems*, 31(1), 41-57.

Meçe, E.K.; Sheme, E.; Trandafili, E.; Juiz, C.; Gómez, B.; Colomo-Palacios, R. (2020) Governing IT in HEIs: Systematic Mapping Review. Bus. *Syst. Res. J*., 11, 93-109.

Montenegro, Carlos W. & Flores, Denys A. 2015. An integrated model for ICT Governance and management at the Council for evaluation, accreditation and quality assurance of higher education institutions in Ecuador (CEAACES). In: *International Conference on Computing, Communication and Security (ICCCS). IEEE*, 2015. 1-9.

Noor Azizi Ismail (2008). Information technology governance, funding and structure: A case analysis of a public university in Malaysia. *Campus-Wide Information Systems.*

Pawan, Elvis, Utami, Ema, Yunita, Selviana, & al. (2019). Measurement of Maturity Level Higher Education Governance Using Balanced Scorecard (BSC) and COBIT 4.1. In *International Conference on Information and Communications Technology (ICOIACT). IEEE*, 2019. 948-953.

Putz, Rosane B. Zaneti, RASOTO, Vanessa I., & ISHIKAWA, Edison. Brazilian (2017). Federal universities information technology governance: An analysys of the strategic alignment dimension. In: 2017 *12th Iberian Conference on Information Systems and Technologies (CISTI).* IEEE, 2017. 1-7.

R. Peterson, “Crafting information technology governance,” *Inf. Syst. Manag.*, 2004.

Ribeiro Jorge & Gomes Rui (2009). IT governance using COBIT implemented in a high public educational institution: a case study. In: *Proceedings of the 3rd international conference on European computing conference.* 41-52.

Reschiwati, WiwikPratiwi, Suratman, Adji, Ibrahim, Ida Musdafia, & al. (2021). Implementation of Good University Governance in Private Higher Education Institutions in Indonesia. *Journal of Southwest JiaotongUniversity,* 56 (3).

Scalabrin Bianchi, Isaías, Dinis Sousa, Rui, & Pereira, Ruben (2021). Information technology governance for higher education institutions: A multi-country study. In*: Informatics. Multidisciplinary Digital Publishing Institute*, p. 26.

Setiyawan, Dodi. A Proposed Model of IT Governance within Cloud Computing and Data Management in Higher Education. *International Journal of Advanced Engineering Research and Science*, 2019, vol. 6, no 10, p. 2349-6495

W. Van Grembergen, S. De Haes, and E. Guldentops, 2011.Structures, Processes and Relational Mechanisms for IT Governance, in *Strategies for Information Technology Governance.*

Wu, S. P.-J., Straub, D. W., & Liang, T.-P. (2015). How information technology governance mechanisms and strategic alignment influence organizational performance: Insights from a matched survey of business and its managers. *Management Information Systems Quarterly*, 39(2), 497-518.

W. Peter and R. Jeanne, (2005). “A Matrixed Approach to Designing IT Governance,” *MIT Sloan Manag. Rev.*

Yanosky, Ronald & CARUSO, Judith Borreson (2008). Process and politics: IT governance in higher education. *ECAR Key Findings.*

EDUCATIONAL CHALLENGES WITH CHATGPT: INTEGRATE OR MARGINALIZE?

Vitor Gonçalves and Bruno F. Gonçalves
CIEB, Instituto Politécnico de Bragança
Campus de Santa Apolónia, 5300-856 Bragança, Portugal

ABSTRACT

Artificial intelligence language models have revolutionized the way education is accomplished, namely the teaching-learning processes. Currently there is a very broad debate about these models that is not only in the academic and scientific communities, but also in the civil community. It is discussed and reflected, essentially, about ChatGPT as a model that can contribute to innovation in the educational context, but also about the ways in which this technology can condition learning and calls into question the functions of teachers and even the current school. It is in this sense that this research aims, by conducting a systematic review of the literature, to identify the challenges that language models, namely the ChatGPT bring to the educational context. The results point to the importance of integrating these models in the teaching-learning process, but also to the existence of a set of challenges for the school as an institution, but also for teachers and students as central actors in the educational process. In addition to identifying the challenges, some suggestions are also presented that may contribute to help solve some of the challenges identified.

KEYWORDS

Artificial Intelligence, ChatGPT, Education, Innovation, Teaching-Learning Process, Digital Technologies

1. ARTIFICIAL INTELLIGENCE IN EDUCATION

If there is one area that constantly needs to modernize, reinvent and innovate, it is education, given the responsibility it has in endowing and training individuals with knowledge and skills useful to their development, but also to the development of societies. This has been the case since prehistoric times (ancient Greece and Rome, Middle Ages, modernity (Enlightenment, French Revolution, Industrial Revolution and the Information Age) and now post-modernity (Industry 4.0, 4th Industrial Revolution or Education 4.0). Education 4.0 designates the educational approach and set of strategies that would be desirable to address the current needs. of the so-called "Fourth Industrial Revolution". This term presented by Schwab (2015), has emerged to describe this whole generation of technological advances, namely, the Internet of Things, Big Data, Robotics, Artificial Intelligence, among others.

Considering the importance that technologies 4.0 have for innovation and improvement of the quality of teaching-learning processes and for the acquisition of skills by both students and teachers, it is essential to conduct research involving artificial intelligence language models in the educational context, namely, the challenges inherent in their use.

The use of artificial intelligence in education has had a great impact, including greater efficiency, global learning, personalized/personalized learning, smarter content, and greater effectiveness and efficiency in education administration, among others (Timms, 2016). The use of artificial intelligence in education is rapidly expanding (Roos, 2018) and has been incorporated into administration, instruction or teaching, and learning (Chassignol et al., 2018). Virtual advising without human intervention, assessment managers to design exams, personalized recording and evaluation of academic performance, strengthening personalized education through crowdsourcing, among others.

Artificial intelligence language models, like everything in education, have their advantages and disadvantages for the teaching-learning process. Some of the advantages have to do with being able to customize the educational process for each student, ensuring that they learn at their own pace and according to their needs. The guarantee of autonomous learning seems to be another advantage, since IA models can not only teach students but also assess their learning, without the need for a teacher to be present. Another advantage is improved efficiency, allowing the teacher to focus on more complex tasks, while AI can handle more routine and repetitive tasks. Recording and analyzing data on student performance is another advantage, as it enables teachers to identify areas that need improvement. Also in a recent research, Nalbant (2021) identifies a number of advantages of artificial intelligence for education, namely: access to information, distance learning, personalization in education, global knowledge, student attendance tracking, digitalization and environmental protection, removing barriers, error reduction, taking on challenging explorations, medical applications, uninterrupted operation for long periods and the daily applications. In addition to the disadvantages, there also seem to be some disadvantages with, for example, the absence of humanity in teaching and learning, as there is no meaningful human interaction. The lack of empathy, which leads to a lack of expression of students' emotions and feelings, which can affect the quality of teaching and learning. Dependence on technology can also be problematic if the technology fails or becomes obsolete. To these disadvantages we can add others (Nalbant, 2021): technology addiction, negative impact on social life, negative impact on health, probability of causing unemployment, lack of creativity, income distribution imbalance, high cost, unethical and the lack of experience.

Obviously, for each artificial intelligence language model there is a set of associated advantages and disadvantages. There are several authors who reflect and discuss these matters (Alshurafat, 2023; Baidoo-Anu & Owusu Ansah, 2023; Celik et al., 2022; Deng & Lin, 2022; Lund & Wang, 2023; Pfeffer et al., n.d.; Segrave et al., 2005). However, since this is not the focus of this research, we have chosen not to talk about them in this study.

The use of large language models in education has been identified as a potential area of interest due to the diverse range of applications they offer (Kasneci et al., 2023). The use of chatbot technology in education is one of the most important approaches to enhancing and promoting a more personalized learning experience (Cunningham-Nelson et al., 2019). Chatbots are conversational or interactive agents that provide instant response to the user (Okonkwo & Ade-Ibijola, 2020; Smutny & Schreiberova, 2020). Chatbots can instantly provide students with standardized details, such as course contents (Cunningham-Nelson et al., 2019) practice questions and answers (Ranoliya et al., 2017; Sinha et al., 2020), evaluation criteria (Benotti et al., 2017; Durall & Kapros, 2020), assignment due dates, advice (Ismail & Ade-Ibijola, 2019), campus path direction (Mabunda & Ade-Ibijola, 2019), and study materials. From the perspective of learning opportunities, these models can enable diverse types of experiences for: elementary school students, middle and high school students, university students, group and distance learning, empowering students with disabilities, and for professional training (Kasneci et al., 2023). Effectively, this type of models can contribute to the teaching-learning process for several areas and study cycles in education and, consequently, to the development of the various educational agents. However, there are implications that are important to bear in mind, such as security, reliability and privacy, manipulation through the spread of misinformation, prejudice and social discrimination, and the absence of personalized feedback.

An example of a chatbot probably more recent at this moment is ChatGPT that seems to be bringing to education several types of implications that are currently under debate in the scientific community, but also in civil society. Students, parents, employees, directors of educational institutions, parents, higher education institutions, politicians and ordinary citizens reflect and debate about this kind of models in the teaching-learning process: What to do? How to deal with the situation? Do we ban it or integrate it into the teaching-learning process? Is it possible to forbid? Is it possible to integrate? If yes, in what way? There are countless questions involving the theme. It is important to note that ChatGPT is a specific type of chatbot known as a "language model", which uses artificial intelligence and machine learning to generate responses based on context and previous conversation history. This model has been trained on a massive amount of data, allowing it to understand and generate human-like text with remarkable accuracy (Aljanabi et al., 2023). Chatbot technology is a good innovation with capabilities to improve not only teaching and learning, but all other aspects of education (Okonkwo & Ade-Ibijola, 2021), and therefore presents itself as a very important area of research today.

2. METHOD

This research is supported by a systematic literature review that aims to identify some of the challenges that artificial intelligence language models such as ChatGPT can bring to the educational context. Presenting some suggestions to overcome these obstacles is the second objective of the research.

The systematic literature review was carried out taking into account the definition of the following set of criteria (keywords): (i)Type of documents: scientific articles; (ii) Language: Portuguese and English; (iii) Keywords: "challenges of artificial intelligence language models in education"; "challenges of ChatGPT in education"; "Problems and opportunities of ChatGPG for education"; "Education with ChatGPT". It should be noted that, in addition to being searched in English, the keywords were also translated into Portuguese and searched in the following databases: (iv) Scielo, Springer, Web of Science, Scopus and Google Scholar; (v) Period: last six years (January 1, 2018 to February 1, 2023).

In addition to formulating the review question and research criteria, the literature review took into account the selection and evaluation of studies, data extraction, and also the analysis, interpretation, and presentation of results.

This study is further supported by exploratory research because the researchers, supported by the set of previously defined criteria, identify some of the challenges of ChatGPT for the educational context. Data of qualitative nature were recorded in the researcher's diary according to previously defined criteria. Subsequently, they were processed, analyzed, and categorized in Microsoft Excel with the aim of identifying the various directions of response. The following table identifies only the articles selected for the scope of the research:

Table 1. Articles from the systematic literature review

Authors	Year	Title
Aljanabi, M., Ghazi, M., Ali, A. H., & Abed, S. A.	2023	ChatGpt: Open Possibilities. Iraqi Journal For Computer Science and Mathematics
Baltar, R., & Baltar, C. S.	2023	Professores serão substituídos pela inteligência artificial?
Carmo, C. R. S., Carmo, R. D. O. S., & de Melo, G. D.	2022	A inteligência artificial e os desafios na avaliação da escrita académica.
Celik, I., Dindar, M., Muukkonen, H., & Järvelä, S.	2022	The promises and challenges of artificial intelligence for teachers: A systematic review of research.
Chen, X., Zou, D., Xie, H., Cheng, G., & Liu, C.	2022	Two decades of artificial intelligence in education
de Farias, S. A.	2023	Pânico na Academia! Inteligência Artificial na Construção de Textos Científicos Com o Uso do ChatGPT
García-Peñalvo, F. J.	2023	The perception of Artificial Intelligence in educational contexts after the launch of ChatGPT: Disruption or Panic?
Kasneci, E., Seßler, K., Küchemann, S., Bannert, M., Dementieva, D., Fischer, F., ... & Kasneci, G.	2023	ChatGPT for Good? On Opportunities and Challenges of Large Language
King, M. R., & chatGPT.	2023	A Conversation on Artificial Intelligence, Chatbots, and Plagiarism in Higher Education
Pfeffer, O. P., Sailer, M., Schmidt, A., Seidel, T., Stadler, M., Weller, J., ... & Kasneci, G.	2022	ChatGPT for Good? On Opportunities and Challenges of Large Language Models for Education.
RudolphA, J.	2022	ChatGPT: Bullshit spewer or the end of traditional assessments in higher education?
Susnjak, T.	2022	The End of Online Exam Integrity?
Thorp, H. H.	2023	ChatGPT is fun, but not an author
van Dis, E. A., Bollen, J., Zuidema, W., van Rooij, R., & Bockting, C. L.	2023	ChatGPT: five priorities for research
Velásquez, F. R.	2023	O ChatGPT na pesquisa em Humanidades Digitais: Oportunidades, críticas e desafios
Zhai, X.	2023	ChatGPT for Next Generation Science Learning

As can be seen in the previous table, of the total 23 articles identified, 16 were selected and, therefore, 7 were excluded for not meeting the requirements defined for conducting the systematic literature review.

3. DISCUSSION OF RESULTS

The emergence of artificial intelligence language models has had a huge impact on the way we look at education at all stages - planning, implementation, monitoring and evaluation. Models like the ChatGPT have arrived and set whole communities thinking and arguing about the topic to find the most appropriate answers possible. Integrating technologies into the teaching-learning process has never been an easy task, especially to get them used for educational purposes. The biggest challenge is even to ensure control and monitoring in the use of these technologies and to create limits for their use in the educational context. The ChatGPT as a technology that offers access to (un)information can effectively pose a problem for the stakeholders in the teaching-learning process. AI is integrated into several instructional technologies such as chatbots (Clark, 2020).

A recent study provides a structured understanding of the use of chatbots in educational processes, including teaching and learning administrative tasks, student assessment, and research and development (Durall & Kapros, 2020; Hien et al., 2018; Ho et al., 2018; Medeiros et al., 2018; Sinha et al., 2020). The use of chatbots has several implications in the various vital areas of education. The main challenges faced by the use of chatbots for educational activities have been elucidated in accordance with the reports of (Cunningham-Nelson et al., 2019; Huang et al., 2020; Rapp et al., 2021; Ruane et al., 2019). Some challenges of ChatGPT seem to be related to the following: academic writing, ChatGpt as a search engine, coding, detecting security vulnerabilities and social media. Regarding academic writing "ChatGPT-3 is not a substitute for human intelligence and creativity. It should be used as a tool to assist with academic writing, rather than to replace it" (Aljanabi et al., 2023). In this sense, as is logical, the student cannot use ChatGPT to write papers, activities, articles and other elements for him, but rather assist him in this writing. Even in the writing aid, the output generated by ChatGPT-3 must be checked and reviewed by the user, as it is not always 100% accurate (Nguyen, 2023). One of the main advantages of using ChatGPT as a search engine is its ability to understand and answer natural language questions. However, one of the main limitations is its cost and accessibility (Aljanabi et al., 2023), as ChatGPT is currently only available to a select group of developers and researchers, which limits its potential user base (Aljanabi et al., 2023). Also the cost of using ChatGPT may be prohibitive for some users, making it less accessible to small businesses and individuals (Aljanabi et al., 2023). In this sense, both accessibility and the associated costs seem to be two important challenges that education has to overcome, since education is the full representation (or at least it should be) of the democratization of knowledge and free access to knowledge. When it comes to coding is that ChatGPT is not yet able to fully understand the nuances of programming languages and therefore may not always provide the most accurate or useful information (Aljanabi et al., 2023). This is an emerging challenge for educational context, in that it requires teachers and students to validate and/or corroborate with another source whether or not the output is actually correct. Will they corroborate the information or will they facilitate it? At the level of security vulnerability detection, it appears that it is not possible to deal with certain types of queries, such as reverse engineering or malware analysis (Hammad, 2023). Here too, it is important to understand that, especially in an educational context, special attention should be given to security issues, as they are a daily challenge that schools have to deal with every day. Protecting our children, young people and adolescents has to be a priority. Finally, another challenge is that ChatGPT is not yet able to fully understand the nuances of human language and social media communication and therefore may not always provide the most accurate or useful information (Aljanabi et al., 2023). As previously mentioned, it is also important to corroborate sources and not copy and paste information that is uniquely and absolutely correct.

In addition to these challenges, there are others of equal importance, such as the danger of plagiarism. This is the problem that comes to mind when talking about ChatGPT. Just as a journalist can produce an article with ChatGPT support, a student can also carry out a work in its entirety using this technology. Incidentally, fearing that students will start cheating, the city of New York in the United States of America has banned the use of the tool on all devices and networks in public schools. However, in addition to plagiarism, there is the assumption that these automated responses may expose students to texts with sexism, racism and other discriminatory manifestations. An example is the case presented by Professor Darren Hick, from Furman University in the United States, who reported that one of his students had used the chatbot to write a 500-word essay on the 18th century philosopher, David Hume, and his work "The horror paradox" (Mitchell, 2022). A student of the Master's degree in Biomedical Engineering and Biophysics at the Faculty of Sciences of the University of Lisbon (Portugal) used the GPT Chat to answer a question from an exam in the discipline "Medical Devices I" (Araújo, 2023). is that the professor himself considers that the tool should not be banned, since "Blocking GPT Chat by universities does not solve the danger of fraud". He also suggests that the alternative involves developing a solution that easily detects plagiarism and resorts to artificial intelligence (Araújo, 2023). These obviously not being the only cases, they clearly demonstrate the attempt to use this technology by students to cheat. This teacher was able to detect the case of plagiarism, but is it possible to detect all of them? Do teachers, for the purpose of evaluating a given work, have time to confirm line by line, paragraph by paragraph, page by page whether there is plagiarism in each student's work? Or should we go for the solution suggested by the professor - Similarity Detector Software? How can we ensure that the work is not partially or fully performed by ChatGPT? Can we do anything about this practice? What? So, afraid that students will start cheating, the city of New York in the United States of America has banned the use of the tool on all devices and networks in public schools. The French University of Sciences Po has also banned the use of GPT Chat. But is banning any good? Ensuring the authenticity of a piece of work is indeed a huge challenge in education, but maybe it's really time to face this not as a problem, but as an opportunity for change. Perhaps it is really important to reflect, debate and discuss in the academic and scientific communities, but even in civil society, if education instead of being afraid, it should not take the opportunity to rethink new teaching methods, but also assessments. Also Zhai (2022) points to the challenge of rethinking the evaluation. Maybe the solution has to come through here!

However, in addition to the huge risk of plagiarism that can be tackled through similarity detector software, but also through other emerging technologies, Chat GPT brings other challenges for education. Other studies also identifies some important challenges such as, for example, issues related to bias and fairness in which large language models can perpetuate and amplify existing biases and unfairness in society, which can negatively impact teaching and learning processes (Kasneci et al., 2023). Also learners may rely too heavily on the model. This confidence, without questioning, without corroborating with another source, without contrasting and comparing, can lead to the presentation of false information. The effortlessly generated information could negatively impact their critical thinking and problem-solving skills (Kasneci et al., 2023). It is important to be aware of the limitations of large language models and use them only as a tool to support and enhance learning (Pavlik, 2023). Another challenge is also the dependence of teachers on this type of models. These models can provide accurate and relevant information, but they cannot replace the creativity, critical thinking, and problem-solving skills that are developed through human instruction. It is therefore important for teachers to use these models as a supplement to their instruction, rather than a replacement (Kasneci et al., 2023). A lack of understanding and expertise that has to do with the fact that many educators and educational institutions may not have the knowledge or expertise to effectively integrate new technologies in their teaching (Redecker, 2017). Educational theory has long since suggested ways of integrating novel tools into educational practice (Salomon, 1993). However, it is recognized that the constant integration of technologies in the teaching-learning context is not easy, as it requires constant updating on the part of the teacher and digital skills that are absolutely fundamental. As with any other technological innovation, it is very important to effectively use them to supplement or enhance specific learning processes. generated answers is another challenge, as It is becoming increasingly difficult to distinguish whether a text is machine- or human-generated, presenting an additional major challenge to teachers and educators (Cotton et al., 2023; Dehouche, 2021; Elkins & Chun, 2020; Gao et al., 2022). This situation is also in line with what was discussed above. Also the maintenance of large language models could be a financial burden for schools and educational institutions, especially those with limited budgets (Kasneci et al., 2023). This is really a huge challenge, as educational institutions have fewer and fewer resources. The use of large language models in education raises concerns about data privacy and security, as student data is often sensitive and personal (Kasneci et al., 2023). Obviously, this raises several questions related to data confidentiality and access authorizations. Sustainable usage has to do with high energy consumption. Hence, energy-efficient hardware and shared (e.g., cloud) infrastructure based on renewable energy are crucial for their environmentally sustainable operation and scaling needed in the context of education (Kasneci et al., 2023). At a time when we live in an energy and environmental crisis, it is fundamental to be concerned about these two matters, which also constitute a great challenge Saving environmental concerns are also a great challenge in this matter. With regard to the cost to verify information and maintain integrity it is important to verify the information provided by the model by consulting external authoritative sources to ensure accuracy and integrity (Kasneci et al., 2023). Finally, the lack of adaptability, since large language models are not able to adapt to the diverse needs of students and teachers, and may not be able to provide the level of personalization required for effective learning (Kasneci et al., 2023). Eventually, with later models, this adaptability tends to increase, but it is still considered a challenge for education today.

Obviously there are many challenges and many more will arise in the near future as models like ChatGPT are here to stay. The important thing is to integrate these models into the teaching-learning process and not put them aside or marginalize them. When we refer to integrating these models, we don't mean that it has to be the only tool for acquiring skills, but rather as a tool that can effectively contribute to help in the learning process of both students and teachers: for students in terms of learning and for teachers in terms of preparing and teaching their classes. There are three vectors that, supported by the literature and our experience in ChatGPT, we have identified and that seem fundamental to contribute to overcome some of the challenges: rethink, train and raise awareness.

Rethink the extent to which we should be disruptive, not be afraid to be bold and take decisions to improve the quality of education. It is important to rethink whether the current educational model responds to reality, whether it is current and whether it responds to the needs of students and the labor market itself, in short, companies, organizations. In this line, it is necessary to reflect on the methodologies and pedagogies to be adopted in the context of the classroom, whether face-to-face or online. But it is also necessary to go further. We probably have to think about the way in which educational content is produced, how it is disseminated and how it is taught. The teaching and assessment processes are also at stake at the moment, and therefore must be deeply debated and questioned.

Train because teachers who have skills in these technologies can contribute more to the dynamism of learning itself, to help motivate other colleagues. However, it is not enough for students to have skills, it is essential that teachers are able to keep up with this change and, likewise, use these technologies in an innovative way. Training should be not only in the technical and digital field, but also and especially in careful preparation in identifying and selecting reliable, true and useful information. In a world where we are no longer able to distinguish between what is true and what is false, models such as the GPT Chat have created even more problems in this regard. In this sense, it is necessary to invest heavily in training educational actors in accessing, analyzing, processing, categorizing and assigning information.

Raise awareness because it is very important to raise everyone's awareness of the impact that these models have on education, but also on citizens' daily lives. Raising awareness of the benefits of these environments, but also of their previously mentioned disadvantages The inherent challenges that these models bring to education are also central aspects that deserve special attention. The need for training in the area of information and in the digital area, the importance of technologies that guarantee that a given product was designed by the author and not by technology and the change in the current paradigm in education are aspects that educational actors must reflect on, but for that this reflection has to be raised, they have to be sensitized to it.

4. CONCLUSIONS

The research allowed to identify a set of challenges that models such as ChatGPT can create in an educational context. Some of these challenges are possible to identify at present, but others will certainly arise in the future with the evolution of technology, particularly in the area of artificial intelligence. We also identified some ways to overcome these challenges that can be adopted partially or totally for education.

The title of this article presents us with the question of whether to integrate or marginalize ChatGPT. The answer is very simple: we should seek to integrate this and similar technologies in the teaching-learning process, regardless of the challenges associated with this process, implications or even fears that as professionals and researchers in education we may have. Education needs disruption, modernization of practices and methodologies that are innovative and closer to the students. It is necessary to try to reinvent the role of the school, but also the role of the teacher. It is impossible that technologies like ChatGPT continue to emerge and the teacher continues to be the same, with the same functions, changing absolutely nothing. It is necessary to act quickly, to help teachers to change their practices, but also the students.

We all have to be responsible for the change that starts with the integration of these technologies in the teaching-learning processes. Putting these technologies aside, pretending they don't exist, or simply banning them is no good, is not an option, and is impossible in a digital society. Rethink, train and raise awareness seem to be three ways that can contribute to help overcome the challenges. Rethink in that we must be disruptive, not be afraid of fear, and innovate daily with these technologies. Train, because teachers with skills in these technologies can contribute to improving the quality of the educational process, but also to increase the literacy of students in different subjects and in the digital domain. Raise awareness since it is important to aggregate efforts for the importance of these technologies in education, for the dynamics of learning, for the approach to students, but also to the educational reality in a digital society.

Are we ready to revolutionize education? Are we ready or are we not ready? We have only one alternative: to be ready!

ACKNOWLEDGEMENT

This work has been supported by FCT – Fundação para a Ciência e Tecnologia within the Project Scope: UIDB/05777/2020.

REFERENCES

Aljanabi, M. et al., (2023). ChatGpt: Open Possibilities. Iraqi Journal for Computer Science and Mathematics, 4(1), 62–64.

Alshurafat, H., (2023). The Usefulness and Challenges of Chatbots for Accounting Professionals: Application on ChatGPT. Available at SSRN 4345921.

Araújo, R., (2023). Professor descobriu que aluno usou ChatGPT num exame, mas diz que ferramenta não deve ser banida. TSF Rádio Notícias.

Baidoo-Anu, D., and Owusu-Ansah, L., (2023). Education in the Era of Generative Artificial Intelligence (AI): Understanding the Potential Benefits of ChatGPT in Promoting Teaching and Learning. Available at SSRN 4337484.

Benotti, L. et al. (2017). A tool for introducing computer science with automatic formative assessment. IEEE Transactions on Learning Technologies, 11(2), 179–192.

Celik, I. et al. (2022). The promises and challenges of artificial intelligence for teachers: A systematic review of research. TechTrends, 66(4), 616–630.

Chassignol, M. et al., (2018). Artificial Intelligence trends in education: a narrative overview. Procedia Computer Science, 136, 16–24.

Clark, D., (2020). Artificial intelligence for learning: How to use AI to support employee development. Kogan Page Publishers.

Cotton, D. et al., (2023). Chatting and Cheating. Ensuring academic integrity in the era of ChatGPT.

Cunningham-Nelson, S. et al, (2019). A review of chatbots in education: practical steps forward. 30th Annual Conference for the Australasian Association for Engineering Education (AAEE 2019): Educators Becoming Agents of Change: Innovate, Integrate, Motivate, 299–306.

Dehouche, N., (2021). Plagiarism in the age of massive Generative Pre-trained Transformers (GPT-3). Ethics in Science and Environmental Politics, 21, 17–23.

Deng, J. and Lin, Y., (2022). The Benefits and Challenges of ChatGPT: An Overview. Frontiers in Computing and Intelligent Systems, 2(2), 81–83.

Durall, E. and Kapros, E. (2020). Co-design for a competency self-assessment chatbot and survey in science education. Learning and Collaboration Technologies. Human and Technology Ecosystems: 7th International Conference, LCT 2020, Held as Part of the 22nd HCI International Conference, HCII 2020, Copenhagen, Denmark, July 19–24, 2020, Proceedings, Part II 22, 13–24.

Elkins, K. and Chun, J. (2020). Can GPT-3 pass a Writer's turing test? Journal of Cultural Analytics, 5(2).

Gao, C. et al., 2022. Comparing scientific abstracts generated by ChatGPT to original abstracts using an artificial intelligence output detector, plagiarism detector, and blinded human reviewers. BioRxiv, 2012–2022.

Hammad, M., (2023). The Impact of Artificial Intelligence (AI) Programs on Writing Scientific Research. Annals of Biomedical Engineering, 1–2.

Hien, H. et al., (2018). Intelligent assistants in higher-education environments: the FIT-EBot, a chatbot for administrative and learning support. Proceedings of the 9th International Symposium on Information and Communication Technology, 69–76.

Ho, C. et al., (2018). Developing a chatbot for college student programme advisement. 2018 International Symposium on Educational Technology (ISET), 52–56.

Huang, M. et al., (2020). Challenges in building intelligent open-domain dialog systems. ACM Transactions on Information Systems (TOIS), 38(3), 1–32.

Ismail, M. and Ade-Ibijola, A., (2019). Lecturer's apprentice: A chatbot for assisting novice programmers. 2019 International Multidisciplinary Information Technology and Engineering Conference (IMITEC), 1–8.

Kasneci, E. et al., (2023). ChatGPT for Good? On Opportunities and Challenges of Large Language Models for Education.

Lund, B. and Wang, T., (2023). Chatting about ChatGPT: how may AI and GPT impact academia and libraries? Library Hi Tech News, ahead-of-p(ahead-of-print). https://doi.org/10.1108/LHTN-01-2023-0009

Mabunda, K. and Ade-Ibijola, A., (2019). Pathbot: An intelligent chatbot for guiding visitors and locating venues. 2019 6th International Conference on Soft Computing & Machine Intelligence (ISCMI), 160–168.

Medeiros, R. et al., (2018). A systematic literature review on teaching and learning introductory programming in higher education. IEEE Transactions on Education, 62(2), 77–90.

Mitchell, A., (2022). Professor catches student cheating with ChatGPT:'I feel abject terror.' New York Post. https://nypost. com/2022/12/26/students-using-chatgpt-tocheat ….

Nalbant, K., (2021). The importance of artificial intelligence in education: a short review. Journal of Review in Science and Engineering, 2021, 1–15.

Nguyen, M.-H., (2023). Academic writing and AI: Day-1 experiment. Center for Open Science2023.

Okonkwo, C. and Ade-Ibijola, A., (2020). Python-Bot: A chatbot for teaching python programming. Engineering Letters, 29(1).

Okonkwo, C. and Ade-Ibijola, A., (2021). Chatbots applications in education: A systematic review. Computers and Education: Artificial Intelligence, 2, 100033.

Pavlik, J., (2023). Collaborating With ChatGPT: Considering the Implications of Generative Artificial Intelligence for Journalism and Media Education. Journalism & Mass Communication Educator, 10776958221149576.

Pfeffer, O. et al. (n.d.). ChatGPT for Good? On Opportunities and Challenges of Large Language Models for Education.

Ranoliya, B. et al., (2017). Chatbot for university related FAQs. 2017 International Conference on Advances in Computing, Communications and Informatics (ICACCI), 1525–1530.

Rapp, A. et al., (2021). The human side of human-chatbot interaction: A systematic literature review of ten years of research on text-based chatbots. International Journal of Human-Computer Studies, 151, 102630.

Redecker, C., (2017). European framework for the digital competence of educators: DigCompEdu. Joint Research Centre (Seville site).

Roos, S., (2018). Chatbots in education: A passing trend or a valuable pedagogical tool?

Ruane, E., (2019). Conversational AI: Social and Ethical Considerations. AICS, 104–115.

Salomon, G., (1993). Distributed cognition. Pychological and Educational.

Schwab, K., (2015). The Fourth Industrial Revolution: What It Means and How to Respond? Foreign Affairs.

Segrave, S., (2005). The power of the 6 three model for enhancing academic teachers' capacities for effective online teaching and learning: Benefits, initiatives and future directions. Australasian Journal of Educational Technology, 21(1).

Sinha, S. et al., (2020). An educational Chatbot for answering queries. Emerging Technology in Modelling and Graphics: Proceedings of IEM Graph 2018, 55–60.

Smutny, P. and Schreiberova, P., (2020). Chatbots for learning: A review of educational chatbots for the Facebook Messenger. Computers & Education, 151, 103862.

Timms, M., (2016). Letting artificial intelligence in education out of the box: educational cobots and smart classrooms. International Journal of Artificial Intelligence in Education, 26, 701–712.

Zhai, X., (2022). ChatGPT user experience: Implications for education. Available at SSRN 4312418.

ONLINE VS OFFLINE ATTENDANCE RATES: ARE THEY CORRELATED AND HOW DO THEY COMPARE?

Victor K. Y. Chan
Macao Polytechnic University
Rua de Luís Gonzaga Gomes, Macao, China

ABSTRACT

The paper empirically studied the interplay between students' online and offline attendance rates and whether students' online attendance rate was higher or lower than their offline attendance rate. A data sample of students' online and offline lecture attendance rates was collected from the author's university in the academic year 2022/2023, covering four senior undergraduate courses of disciplines ranging from software engineering to management. All these four courses were delivered partially online and partially offline due to the then lockdowns and other expedient controls amidst the COVID-19 pandemic. No evidence was found to suggest that a student's online attendance rate in a course was correlated with his/her offline attendance rate in that course. Also, students' mean online attendance rate was found to be higher than their mean offline attendance rate. This finding offers guidance to universities, educators and students on the upside of the online lectures in terms of attendance.

KEYWORDS

Online attendance rates, offline attendance rates, university courses, correlation, paired sample *t*-test.

1. INTRODUCTION

Ever since the advent of online lectures or classes, related researches have been focusing on the pedagogical efficacy of online lectures (probably vis-à-vis that of traditional, offline lectures) or almost equivalently the relationship between online lecture (probably versus traditional, offline lecture) attendance and academic performance (Michinov *et al.*, 2011; Galyon *et al.*, 2016; Kauffman *et al.*, 2018; Nieuwoudt, 2020; Diab-Bahman *et al.*, 2022). Some other studies centred around the advantages and disadvantages of online lectures versus offline ones. In particular, advantages encompass convenience (Jensen, 2011; Almahasees, Mohsen and Amin, 2021), more engagement in quantitative analysis (Dumford and Miller, 2018), flexible learning environment (Nieuwoudt, 2020; Almahasees, Mohsen and Amin, 2021; Bdair, 2021), academic achievement (Bdair, 2021), student-centred learning (Bdair, 2021), self-learning and low costs (Almahasees, Mohsen and Amin, 2021). Disadvantages include difficulty in maintaining attention in the less structured environment of online lectures in comparison with offline counterparts (Jensen, 2011), demotivation of students in terms of their learning effort (Jensen, 2011; Almahasees, Mohsen and Amin, 2021), less engagement in collaborative learning, student-faculty interactions and discussions with diverse others (Dumford and Miller, 2018), less exposure to effective teaching practices (Dumford and Miller, 2018), lower quality of interactions between students and between students and faculty (Dumford and Miller, 2018; Almahasees, Mohsen and Amin, 2021; Bdair, 2021), inadequacy of behavioural control, online lecture training, Internet connectivity and platform standardization (Bdair, 2021), academic integrity compromise (Bdair, 2021), learning environment trading off engagement, teamwork and social interaction and overloading students with assignments (Bdair, 2021), family burden in respect of the demand for computers and Internet access (Bdair, 2021), technical and Internet issues, data privacy and security (Almahasees, Mohsen and Amin, 2021). Dependent on the settings, these studies, among many others, rendered contradictory findings, for example, some favouring the online modality's superiority over the offline one whilst some others inclining towards the converse.

Most importantly, there seems to be an issue that has so far been overlooked. To the best of the author's knowledge, there have not been extensive studies, if any at all, on the students' attendance behaviours in online versus offline lectures. For example, are students on average more ardent in taking part in online lectures than offline lectures or vice versa? Do students participating much in online lectures also do so in offline counterparts and vice versa? Do students meeting any attendance rate threshold in the online lectures also satisfy the same in offline lectures?

This article seeks to fill these gaps by answering some of these questions. In other words, it delves into the comparative behavioural or decisional patterns of students as to what their behaviours and decisions to attend or skip online versus offline lectures are. This opens up a new frontier of research that hitherto remains nascent among universities, their management, educators or even researchers.

In fact, it might be a matter of chance for universities and other educational institutions around the globe to confront the COVID-19 pandemic and thus to adopt the recourse of online lecturing over the past few years. In particular, in various places worldwide, fickle lockdowns and other expedient controls were implemented by governments such that lectures varied among multiple forms spanning online and offline lectures, minimal contact lectures in "mega" venues with sufficient personal/social distancing and so forth. Even the same course lasting a few months might administer lectures in omnibus forms at different points in time to comply with lockdowns and expedient controls promulgated by governments. This occasion enabled this study to compare online and offline attendance in the same set of courses without being confounded by the distinct natures of various courses unlike the cases in which lectures of natively online courses were to be compared with those of natively offline courses. It was in such force majeure that this study was realized.

2. THE PRESENT STUDY

2.1 Materials

For the study, an empirical data sample was collected from the Faculty of Business at the author's university in the city of Macao (Special Administrative Region), China during the first semester of the academic year 2022/2023, the semester's lecturing being scheduled for the period from 19 August 2022 to 24 November 2022. The sample comprised the courses entitled "System Analysis and Design" and "E-commerce Project Management" and two courses entitled "Operations Management". Table 1 delineates some distinguishing features of these four courses.

Table 1. Distinguishing features of the four courses for the present study

Course title	Language of instruction	Regular weekly lecture frequency	Programme specialization, year of undergraduate study	Number of students
System Analysis and Design	English	Two 1.5-hour lectures	E-commerce, 3	20
E-commerce Project Management	English	Two 1.5-hour lectures	E-commerce, 3	20
Operations Management	English	Two 1.5-hour lectures	Management, 3	19
Operations Management	Chinese/ English (bilingual)	Two 1.5-hour lectures	Management, 3	19

All these courses were meant to deliver offline courses in physical lecture rooms or computer laboratories. Regrettably, in mid-June 2022, Macao was beset by a bout of the pandemic. As a precaution in the aftermath, the author's university imposed a transient mandate of online lectures in place of the original offline counterparts. Thereby, approximately a quarter of all the lectures for these four courses were conducted online, the rest reverting to the offline paradigm. Therefore, these courses in that semester were among the rare cases in which both the online and offline modalities coexisted.

At the author's public university, it was a "statutory" requirement that the attendance be taken for each student in each lecture. This was done through a specialized attendance system of the university, which served as the sole source of such data for this study. The online lectures were in the form of Zoom (*Zoom Online Video Conferencing | Zoom*, no date) sessions with the professor-in-charge being the host and students being invited to the Zoom sessions over the Internet by means of an "Assignment" on the university's Canvas learning management system (*Canvas LMS | Learning Management System | Instructure*, no date). This way of inviting students to attend online lectures had been practised for around two years of the pandemic for quite some courses prior to the above four courses, so all students should be so well versed in the operation and technology that no related hindrance could be assumed. Locally in Macao, there were widespread ownership and/or use of such information and communications technology (ICT) as notebook/laptop computers, desktop computers, mobile phones and Internet connectivity (Xie *et al.*, 2020) through which Zoom sessions were accessible. Taking into account students' familiarity with online courses and the availability of ICT equipment for that purpose, it was students' behavioural and decisional choice to opt for presence at or absence from any online lectures at will. In other words, the attendance records collected were the fidelity of the students' behaviours and decisions without being confounded by such extraneous factors as operational and technological barriers, ICT equipment availability, etc.

In addition, given attendance taking being "statutory", the process was rigorous and formal in each lecture. For online lectures on Zoom, the professor-in-charge called the name of each student and waited for the student to respond by voice before recording the student's presence on the Internet/Intranet-based, specialized attendance system, which stored such records immediately on the university's server(s), guaranteeing the technological data integrity through a series of security measures of the system. In case of suspected impersonation, the professor might choose to verify the student's live picture on Zoom against his/her official picture on the attendance system. Even if no camera was on the student's side, the professor might still challenge the student with questions for identification purposes. For offline lectures, the professor called the name of each student before recording his/her presence on the attendance system and could simply tally the student's physical outlook with his/her official picture on the system to minimize chance of impersonation. Hence, the credibility of the attendance records underlying this study was beyond reasonable doubt.

2.2 Methods

An online attendance rate (r_n) and an offline attendance rate (r_f) were calculated for each student in each course as follows:

$$r_n = \frac{\text{Total number of online lectures attended by the student for the course}}{\text{Total number of online lectures for the course}}$$

and

$$r_f = \frac{\text{Total number of offline lectures attended by the student for the course}}{\text{Total number of offline lectures for the course}}.$$

A Pearson correlation coefficient ρ was computed between the online attendance rates r_n and the offline attendance rates r_f across all cases in the above data sample. The higher such a correlation coefficient, the stronger and positive the association between the online attendance rate r_n and the offline attendance rate r_f is. The corresponding *z*-statistic to test (two-tailed) the null hypotheses H_0: $\rho = 0$ was also calculated in order to verify with statistical significance whether the online attendance rate r_n and the offline attendance rate r_f were actually statistically correlated based on all cases in the above data sample. To visualize the probable correlation, a scatterplot of the online attendance rate r_n against the offline attendance rate r_f was plotted. As usual, a cluster of points in the vicinity of a straight would indicate a high correlation. Any point(s) apparently deviating from the straight line could also be identified.

To determine whether the online attendance rate r_n differed from the offline attendance rate r_f in each case across the population, a paired sample *t*-test was conducted. The test results would reveal whether the population mean difference ($r_n - r_f$) between the online attendance rate r_n and the offline attendance rate r_f in each case across the population differed from zero with statistical significance whereas r_n and r_f in each case of the data sample constituted the support for such results. Collaterally, the 95% confidence interval (CI) for the population mean difference ($r_n - r_f$) between the online attendance rate r_n and the offline attendance rate r_f was derived.

3. RESULTS

Table 2 enumerates the Pearson correlation coefficients ρ and its 95% confidence interval (CI) together with the corresponding *z*-statistic elucidated in section 2.2 above and its *p*-value (two-tailed) whereas Figure 1 is the scatterplot of the online attendance rate r_n against the offline attendance rate r_f.

Table 2. ρ and its 95% confidence interval (CI) together with the corresponding z-statistic and its p-values (two-tailed)

Pearson Correlation coefficient	95% CI	z (p-value)
$\rho = 0.104$	-0.1213 to 0.3191	0.091 (0.363)

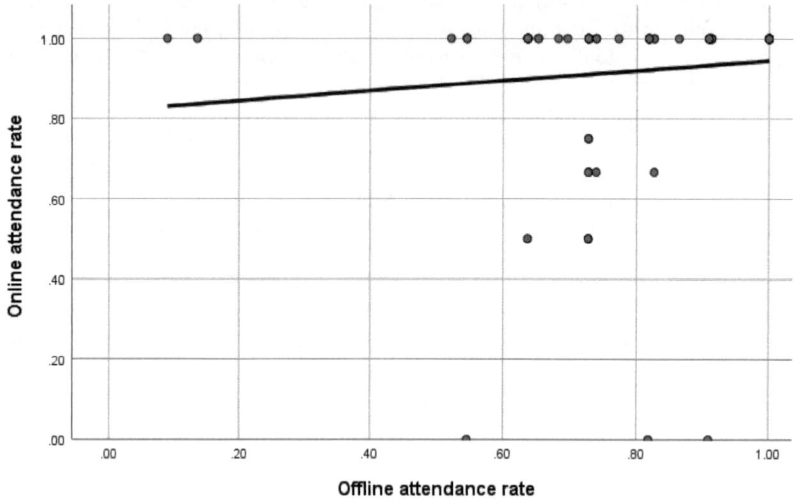

Figure 1. The scatterplot of the online attendance rate (r_n) against the offline attendance rate (r_f)

The Pearson correlation coefficient ρ for this study's data sample was positive. However, its CI spanned the zero-point, and there was no evidence to conclude with statistical significance that the ρ for the population differed from zero ($p > 0.05$). In a sense, there was no evidence to conclude with statistical significance that in the population, the online and offline attendance rates of a student in a particular course were statistically correlated or associated with each other though the online and offline attendance rates in the data sample exhibited weak, positive correlation or association. This finding concurred with Figure 1, in which most points deviated from the straight line.

Figure 2 illustrates the differences ($r_n - r_f$) between the online attendance rates r_n and the offline attendance rates r_f in all cases. Apparently, in the overwhelming majority of all cases, the differences are positive, so the online attendance rates r_n very likely exceed the offline attendance rates r_f on average.

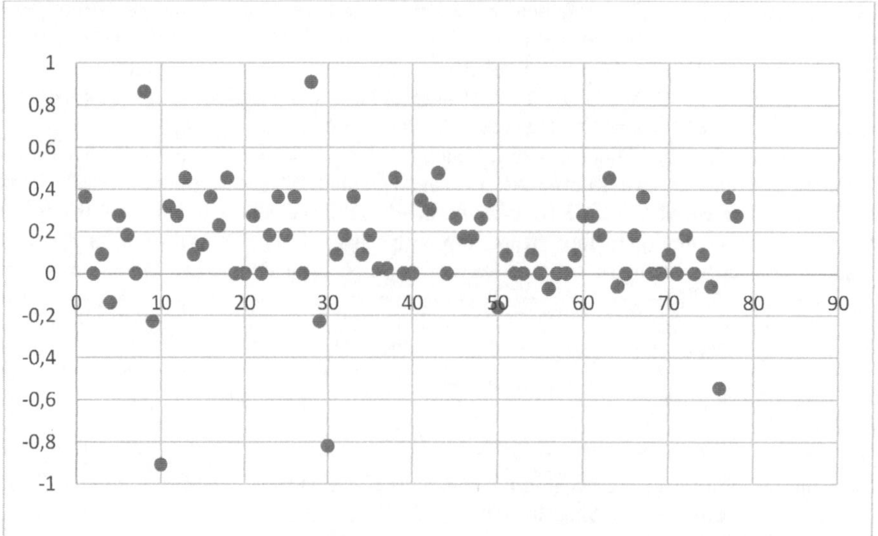

Figure 2. Differences ($r_n - r_f$) between the online attendance rates r_n and the offline attendance rates r_f in all cases

To statistically ascertain the online attendance rates r_n exceeding the offline attendance rates r_f on average, Table 3 depicts the sample mean difference $(r_n - r_f)$ between the online attendance rates r_n and the offline attendance rates r_f in each case, the 95% CI for the population mean difference, the paired sample t-statistic to test (two-tailed) whether the population mean difference between the online attendance rate r_n and the offline attendance rate r_f in each case differed from zero, the corresponding p-value and the corresponding degrees of freedom

Table 3. Sample mean difference $r_n - r_f$, the 95% confidence interval (CI) for the population mean difference, the paired sample t-statistic, the corresponding p-value (two-tailed) and degrees of freedom

Sample mean difference $r_n - r_f$	95% CI for the population mean difference	t (p-value), df
0.12682	0.06527 to 0.18837	4.103 (0.000**), 77

** $p < 0.01$
df = degrees of freedom

The sample mean difference $(r_n - r_f)$ between the online attendance rates r_n and the offline attendance rates r_f in each case and its CI were found positive. The t-test (two-tailed) revealed that the population mean difference $(r_n - r_f)$ between the online attendance rate r_n and the offline attendance rate r_f in each case differed from zero ($p < 0.01$). Based on further straightforward, statistical reasoning, the t-test (one-tailed) deduced that the population mean difference $(r_n - r_f)$ between the online attendance rate r_n and the offline attendance rate r_f in each case was greater than zero ($p < 0.005$).

4. CONCLUSION

The study did not lend statistically significant support to any correlation or association between the online and offline attendance rates of a student in a particular course. Subject to further investigation, the reason might be that the factors inducing students to attend or skip were different between online and offline lectures. Students of disparate backgrounds were enticed into or from online and offline lectures by such diverse factors to uneven extents, so, for example, convenience being an advantage of online lectures (Jensen, 2011; Almahasees, Mohsen and Amin, 2021) might be a factor to boost a busy part-time student's online attendance compared with his/her offline attendance but might not exercise such an effect on a less busy full-time student. This way, the comparative pattern of online versus offline attendance might be inconsistent across students, rendering the aforesaid correlation or association devoid of statistical significance. Other advantages of online lecture than convenience might also be such factors as more engagement in quantitative analysis (Dumford and Miller, 2018), flexible learning environment (Nieuwoudt, 2020; Almahasees, Mohsen and Amin, 2021; Bdair, 2021), academic achievement (Bdair, 2021), student-centred learning (Bdair, 2021), self-learning and low costs (Almahasees, Mohsen and Amin, 2021). Likewise disadvantages of online lectures could also be other factors, for example, difficulty in maintaining attention in the less structured environment of online lectures in comparison with offline counterparts (Jensen, 2011), demotivation of students in terms of their learning effort (Jensen, 2011; Almahasees, Mohsen and Amin, 2021), less engagement in collaborative learning, student-faculty interactions and discussions with diverse others (Dumford and Miller, 2018), less exposure to effective teaching practices (Dumford and Miller, 2018), lower quality of interactions between students and between students and faculty (Dumford and Miller, 2018; Almahasees, Mohsen and Amin, 2021; Bdair, 2021), inadequacy of behavioural control, online lecture training, Internet connectivity and platform standardization (Bdair, 2021), academic integrity compromise (Bdair, 2021), learning environment trading off engagement, teamwork and social interaction and overloading students with assignments (Bdair, 2021), family burden in respect of the demand for computers and Internet access (Bdair, 2021), technical and Internet issues, data privacy and security (Almahasees, Mohsen and Amin, 2021). Like the factor of convenience, all these advantages and disadvantages might also be factors contributing to the lack of statistical significance of the aforesaid correlation or association.

Nonetheless, this study verified with statistical significance that the population mean difference between the online attendance rate and the offline attendance rate for each case was positive. Again, subject to further investigation, the underlying disparity might be because the aforementioned advantages of the online lectures outstripped their disadvantages for the majority of students.

This study is not without its critics. First, the total number of courses and thus students covered by the data sample were only moderately sufficient due to the transiency of the pandemic. Second, despite the stringent practice of the author's university, impersonation and other attendance frauds might still have interfered with the findings.

Having said this, this study affirms that online attendance rates are in general higher than offline attendance rates, all other things being equal. This finding offers guidance to universities, educators and students on the upside of online lectures in term of attendance.

ACKNOWLEDGEMENT

The author is indebted to his university and students for their being the "field" of this field study.

REFERENCES

Almahasees, Z., Mohsen, K. and Amin, M.O. (2021) 'Faculty's and Students' Perceptions of Online Learning During COVID-19', *Frontiers in Education*, 6(May), pp. 1–10. Available at: https://doi.org/10.3389/feduc.2021.638470.

Bdair, I.A. (2021) 'Nursing students' and faculty members' perspectives about online learning during COVID-19 pandemic: A qualitative study', *Teaching and Learning in Nursing*, 16(3), pp. 220–226. Available at: https://doi.org/10.1016/j.teln.2021.02.008.

Canvas LMS | Learning Management System | Instructure (no date). Available at: https://www.instructure.com/en-au/canvas (Accessed: 17 March 2023).

Diab-Bahman, R. *et al.* (2022) 'The effect of attendance on student performance: implications of using virtual learning on overall performance', *Journal of Applied Research in Higher Education*, 14(3), pp. 1175–1192. Available at: https://doi.org/10.1108/JARHE-04-2021-0135.

Dumford, A.D. and Miller, A.L. (2018) 'Online learning in higher education: exploring advantages and disadvantages for engagement', *Journal of Computing in Higher Education*, 30(3), pp. 452–465. Available at: https://doi.org/10.1007/s12528-018-9179-z.

Galyon, C.E. *et al.* (2016) 'Comparison of group cohesion, class participation, and exam performance in live and online classes', *Social Psychology of Education*, 19(1), pp. 61–76. Available at: https://doi.org/10.1007/s11218-015-9321-y.

Jensen, S.A. (2011) 'In-Class Versus Online Video Lectures: Similar Learning Outcomes, but a Preference for In-Class', *Teaching of Psychology*, 38(4), pp. 298–302. Available at: https://doi.org/10.1177/0098628311421336.

Kauffman, C.A. *et al.* (2018) 'Relationship between classroom attendance and examination performance in a second-year medical pathophysiology class', *Advances in Physiology Education*, 42(4), pp. 593–598. Available at: https://doi.org/10.1152/advan.00123.2018.

Michinov, N. *et al.* (2011) 'Procrastination, participation, and performance in online learning environments', *Computers and Education*, 56(1), pp. 243–252. Available at: https://doi.org/10.1016/j.compedu.2010.07.025.

Nieuwoudt, J.E. (2020) 'Investigating synchronous and asynchronous class attendance as predictors of academic success in online education', *Australasian Journal of Educational Technology*, 36(3), pp. 15–25. Available at: https://doi.org/10.14742/AJET.5137.

Xie, Y.J. *et al.* (2020) 'Relationships between the usage of televisions, computers, and mobile phones and the quality of sleep in a Chinese population: Community-based cross-sectional study', *Journal of Medical Internet Research*, 22(7). Available at: https://doi.org/10.2196/18095.

Zoom Online Video Conferencing | Zoom (no date). Available at: https://explore.zoom.us/en/lp/video-conferencing/?creative=649288063628&keyword=webconferenceplatforms&matchtype=e&network=s&device=c&zcid=6095&utm_id=19735288070_649288063628&gclid=Cj0KCQjwn9CgBhDjARIsAD15h0B9FdK_h6JAkHpGIK60d6ylhxW1MraG0S380gFnavqaAEo (Accessed: 17 March 2023).

VIRTUAL TEACHER PROFESSIONAL DEVELOPMENT IN NEPAL DURING COVID-19

Binod Prasad Pant, Bal Chandra Luitel and Niroj Dahal
Kathmandu University School of Education, Department of STEAM Education, Nepal

ABSTRACT

This paper aims to share the professional development of teachers through virtual mode during COVID-19. The paper is a part of PhD research conducted by the first author, the second author is the research supervisor, and the third author is the critical friend of the first author. Participatory Action Research (PAR) was conducted in one of the rural schools of Nepal. During the time of fieldwork, the school was physically closed due to COVID-19. Even if the school was closed, the first author continued professional development sessions virtually for teachers for about one year. It was found that teachers engaged in developing several information technology-related skills which are useful for teaching. At the same time, teachers also developed projects for basic-level students, which became useful during the school closure.

KEYWORDS

Professional development, Virtual mode, E-learning, Participatory action research, Professional development

1. BACKGROUND OF THE STUDY

At first, the overall purpose of the project is discussed, and later, the research (which is one component of the entire project) issue is discussed. The entire project, named as NORHED Rupantaran at Kathmandu University School of Education, aims to accelerate changes in the quality of basic education teaching and learning in Nepal through creative, transformative, and contextualized pedagogical strategies in accordance with the recently adopted ideas of Global Goals for Sustainable Development. In addition to this, the project also aims to establish educational programs in two partner universities (Kathmandu University and Tribhuvan University) that are innovative, transformative, and contextualized as a result of the project. While working in/for the community by the team of university academia with community people, it was also aimed to strengthen the motivation and capacity of a variety of stakeholders both at the local level and at the higher education level. More specifically, the following strategies have been accomplished: a) conducting experiments with novel methods in schools located in a variety of environments in Nepal with the goal of fostering improvements in education, health, sustainability, productivity, and opportunities for earning a living; b) providing intensive training for teachers as well as training in leadership for principals in order to support the professional development of teachers and principals. Inclusion of marginalized groups in all aspects of the project, capacity building at Tribhuvan University and Kathmandu University to conduct rigorous research that informs policy development, and sustained engagement with key stakeholders to scale promising models for improving teaching and learning outcomes in Nepal are some of the goals of this initiative. This serves as a model and involves teachers and students in meaningful and sustainable efforts to improve water and sanitation in schools and local communities. These initiatives aim to improve water and sanitation in schools and local communities. This method has a positive impact on the outcomes of education and health initiatives and serves as a model for efficient social entrepreneurship. Furthermore, this method clearly has a multiplier effect. The research determined that there are two categories of schools: action and reference. The action school is analogous to the "treatment" site, where most of the project's efforts were concentrated on the action school. The four reference schools would receive a delayed intervention based on the action school's outcomes.

The area of research interest of the first author is pedagogical innovations, which is one of the focuses of the projects that could benefit from intervention. In the fieldwork at the action school, the first author immediately began to think about how he could collaborate with fellow researchers in producing new information. During the first few months of fieldwork, he found it quite challenging to cultivate a vibrant atmosphere that encouraged the participation of educators. Eventually, it began "coping" with the situation by being more open to receiving and implementing fresh concepts.

After a few months of conducting fieldwork in a face-to-face manner, the school was forced to close because of COVID-19, and we were unable to meet in person in the school. Once the school had been closed for a few months, we transitioned to holding our teacher professional development seminars online. Within the scope of this discussion, this paper presents one aspect of the fieldwork done by the first author. In this context, this article explains how the first author engaged with schoolteachers in order to organize learning and sharing opportunities for participants.

2. METHODOLOGY

This article showcases one of the results of the first author's current doctoral research. The second author is the research supervisor of the first author, while the third author is the critical friend. The paper is based solely on the first author's experiences as a participatory action researcher. During the action school fieldwork, the first author interacted with sixteen schoolteachers.

We utilized PAR as our study method. Chevalier and Buckles (2019) argued that the three crucial components of the PAR model—participation (living in society), action (experiences), and research (knowledge creation)—should be properly integrated. In this context, the first author collaborated with schoolteachers (i.e., participation) through a series of virtual workshops (actions) to construct the knowledge (research) that teachers' professional development can be continued through the virtual mode, despite the challenging environment. Participatory Action Research (PAR), a collaborative cyclic process of planning, implementation, observation, reflection, and further planning, is required (Herr & Anderson, 2014). Hence, we employed PAR to foster the ICT integration skills of the participants of the research site. The following procedures were utilized during fieldwork.

Needs Identification: At first, the first author discussed with schoolteachers after the school closure due to COVID-19. I (first author) was taking the initiative to continue ongoing teacher professional sessions in any form. I had a telephone conversation with a few teachers. Few schoolteachers who were relatively young and had access to an internet connection shared that they were available online. But, other teachers who were comparatively seniors did not show interest in joining in the virtual mode. After several communications via telephone, we agreed to meet once a week for two hours. We also agreed on the modality of the virtual session. The modality of the virtual session was sharing teachers' resources and initiatives during COVID.

Actions: In this phase, we organized almost 20 virtual sessions, each one and a half hours to two hours. In the first 8 sessions, we focused on enhancing IT skills (such as preparing PowerPoint slides, exploring authentic materials from different learning sites, and using readymade online resources for students). In the next 12 days, we focused on developing projects in different subjects so that these projects could be given to students to ensure their learning during their stay at home. We also invited some other persons (such as the local education officer of the municipality) on some days to showcase our ongoing activities and to provide them lessons on how teacher professional devilment can be done in the virtual mode.

Reflection: We used action and reflection iteratively as per the principle of PAR. At the end of each virtual session, we used to discuss the usefulness of the session. In some sessions, schoolteachers seem very active when the content is needy, such as developing PowerPoint slides. But, in some sessions, we observed that teachers were not paying attention as the contents were not interesting. For example, I discussed some theoretical approaches to becoming an effective online teacher. But, I realized that my lecture was not attracting them. Such a session helped me change the session's content and modality. Since this is PAR, I always invited them to discuss the training modality and participate in decision-making while choosing the workshop topics and modality.

Each session was recorded with the participant's consent. After completing each session, I developed a reflective summary of the class and went through the recording to observe teachers' engagement during the virtual workshops. I also interviewed three times for 4 participants to document their experiences as a part of the virtual workshop. At the same time, I remained in touch with my research supervisor and critical friend. The discussion with my research supervisor and a critical friend helped me to amend a few activities and explore better alternatives during the fieldwork. For example, in one of the sessions, my research supervisor observed my session and suggested not to deliver the readymade projects to the teachers but invite them to develop the program and share in the session. My critical friend supported me in designing activities, especially in IT. The critical friend also observed my sessions and provided observations regarding my delivery style. I improved the design of virtual sessions to keep teachers' activities at the center.

3. RESULTS AND DISCUSSIONS

This session demonstrates the major results and discussion. Based on the interviews with schoolteachers, observation notes of the session, and reflective diary of the first author. The following two major themes were developed from the fieldwork.

3.1 Teachers' Competencies in IT Skills

One of the major results obtained from the study is the enhancement of teachers' competencies in IT skills. In this regard, almost all teachers reported that their skills had improved notably. I observed that the progress in developing competencies is not uniform. But, teachers have made remarkable progress. In this regard, one teacher said;

> *"I learned both English and Nepali typing. I can now type and print exam questions. Most importantly, I'm ready to use slides to teach. At first, I wasn't sure about how to use ICT tools. Now, I'm confident that I can use them. And we learned a lot about each other."*

On another day of the virtual session, I shared the procedure of making an assessment tool using google forms. The teachers faced cognitive overload due to the first-time exposure to online tools. Nevertheless, one female teacher developed and shared the Google form in the virtual session. When she shared the Google form with her peers, other friends were motivated and realized they could also make such forms. The form was on exploring students' prerequire knowledge of science. She developed this form to test the students' existing status on the subject matter so that she could plan the lesson connecting with the prerequire knowledge.

Next week, we demonstrated the collaborative use of google docs with the teachers. We asked them to write their experiences during COVID-19. The teachers were amazed to see that all collaborators could work on the same document simultaneously. When I conducted an interview, one teacher shared,

> *"It was just like magic in which we all could read, write, and comment on the document at the same time. I think I will try to use it in the computer lab once the school opens."*

Teachers also faced different difficulties in this journey. Regarding the difficulties, one teacher shared,

> *It was enjoyable and challenging. Online learning was not my cup of tea. Due to an unreliable internet connection, there was a communication barrier that hindered my ability to learn. We learned how to develop presentation slides for teaching purposes. Importantly, we learned how to use Google Meet to conduct online classes. Also, we were able to connect with the students. But, few students could not enroll due to a lack of technology access.*

In the following weeks, I talked about my experiences using presentation slides for online instruction. I used PowerPoint slides to discuss inquiry-based teaching and learning. I provided examples of asking the questions by illuminating images on the slides to estimate the students' comprehension. On the next day, they also worked on slide creation on a mobile set. I discussed incorporating Google Earth into teaching and learning activities next week. We took a virtual tour of Nepal, using it to teach social studies in various classroom settings. This event became very interesting to teachers.

During the interview, one teacher said,

> *It was overjoyed to use presentations to teach. I have always had a strong desire to learn about technology. I was interested in finding out how to use the presentation to teach. I gained more excellent ideas about creating slides for instruction and using them in the classroom. It produced excellent results. I believe I can now adapt my traditional teaching to a changing world due to technology. Using technology in the school increased students' engagement, as per my experience.*

I realized that she was self-motivated to learn technological skills for education after our conversation. She concurred with the result that using technology in the classroom significantly increases student involvement and learning enjoyment, as Heafner (2004) argued.

In a similar line, I requested another teacher who was heavily involved in the virtual session. He shared his experience in this way.

> *I learned how to create slides with internet-sourced images using a mobile phone. I also attempted to insert a video but was unable to. I finally tried to teach sixth graders about our society in the classroom for the first time. I utilized the question-asking method I learned from you during our online meeting. I found the classroom to be productive since every kid was attentive to the photos and engaged in the discussion. I asked the students to discuss their image interpretation, and they responded enthusiastically.*

According to his perspective, such authentic activities are particularly valuable for learners as part of engaged activities, as discussed by Vygotsky and Cole (1978), and they are highly relevant to teaching and learning, allowing teachers to attain students' ZPD. It provides substantial support that learning is a social process involving interactions with peers and more proficient adults, as argued by Vygotsky's ZPD theory.

The usage of presentation slides with pertinent images has aided in the comprehension of classroom questions. It has produced a classroom climate conducive to learning. Frey and Birnbaum (2002) also showed that PowerPoint presentations helped students maintain their focus throughout lectures. The TPACK knowledge was exhibited by harmonizing PowerPoint and inquiry-based learning to teach cultural components of the social studies curriculum.

3.2 Teachers' Engagement in Project Development

Another remarkable development is teachers' engagement in project-based task development. During the school's closure due to COVID-19, the local government and parents were also advocating for the development of projects for the students so that students could learn through such projects conducted at their homes. So, I invested 12 virtual sessions for project development activities. Teachers developed sample projects, shared them in the virtual sessions, received feedback from their peers, and finalized the projects. Once they finalized the projects, they gave them projects to their students.

One project developed by schoolteachers is:

Project: My Food Chart
Name of the Teacher:
Subject: <u>Science/Math/English</u>
Name of the Project: <u>My Food Chart</u>
Project Period: <u>1 week</u>

Students Name:
Due Date:
Date of submission:

1. **Learning Goals**
    1. Make learners familiar with the different healthy and unhealthy food.
    2. Integrate the content with the real context of the learner's day-to-day life.
    3. Make learners familiar with the connection among different disciplines.
    4. Develop creative writing skills.

2. **Project's Outputs (Products)**
    1. Drawing of the food items on chart paper.
    2. Classification of different food items in a chart paper.
    3. Addition of the price of different items of food in A4 paper.
    4. Creative Writing about favorite food on the paper.

3. **Major Steps and Timeline**
    1. Prepare a chart paper and A4 paper.
    2. Prepare a list of food items for three days with their prices on paper.
    3. Show the addition of those prices of the food items on the same paper.
    4. Draw the food items on a Chart Paper. It needs to be drawn under the different categories according to the nutrients they provide. (For e.g.: potato gives energy, so it needs to be drawn under energy-giving food and so on…)
    5. Write about your favorite food in a paragraph.

4. Students' Learning Assessment

Table 1. Students' Learning Assessment

Measuring Range → Criteria ▼	3 (Excellent)	2(Good)	1(Developing)
Drawing of food items			
Usage of words (Vocabulary)			
Addition			

Note: teachers should make specific criteria as per the context under "excellent," "good" and "developing" measuring range.

In a similar way, teachers developed projects and distributed them to students. Students completed projects as per the instructions and submitted them to teachers in person. As the teachers and students were from the same community, meeting students and teachers in small groups was possible despite the school closure. One of the participating teachers shared.

> *In the virtual session, I did not only learn ICT skills, but I also learned how to develop the project. The project I developed was more interdisciplinary. It helps my students learn different subjects like mathematics and science, and language.*

This reflection of the teacher shows that the teacher has developed skills in developing projects for the students. Project-based learning has been clearly mentioned in the recently developed curriculum at the school level in Nepal. It is a good time for teachers to learn how to develop the project and implement them in their school setting.

During the interview, another participant shared that.

> *The projects we developed in our virtual session helped greatly during this COVID time. At least our students got engaged and learned something even during school closure.*

As a lead researcher, I observed and experienced that this activity was timely as teachers got opportunities to implement and refine them. Almost all teacher admired their involvement in the virtual sessions.

Nepal's 2019 national curriculum framework identifies different uses of ICT in education: as a means of communication in school administration, as a tool to facilitate teaching and learning, as a medium to improve the competence of stakeholders, including teachers, and as a separate subject to be incorporated into the curriculum to improve students' life skills. As one of the guiding principles for managing learning outcomes in curriculum creation, the ability required to use ICT has been prioritized. The teachers' competencies framework of Nepal (MoE, 2016) specifies eight competencies for managing and selecting teachers at school levels: content knowledge, pedagogical knowledge, knowledge of learners, learning environment and classroom management, communication and collaboration, continuous learning and professional development, legal-based and professional conduct, and information and communication technology.

The Ministry of Education's 2013 ICT in Education Master Plan mandated the use of ICT in our profession, and we are required to hone our skills for ICT integration. I participated in many trainings to improve my material and pedagogical expertise. Nonetheless, I sensed a glaring lack of expertise regarding the appropriate use of technology in my pedagogy. I participated in multiple TPDs for ICT integration. The curricula emphasized technology (Microsoft Word, Excel, and PowerPoint) without providing considerable assistance to increase students' learning. I felt much more emphasis was needed on ICT pedagogy (McCarney, 2004).

The UNESCO ICT Competence Framework (2018) emphasized that teachers should have ICT competencies to integrate ICT into their profession and achieve equity and quality education. In the context of Nepal, reaffirming ICT as one of eight domains of a teacher competency framework, as outlined in the National Curriculum Framework. The TPACK framework (Koehler et al., 2013) articulates the harmonious interaction between three bodies of knowledge (Technology, Content, and Pedagogy) for effective technology integration. It shows that ICT has become a cross-cutting issue in Nepal's teacher education. Due to COVID-19, the discourses on using ICT in emergencies have become pertinent for continuing education activities, even if such conditions arise in the future. The development of projects using different technical tools and making them accessible through different learning management systems can be further explored. In this context, this research study is beneficial for teachers and educators interested in using instructional practices using ICT.

4. CONCLUSIONS

The teacher development workshop through virtual mode during COVID-19 became effective. Participatory Action Research (PAR) became helpful from the beginning of the study to the reflection of the entire process. Teachers developed several IT skills during the virtual session and developed several projects. The teacher-made projects became useful in times of emergency. Teachers reported that students were involved in learning through projects and different IT tools. However, different challenges were faced during the fieldwork. A few challenges were internet connections and a few teachers' unwillingness to participate virtually.

REFERENCES

Chevalier, J. M., and Buckles, D. J. 2019. Participatory action research: Theory and methods for engaged inquiry (2nd ed.). London: Routledge.

Frey, B. A., and Birnbaum, D. J. 2002. *Learners' perceptions on the value of PowerPoint.* Harvard University Press.

Heafner, T. 2004. Using technology to motivate students to learn social studies. *Contemporary issues in technology and teacher education*, *4*(1), pp. 42-53.

Herr, K., & Anderson, G. L. 2014. *The action research dissertation: A guide for students and faculty*. Sage. http://www.jstor.org/stable/24636917

Koehler, M. J., Mishra, P., and Cain, W. 2013. What is technological pedagogical content knowledge (TPACK)? *The Journal of Education, 193*(3), pp. 13-19. Lectures. https://files.eric.ed.gov/fulltext/ED467192.pdf

McCarney, J. 2004. Effective models of staff development in ICT. *European Journal of Teacher Education, 27*(1), pp. 61-72. https://doi.org/10.1080/0261976042000211801

Ministry of Education [MOE]. 2016. *Teacher Competency Framework. National Center for educational development.* Sanothimi, Bhaktapur. https://cehrd.gov.np/

UNESCO. 2018. *UNESCO ICT competency framework for teachers*. The United Nations.

Vygotsky, L. S., and Cole, M. (1978). *Mind in society: Development of higher psychological processes*. Harvard University Press.

PREDICTING DROP-OUT INTENTIONS FROM STUDENTS' PERCEPTIONS OF ACADEMIC FEEDBACK AND PROCRASTINATION TENDENCIES. 'CAN PERSONALIZED FEEDBACK SOMETIMES BE TOO MUCH OF A GOOD THING?'

Ionut-Dorin Stanciu, Ph.D. and Liliana Ciascai, Ph.D.
[1]*Technical University of Cluj-Napoca, Department of Psychology and Pedagogy*
15 Daicoviciu Street, Cluj-Napoca, Cluj, Romania
[2]*Babes-Bolyai University, Faculty of Psychology and Education Sciences*
7 Sindicatelor Street, Cluj-Napoca, Romania

ABSTRACT

In this research, we investigated a three-way relationship between academic feedback, procrastination tendencies, and the intention of dropping out of school in a sample of 532 undergraduate students. We used a cross-sectional, associative research design, with scales developed based on corroborating themes, obtained in previous interviews with the students, with instruments from the literature. The models were tested using variance-based SEM (PLS) and our measurements showed very good psychometric properties. The study's findings indicate that overall, academic feedback is negatively associated with both procrastination and drop-out intention, but when the personalization aspect of the academic feedback is differentiated from the 'impersonal' academic feedback, it might not help mitigate the intention to drop out. Our findings provide additional support to existing research while opening a new line of inquiry into the interplay between the three main constructs mentioned above, which, to date, are scarcely studied together. Albeit with several conceptual and methodological limitations, educational researchers and educators can benefit from the evidence provided in this research.

KEYWORDS

Academic drop-out, procrastination, academic (teacher-provided) feedback, PLS-SEM

1. INTRODUCTION

While education is increasingly acknowledged as an important factor for societal development, many educational institutions across the World struggle with student retention. Student retention, or, conversely, the phenomenon of academic drop-out is neither new nor ignored. However, its complexity and the multitude of causes and inter-related mechanisms and contributing factors, make it a difficult challenge for educators as well an ever-current field of research. Specifically, academic drop-out can be the result of many factors, varying from personal (e.g., attitudes, self-beliefs, self-regulation strategies, etc.) to institution-related (resources, interactions with the instructors and administrative staff, etc.). In this paper, we perform a quick overview of how academic drop-out is related to two of its most important predictors, i.e., an external factor, academic feedback (as a form of teacher support), and procrastination, as a personal, internal factor, and we present the results of our research which investigated their interrelationships.

2. THEORETICAL BACKGROUND / LITERATURE REVIEW

2.1 Academic drop-out and procrastination

Extant research related drop-out intentions to a variety of causes, ranging from students' perceptions regarding teachers' effectiveness, including the provision of feedback (Nikolaidis et al., 2022) to the effective organisation of learning and putting in the required amount of work, i.e., keeping up to date with academic assignments (Donnelly & Kovacich, 2014). However, maintaining a good work ethic is neither easy, nor to be expected in all people, and in all contexts, and many individuals postpone engaging in certain tasks that are required of them, a phenomenon commonly known as procrastinating. Unsurprisingly, procrastination can be encountered in students at all educational levels (Afzal & Jami, 2018) and has an important negative impact on learning and achievement (Ferrari, 2001).

Solomon and Rothblum describe procrastination as the tendency to delay the initiation or completion of important tasks to the point that it can induce discomfort (1984). Schouwenburg's (Schouwenburg, 1995; Schouwenburg & Lay, 1995) research are suggestive of the pervasive nature of procrastination, both as an everyday behavior, which affects almost everyone, and has an undesirable academic conduct (over 2/3 of the students having reported procrastinating in various academic contexts). More recently, Klingsieck (2013) detailed four main psychological perspectives of procrastination. The first perspective frames procrastination as a personality trait, not unlike conscientiousness. The second one treats procrastination as a process that specifically has to do with motivation and volition. The third perspective is more clinical in nature, and looks for psychopathological correlates and potential underpinnings. Finally, the fourth perspective embeds procrastination in a specific context, linked to external factors, such as the task's or collaborator's characteristics (e.g., teacher's style, behavior, goal alignment with the task's outcome, etc.).

In psychological research, but also in educational research, procrastination is often explained via (disadaptive) coping mechanisms (Hen & Goroshit, 2020; Kim et al., 2016), low self-efficacy (Graff, 2019; Kandemir, 2014), or perfectionist tendencies (Serdar et al., 2021). Mental health problems are important in and of themselves, in and out of school, but, beyond that, they are relevant predictors of both academic success and attrition (Cuijpers et al., 2019). As such, some of their close correlates, even though not classifiable as mental disorders, can also be valid predictors of dropout. In turn, procrastination was linked to emotional suffering and mental disorders, including in school contexts (Grunschel et al., 2013; Karami & Mahmoodi, 2018; Rozental et al., 2014, 2015). It also appears to affect more those with learning disabilities (Goroshit & Hen, 2021) and that is more prevalent in men than in women (Fior et al., 2022).

The (undesirable) academic consequences of procrastination were mostly linked to poor academic performance; for instance, Steel's (2007) meta-analytical review showed consistent evidence of negative associations between procrastination and academic success, and an overall framing of procrastination as a self-regulatory failure, at least in the long run (Hen & Goroshit, 2020). Procrastination was also found to be a constant success-impeding factor in academic writing (Lonka et al., 2019), which could be related to the difficulty of approaching ill-defined problems (Lonka et al., 2019; Yurtseven & Doğan, 2019). Because procrastination is closely related to the concepts of persistence and engagement (Park et al., 2019), it bears a direct relevance to drop-out intentions. For instance, with respect to online learning, procrastination was identified as one of the most salient causes of academic drop-out (Doherty, 2006). In turn, academic feedback was related to self-regulation, assignment completion, and academic achievement (Suamuang et al., 2021).

Narciss (2008) defined feedback as the information provided to the learner following performing an action, which has the potential to influence the learner on various levels, i.e., cognitive, metacognitive, and motivational. The dominant perspective is that personalizing feedback can (and is expected to) improve the learners' performance (Aleven et al., 2017; Narciss, 2008). It is important to note, though, that not all existing research unequivocally supports the significant effect of academic feedback on learning performance. For instance, in their research, H. Wang & Lehman (2021) found that, while personalized feedback may benefit learning by improving motivation, its effect on performance outcomes was not significant, and "[we] speculate that more complex factors may have obscured any performance effect" (H. Wang & Lehman, 2021, p. 553).

Students themselves expect instructors to provide quality feedback, capable of helping them navigate academic requirements. In certain situations, students expect concrete and specific advice with respect to what they should be doing, when, and why. (Bozarth et al., 2004; Henry, 2020). Furthermore, students expect the feedback to be positive (i.e., constructive, supporting) and encouraging (Peterson, 2001), and to address not only the outcome (e.g., the final result, the grade) but also the procedure (Şahin & Yurdugül, 2022). Notably,

Şahin & Yurdugül (2022) found that it is especially learners who display procrastination behavior that expect feedback on the procedural aspects of the task, rather than on the final outcome. Importantly, lack of feedback was found to be a cause for dropping out (Bozarth et al., 2004; Doherty, 2006), and overall, a good (mentoral) relationship between students and instructors is associated with reduced procrastination (Q. Wang et al., 2022).

2.2 The research questions and associated hypotheses in the current study

While the bi-directional relationships between 'drop-out intentions', 'academic feedback', and procrastination have been studied rather extensively, we did not identify quantitative research that explicitly related all three constructs. Therefore, we based our study on the theoretical framework developed by Rumberger and Rotermund, which is a multi-level framework that combines individual and institutional tiers in situating and explaining the behavior in academic contexts. Rumberger and Rotermund's framework avoids the temptation of using either purely psychological frameworks, which, as insightful as they are, situate the problem strictly on an (inter- and/or intra-) individual level or, conversely, shift the emphasis solely on pedagogical approaches, which tend to place too little focus on the personal characteristics.

Against the backdrop presented above, our main research questions were whether or not the 'drop-out intention' is significantly associated with 'academic feedback' and 'procrastination' tendencies (RQ1), and if 'academic feedback' is a valid predictor for 'procrastination' (RQ2) while concomitantly predicting 'drop-out' intentions. Consequently, the working hypotheses for RQ1 assumed that a positive association exists between 'academic feedback' and 'drop-out intention' (H1), on the one hand, and 'procrastination' and 'drop-out intentions' (H2), on the other hand. Finally, our last hypothesis assumed that 'procrastination' mediates between 'academic feedback' and 'drop-out intentions' (H3).

3. METHODS

3.1 Research design and measurements

We used a one-time, cross-sectional associative research design, in which 'drop-out intention' (DRPINT) was the predicted (dependent/outcome) variable, and 'academic feedback' (AF) and 'procrastination' (PROCR) were the predictors (with PROCR conceptualized as mediating between AF and DRPINT).

3.2 Measurements

The respondents provided their answers to three 7-step Likert scales (corresponding to the three main constructs), measuring agreement vs disagreement. We based our scales on themes derived from interviews with a pilot group of 30 participants, but we also observed the dimensions used by existing instruments. Specifically, to develop the procrastination items we consulted Tuckman's Procrastination Scale (1991) and McCloskey's Procrastination Scale (2011; 2015). Similarly, for the drop-out intention, we consulted Schmitt et al.'s (2021) WWH-dropout scale and Vaughn et al.'s (2020) Dropout Risk Inventory. With respect to academic feedback, we found Jan-Willem Strijbos's work on feedback measurement to be the most closely related to our research purposes (Strijbos, 2010; Strijbos et al., 2010, 2021). Nevertheless, all the above-mentioned instruments either included too many dimensions which would have made our model too complex, or included aspects that, however relevant for each of the main construct, departed too much from the themes that resulted from our interviews with the pilot participants. Consequently, we designed our own set of items, inquiring directly into [a] the quality and characteristics of academic feedback (a total 18, of which 11 referred to 'impersonal' traits of the feedback, such as timeliness, support, relevance, clarity, etc., and 7 inquiring into whether or not the feedback considered the learners' personality traits, aims, and specific interests and evolution), [b] participants' procrastination tendencies (6), and [c] their drop-out intentions (11).

3.3 Participants and procedure

The participants were $N = 532$ ($M_{age} = 20.65$, $SD_{age} = 2.02$) students enrolled in undergraduate courses at the Technical University of Cluj-Napoca, out of which $N_{female} = 274$ were women ($M_{agefemale} = 20.37$, $SD_{agefemale} = 1.39$) and $N_{male} = 258$ were men ($M_{agemale} = 20.95$, $SD_{agemale} = 2.49$).

The participants were contacted during the first month of the first academic semester and inquired with respect to their perceived challenges and difficulties and to their experiences and expectations regarding their academic performance and context, including the academic relationship with the instructors. These inquires were a normal part of the interactions with the students, and allowed the researchers to invite potentially interested participants to a series of interviews concerning their procrastination tendencies, expectations and perceptions regarding teacher feedback, and possible drop-out intentions. Thirty participants (16 women, 14 men) accepted to provide answers to the interviews, consequent to which a series of common themes were extracted. These extracted themes, corroborated with the measurements previously used in the literature, lead to the development of the questionnaires used in this study.

The measurements were disseminated using a web-based form during the first semester of study of the academic year, via the e-learning platform formally used by the University. The form contained a starting section dedicated to Informed Consent (including a description and an explanation of the research aims, and making clear the rights to withdraw at any time, without consequences, to request data deletion, and to contact the researchers throughout the research and after), continued with the questionnaires, and ended with a debriefing section (allowing the participants to send questions to researchers and maintain contact, if needed).

3.4 Data processing

The data processing and all associated statistical analyses were performed using R Software (R Core Team, 2017) to ensure the transparency and reproducibility of the study. Less than 3% of data were missing, which our analysis determined were missing at random, and no aberrant patterns of responses (e.g., indicative of responding randomly, linearly, or of unengagement) were identified. Consequently, considering the small number of missing data, a mean imputation was performed for the missing values.

Because our study was vastly exploratory in nature (albeit guided by the above-mentioned hypotheses) we performed an exploratory factor analysis, followed by a variance-based modeling structural equation modeling, aka, partial least squares SEM. With specific regard to the PLS-SEM, we developed and compared two models, i.e., one that aggregated all 'academic feedback' items and one factor for each of 'drop-out intention' and 'procrastination', and a second model that followed the factor structure suggested by the EFA, namely two latent 'academic feedback' factors, and one factor for each of 'drop-out intention' and 'procrastination'.

4. RESULTS

Our measurements[1] showed good internal consistency (as a measure of reliability) and validity, for both models, as shown below in Table 1 for model 1 and Table 2 for model 2. Specifically, alpha and rhoA exceed 0.7, while the average variance extracted (AVE) exceeds 0.5 (with the AVE for PROC borderline at 0.5) and no crossloadings.

Table 1. Reliability and validity indices for Model 1

Scale	Mean	Std. deviation	Raw alpha	Std. alpha	rhoA	AVE	AF	PROCR
DRPINT	2.72	1.39	0.93	0.94	0.94	0.58	0.29	0.38
PROCR	4.58	1.27	0.85	0.85	0.86	0.50	0.25	
AF	4.30	1.10	0.96	0.96	0.96	0.54		

Note: the last row of Table 1, above, presents the internal consistency for academic feedback as a single-factor/unidimensional scale.

[1]the list of items and the results of the EFA could not be included in this paper due to space restrictions but they can be requested from the authors

Table 2. Reliability and validity indices for Model 2

Scale	Mean	Std. deviation	Raw alpha	Std. alpha	rhoA	AVE	AF_PERS	AF_GEN	PROCR
DRPINT	2.72	1.39	0.93	0.94	0.94	0.58	0.16	0.31	0.38
PROCR	4.58	1.27	0.85	0.85	0.86	0.50	0.24	0.24	
AF_GEN	4.58	1.15	0.95	0.95	0.96	0.60	0.75		
AF_PERS	3.57	1.27	0.89	0.89	0.91	0.63			

We first tested the model in which all three constructs (i.e., 'academic feedback', 'procrastination', 'drop-out) were intention') conceptualized as single factors (model 1, shown in Figure 1, below). However, because the EFA suggested that 'academic feedback' items load on two different factors, we also tested a subsequent model, which followed the structure suggested in the EFA.

The first model (Model 1) indicated that both 'academic feedback' and 'procrastination' are valid predictors of 'drop-out intention' (H1 and H2), with 'procrastination' mediating between 'academic feedback' and procrastination' (H3).

Figure 1, below, shows the resulted path model in which all 'academic feedback' items were considered together (loading on a single parent factor).

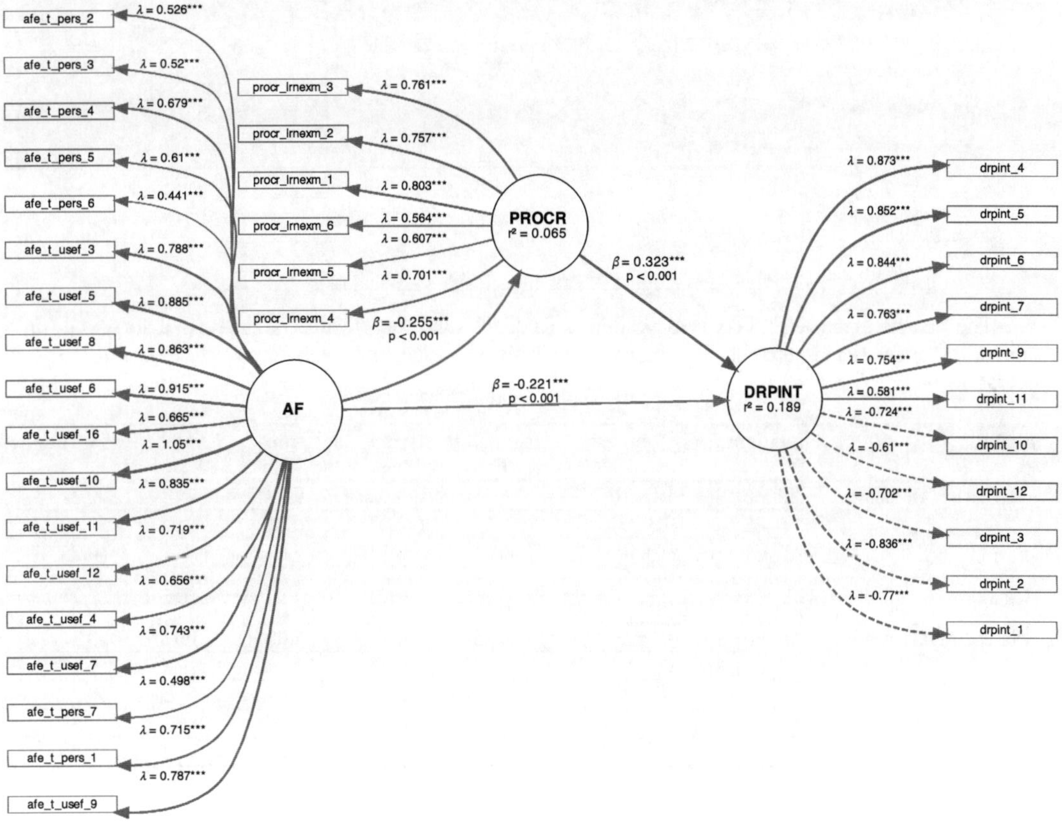

Figure 1. Path model with 1 AF factor

Figure 2, below, shows the path model for the case of two latent factors of 'academic feedback'.

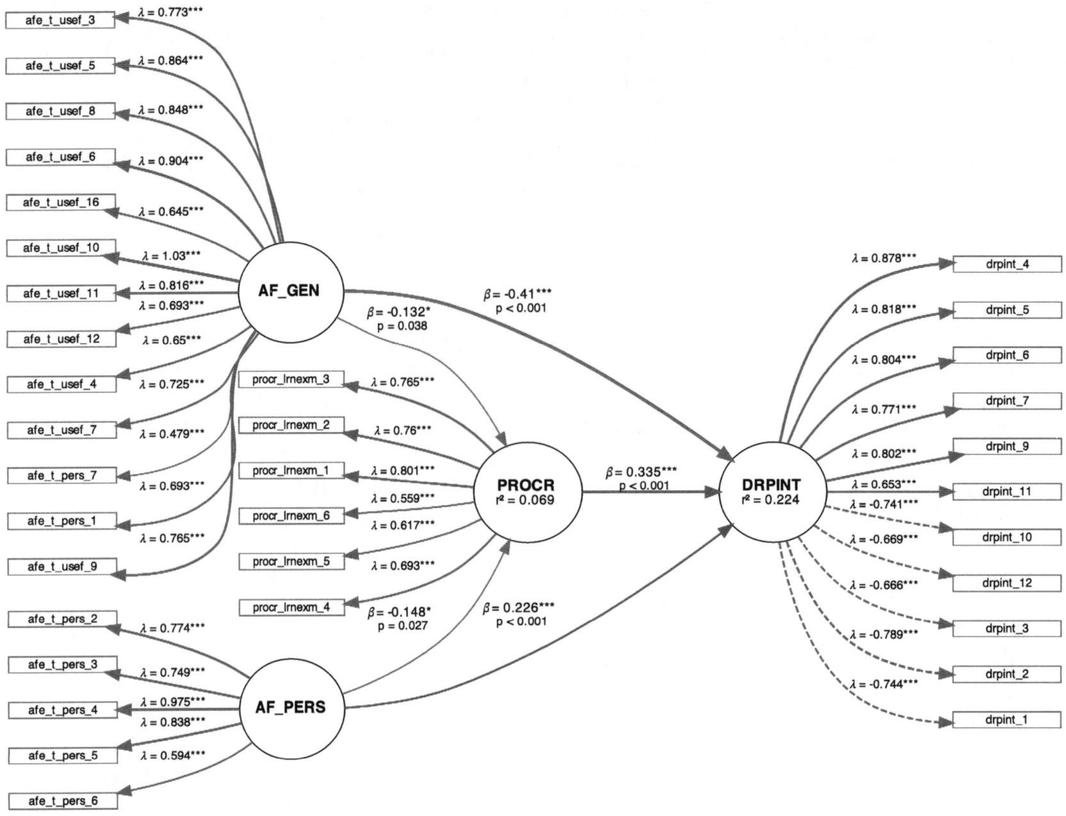

Figure 2. Path model with 2 AF factors

The associations (betas) between 'academic feedback' and 'procrastination', as predictors, and 'drop-out intention' as predicted/outcome, are presented in Table 3, below:

Table 3. Path Estimates

model	relationship	Original Est.	Bootstrap Mean	Bootstrap SD	T Stat.	2.5% CI	97.5% CI
Model 1 (1 AF factor)	AF_PERS -> PROCR	-0.15	-0.15	0.08	-1.78	-0.31	0.02
	AF_PERS -> DRPINT	0.23	0.22	0.06	3.59	0.10	0.35
	AF_GEN -> PROCR	-0.13	-0.14	0.08	-1.76	-0.28	0.01
	AF_GEN -> DRPINT	-0.41	-0.41	0.06	-6.58	-0.53	-0.28
	PROCR -> DRPINT	0.33	0.34	0.04	7.93	0.26	0.42
Model 2 (2 AF factors)	AF_PERS -> PROCR	-0.15	-0.15	0.08	-1.92	-0.31	0.00
	AF_PERS -> DRPINT	0.23	0.23	0.07	3.36	0.10	0.35
	AF_GEN -> PROCR	-0.13	-0.14	0.07	-1.78	-0.28	0.01
	AF_GEN -> DRPINT	-0.41	-0.41	0.06	-6.38	-0.53	-0.29
	PROCR -> DRPINT	0.33	0.34	0.04	7.68	0.26	0.42

AF denotes 'academic feedback'

5. DISCUSSION AND CONCLUSION

The results associated with the first model provided support for all our three hypotheses and explained .22% of the overall variance in the outcome variable ('drop-out intention'). Namely, the effect of AF on DRPINT was β = -0.22 at p < 0.001, and PROCR was also a significant predictor of DRPINT with β = 0.32 at

$p < 0.001$, while, concomitantly, PROCR mediated between AF and DRPINT. However, the most surprising result was provided by model 2, in which AF was conceptualized as two different factors, associated with one, more 'general' (or, in better words, 'impersonal') dimension of feedback and one, more specific dimension, which attracted items inquiring into how much the instructors' feedback was adapted to the learner's personality. Specifically, model 2 shows that the 'personal' AF factor was *positively* associated with the 'drop-out intention', which, at a superficial, face-value approach, appears to speak to the futility of such type of feedback, or of the attempts to personalize academic feedback.

Nevertheless, while surprising, this result is neither unexplainable nor direct evidence of the lack of usefulness of personalizing the academic feedback. For instance, the literature indicates there is some criticism as to what personalized feedback actually means (H. Wang & Lehman, 2021). In this regard, and which we acknowledge as a clear limitation, our 'academic feedback' questionnaire was not nuanced and faceted enough to capture the many intricacies of what 'personalized' 'academic feedback' means (for a lengthy and clarifying discussion about these facets, see H. Wang & Lehman, 2021). One possible explanation is the case when students at risk of dropping out of school are specifically 'targeted' by the instructors, to be provided more personalized feeedback, which (some of) them could interpret; in such a case, these students would report increased levels of personalized instructor feedback. Furthermore, if the reason for this positive association between AF_PERS and DRPINT was a consequence of help-seeking by the students, and is performed in situations of poor academic achievement, then our findings are consistent with those of Suamuang et al. (2021), insofar students can contemplate dropping out as a potential course of action and perceive the personalization of feedback as endorsing their prospects of dropping out.

Overall, our findings support existing research (Grunschel et al., 2013; e.g., Suamuang et al., 2021) while opening new lines of inquiry into the valences of personalized academic feedback. Procrastination is a relatively stable behavioral characteristic, that changes rather insignificantly throughout the semester (Kokoç et al., 2021). It is therefore to be expected that the sooner interventions on deterring procrastination tendencies, the more impactful they would be. Also, our findings provide additional evidence that the feedback provided by teachers and educators does, indeed reduce procrastination, and that, regardless of how it reflects on more distal variables (such as the 'drop-out') is a significant gain in and of itself.

Finally, our research had several methodological and conceptual limitations. First, methodologically, as mentioned above, our questionnaire did not capture the many intricacies of personalized feedback or the sensitive nature of properly balanced AF (Xavier & Meneses, 2021). Second, conceptually, and with relevance for the research design, future development will need to account for learning disabilities, as well as for more precise separation (i.e., operationalization) of the various facets of feedback. In our future work, we intend to refine the instruments used here, perform a thorough CFA and perhaps an exploratory SEM, including doing a thorough analysis of the scales used and reducing their complexity (i.e., maintaining only the minimum number of most relevant items) and investigate in more depth why various types and features of academic feedback may have opposing effects.

ACKNOWLEDGEMENT

This work was partly supported by the project "The Development of Advanced and Applicative Research Competencies in the Logic of STEAM + Health"/POCU/993/6/13/153310, project co-financed by the European Social Fund through The Romanian Operational Programme Human Capital 2014-2020.

REFERENCES

Afzal, S., & Jami, H. (2018). Prevalence of academic procrastination and reasons for academic procrastination in university students. *Journal of Behavioural Sciences*, *28*(1), 51–69.

Aleven, V., McLaughlin, E. A., Glenn, R. A., & Koedinger, K. R. (2017). Instruction Based on Adaptive Learning Technologies. In *Handbook of research on learning and instruction* (Second, pp. 552–560). Routledge.

Bozarth, J., Chapman, D. D., & LaMonica, L. (2004). Preparing for distance learning: Designing an online student orientation course. *Journal of Educational Technology & Society*, *7*(1), 87–106.

Cuijpers, P., Auerbach, R. P., Benjet, C., Bruffaerts, R., Ebert, D., Karyotaki, E., & Kessler, R. C. (2019). The World Health Organization World Mental Health International College Student initiative: An overview. *International Journal of Methods in Psychiatric Research*, *28*(2), e1761. https://doi.org/10.1002/mpr.1761

Doherty, W. (2006). An analysis of multiple factors affecting retention in Web-based community college courses. *The Internet and Higher Education*, *9*(4), 245–255. https://doi.org/10.1016/j.iheduc.2006.08.004

Donnelly, W., & Kovacich, J. (2014). *A Phenomenological Investigation of the Problem of Adult Student Attrition in Community College Online Courses*. *3*(1).

Ferrari, J. R. (2001). Procrastination as self-regulation failure of performance: Effects of cognitive load, self-awareness, and time limits on "working best under pressure." *European Journal of Personality*, *15*(5), 391–406. https://doi.org/10.1002/per.413

Fior, C. A., Polydoro, S. A. J., Pelissoni, A. M. S., Dantas, M. A., Martins, M. J., & Almeida, L. da S. (2022). Impact of Self-Efficacy and Academic Performance in the dropout of higher education students. *Psicologia Escolar e Educacional*, *26*, e235218. https://doi.org/10.1590/2175-35392022235218t

Goroshit, M., & Hen, M. (2021). Academic procrastination and academic performance: Do learning disabilities matter? *Current Psychology*, *40*(5), 2490–2498. https://doi.org/10.1007/s12144-019-00183-3

Graff, M. (2019). Self-Efficacy Beliefs and Academic Procrastination. *North American Journal of Psychology*, *21*(1), 81–100.

Grunschel, C., Patrzek, J., & Fries, S. (2013). Exploring reasons and consequences of academic procrastination: An interview study. *European Journal of Psychology of Education*, *28*(3), 841–861. https://doi.org/10.1007/s10212-012-0143-4

Hen, M., & Goroshit, M. (2020). The effects of decisional and academic procrastination on students' feelings toward academic procrastination. *Current Psychology*, *39*(2), 556–563. https://doi.org/10.1007/s12144-017-9777-3

Henry, M. (2020). Online Student Expectations: A Multifaceted, Student-centred Understanding of Online Education. *Student Success*, *11*(2), 91–98. https://doi.org/10.5204/ssj.1678

Kandemir, M. (2014). The predictors of academic procrastination: Responsibility, attributional styles regarding success/failure, and beliefs in academic self-efficacy. *Education and Science*, *39*(171), 99–114.

Karami, F., & Mahmoodi, M. (2018). The study of the relationship between mental health and academic performance and self-efficacy and academic procrastination of high school students. *Indian Journal of Positive Psychology*, *9*(4), 462–464.

Kim, E., Alhaddab, T. A., Aquino, K. C., & Negi, R. (2016). *Delaying Academic Tasks? Predictors of Academic Procrastination among Asian International Students in American Universities*.

Klingsieck, K. B. (2013). Procrastination: When Good Things Don't Come to Those Who Wait. *European Psychologist*, *18*(1), 24–34. https://doi.org/10.1027/1016-9040/a000138

Kokoç, M., Akçapınar, G., & Hasnine, M. N. (2021). Unfolding students' online assignment submission behavioral patterns using temporal learning analytics. *Educational Technology & Society*, *24*(1), 223–235.

Lonka, K., Ketonen, E., Vekkaila, J., Cerrato Lara, M., & Pyhältö, K. (2019). Doctoral students' writing profiles and their relations to well-being and perceptions of the academic environment. *Higher Education*, *77*(4), 587–602. https://doi.org/10.1007/s10734-018-0290-x

Mccloskey, J. D. (2011). *Finally, My Thesis on Academic Procrastination* [Master's thesis]. Texas at Arlington.

Mccloskey, J., & Scielzo, S. (2015). *Finally!: The Development and Validation of the Academic Procrastination Scale*. https://doi.org/10.13140/RG.2.2.23164.64640

Narciss, S. (2008). Feedback strategies for interactive learning tasks. In *Handbook of research on educational communications and technology* (pp. 125–143). Routledge.

Nikolaidis, P., Ismail, M., Shuib, L., Khan, S., & Dhiman, G. (2022). Predicting Student Attrition in Higher Education through the Determinants of Learning Progress: A Structural Equation Modelling Approach. *Sustainability*, *14*(20), 13584. https://doi.org/10.3390/su142013584

Park, E., Martin, F., & Lambert, R. (2019). *Examining Predictive Factors for Student Succes in a Hybrid Learnig Course*. *20*(2).

Peterson, P. W. (2001). The debate about online learning: Key issues for writing teachers. *Computers and Composition*, *18*(4), 359–370. https://doi.org/10.1016/S8755-4615(01)00068-8

R Core Team. (2017). *R: A language and environment for statistical computing*.

Rozental, A., Forsell, E., Svensson, A., Andersson, G., & Carlbring, P. (2015). Internet-based cognitivebehavior therapy for procrastination: A randomized controlled trial. *Journal of Consulting and Clinical Psychology*, *83*(4), 808–824. https://doi.org/10.1037/ccp0000023

Rozental, A., Forsström, D., Nilsson, S., Rizzo, A., & Carlbring, P. (2014). Group versus Internet-based cognitive-behavioral therapy for procrastination: Study protocol for a randomized controlled trial. *Internet Interventions*, *1*(2), 84–89. https://doi.org/10.1016/j.invent.2014.05.005

Şahin, M., & Yurdugül, H. (2022). Learners' Needs in Online Learning Environments and Third Generation Learning Management Systems (LMS 3.0). *Technology, Knowledge and Learning*, *27*(1), 33–48. https://doi.org/10.1007/s10758-020-09479-x

Schmitt, J., Fini, M. I., Bailer, C., Fritsch, R., & Andrade, D. F. de. (2021). WWH-dropout scale: When, why and how to measure propensity to drop out of undergraduate courses. *Journal of Applied Research in Higher Education*, *13*(2), 540–560. https://doi.org/10.1108/JARHE-01-2020-0019

Schouwenburg, H. C. (1995). *Academic procrastination: Theoretical notions, measurement, and research.*

Schouwenburg, H. C., & Lay, C. H. (1995). Trait procrastination and the big-five factors of personality. *Personality and Individual Differences*, *18*(4), 481–490. https://doi.org/10.1016/0191-8869(94)00176-S

Serdar, E., DemiRel, D. H., & DemiRel, M. (2021). The Relationship between Academic Procrastination, Academic Motivation and Perfectionism: A Study on Teacher Candidates. *The Turkish Online Journal of Educational Technology*, *20*(4).

Solomon, L. J., & Rothblum, E. D. (1984). Academic procrastination: Frequency and cognitive-behavioral correlates. *Journal of Counseling Psychology*, *31*(4), 503. https://doi.org/10.1037/0022-0167.31.4.503

Steel, P. (2007). The nature of procrastination: A meta-analytic and theoretical review of quintessential self-regulatory failure. *Psychological Bulletin*, *133*(1), 65–94. https://doi.org/10.1037/0033-2909.133.1.65

Strijbos, J.-W. (2010). Structural validation of a feedback perceptions questionnaire. *9th International Conference of the Learning Sciences*, *2*, 2.

Strijbos, J.-W., Pat-El, R. J., & Narciss, S. (2010). Validation of a (peer) feedback perceptions questionnaire. *Proceedings of the 7th International Conference on Networked Learning 2010*, 9.

Strijbos, J.-W., Pat-El, R., & Narciss, S. (2021). Structural validity and invariance of the Feedback Perceptions Questionnaire. *Studies in Educational Evaluation*, *68*, 100980. https://doi.org/10.1016/j.stueduc.2021.100980

Suamuang, W., Easter, M. A., & Suksakulchai, S. (2021). Relations Between Instructor Feedback, Self-Regulation, Assignment Completion and Academic Achievement in Thai Higher Learning Institutions. *Malaysian Journal of Learning and Instruction*, *18*. https://doi.org/10.32890/mjli2021.18.1.4

Tuckman, B. W. (1991). The Development and Concurrent Validity of the Procrastination Scale. *Educational and Psychological Measurement*, *51*(2), 473–480. https://doi.org/10.1177/0013164491512022

Vaughn, M. G., Roberts, G., Fall, A.-M., Kremer, K., & Martinez, L. (2020). Preliminary validation of the dropout risk inventory for middle and high school students. *Children and Youth Services Review*, *111*, 104855. https://doi.org/10.1016/j.childyouth.2020.104855

Wang, H., & Lehman, J. D. (2021). Using achievement goal-based personalized motivational feedback to enhance online learning. *Educational Technology Research and Development*, *69*(2), 553–581. https://doi.org/10.1007/s11423-021-09940-3

Wang, Q., Xin, Z., Zhang, H., Du, J., & Wang, M. (2022). The Effect of the Supervisor on Academic Procrastination: The Chain-Mediating Role of Academic Self-Efficacy and Learning Adaptation. *International Journal of Environmental Research and Public Health*, *19*(5), 2621. https://doi.org/10.3390/ijerph19052621

Xavier, M., & Meneses, J. (2021). The Tensions Between Student Dropout and Flexibility in Learning Design: The Voices of Professors in Open Online Higher Education. *The International Review of Research in Open and Distributed Learning*, *22*(4), 72–88. https://doi.org/10.19173/irrodl.v23i1.5652

Yurtseven, N., & Doğan, S. (2019). Structural relationships among academic procrastination, academic motivation, and problem solving skill in prep class college students. *Pegem Eğitim Ve Öğretim Dergisi*, *9*(3), 849–876. https://doi.org/10.14527/pegegog.2019.027

"A PIECE OF CAKE" - SUSTAINABLE EDUCATION PRACTICES WITH TRANSFORMER MODELS

Francisco de Arriba-Pérez, Silvia García-Méndez and Francisco J. González-Castaño
Information Technologies Group, atlanTTic, University of Vigo
School of Telecommunications Engineering, Maxwell street, 36310 Vigo, Spain

ABSTRACT

New advances in Artificial Intelligence (AI), particularly transformer language models, have disruptively changed the current education and training practices. The latter advancements have forced a shift towards a more practical and sustainable learning process with the ultimate objective of creating active, high-quality, and continuous educational services. However, human intervention is still essential to keep control over the outcomes of AI-based solutions. Thus, this work contributes with a fully functional and well-designed Conversational Assistant for Knowledge Enhancement (CAKE), which integrates a transformer model in a chatbot platform with empathetic capabilities. The proposed environment is fully immersive and incorporates gamification. Natural Language Processing techniques are exploited through prompts. The most relevant features of CAKE are its flexibility regarding language, areas and topics covered, and the education level of the end users. Results obtained in two evaluation scenarios endorse the performance of the proposed solution and motivate us to continue this research line toward real-time monitoring of smart learning.

KEYWORDS

Artificial Intelligence, Intelligent Chatbot, Natural Language Processing, Prompt Engineering, Smart and Sustainable Education, Transformer Language Models

1. INTRODUCTION

Digitalization has spread in many fields. However, the impact of artificial intelligence (AI) has been even more notable in education and training (Rudolph et al., 2023). The latter process, when applied to the education sector, is transversal through all levels (elementary and high school, university studies, etc.) and seeks to shift from traditional and theoretical practices to more practical ones, promoting at the same time the sustainability of the learning process (Crompton & Burke, 2018). Accordingly, smart education has attracted the attention of the research community, coinciding with the progress on AI-based solutions and the need for a global framework for high-quality training services (Li et al., 2022). The fundamental change now lies in the real-time monitoring of the learning process thanks to these applications and technologies, including the analysis and detection of the emotional state of the students through sentiment and emotion analyses (Cope et al., 2021).

Since the disruptive transformer models were released, education has been subject to challenges that, when not addressed properly, may result in inadequate pedagogical practices. The educational community, researchers, and policymakers are responsible for providing well-designed applications (Baidoo-Anu & Owusu Ansah, 2023). In this line, transformer models like ChatGPT[1] have caused a huge buzz worldwide. These new AI-based solutions are changing business by creating new market opportunities and eliminating those whose Natural Language Processing (NLP) needs are already covered by them (Baidoo-Anu & Owusu Ansah, 2023). The latter disruptive change also applies to the education field. Among the benefits for students and teachers, the workload reduction (e.g., automatic assessment), the real-time monitoring of student knowledge acquisition, and classroom innovation (e.g., feedback mechanisms) deserve to be mentioned (Cope et al., 2021). However, these systems still require human intervention (Kitamura, 2023). Thus, fully functional and well-designed solutions are essential (Parsazadeh et al., 2018).

[1]Available at https://openai.com/blog/chatgpt, April 2023.

The most popular form of integration of the transformer models in commercial solutions is through chatbots (Brown et al., 2020). They are agents that have the capabilities to establish real-time and human-like conversations with the end users through natural language (text or voice) and are appropriate to our purpose due to their simplicity and quick response time (Tallyn et al., 2018). Since their creation, chatbots have been used in education, entertainment, etc. (Rodrigues et al., 2022) for question-answering and task-solving. In the particular case of the education sector, they contribute to creating an active, high-quality, and continuous training environment (Romero-Rodríguez et al., 2020). The future generation of learners, as digital natives, are willing to incorporate these technologies to get an immersive learning experience (Pedro et al., 2018).

Taking advantage of transformers and the chatbot paradigm, this work contributes with a multi-platform and language application that integrates sustainable learning practices, CAKE (Conversational Assistant for Knowledge Enhancement). The main advantage is its flexibility regarding language, areas and topics covered, and the education level of the end users, thanks to NLP techniques applied to prompt engineering.

The rest of the paper is organized as follows. Section 2 reviews relevant works on innovative educational applications. Section 3 describes the proposed solution, while Section 4 provides relevant details about implementation and experimental results. Finally, Section 5 concludes the article and proposes future research.

2. RELATED WORK

In the era of transformer models, education and training have derived towards smart education (Li et al., 2022). Consequently, there exists a wide variety of technology-based solutions applied in the classrooms. A relevant feature of these systems is that they are ubiquitous. Thus, users apply them both inside and outside of these environments (*e.g.*, in remote learning). Notwithstanding its easy, flexible, and ubiquitous nature, AI-based solutions remove significant barriers, most importantly, money and time, without neglecting the learner's needs (Kumar Basak et al., 2018). It is the case with popular apps for educational purposes (Crompton & Burke, 2018). In fact, educational apps rank second in popularity on Google Play (Singh & Suri, 2022). Since the learning process is composed of high-difficulty tasks like comprehension and problem-solving (Malik et al., 2020), these kinds of systems become essential to support both teachers and students. In this line, Intelligent tutoring systems (ITS) are educational applications based on AI which provide intelligent support for students in an adaptive and personalized manner, aka personalized adaptive learning (PAL) (Peng et al., 2019; Zawacki-Richter et al., 2019). Among the applications of ITS, one of the most popular consists of automatically providing feedback to the students, for example, about their writing skills (Garcia-Gorrostieta et al., 2018). Furthermore, when combined with traditional educational and training approaches, technology-enhanced learning ensures promising learning outcomes and user satisfaction (Mota et al., 2018). Note that the use of multi-modal sources of information is highly appropriate in smart education (Li et al., 2022).

AI-based solutions provide relevant benefits for the whole educational community (*e.g.*, workload reduction and real-time training for teachers and students, respectively) (Cope et al., 2021). Accordingly, ChatGPT has become a common NLP practice in education (Baidoo-Anu & Owusu Ansah, 2023). Apart from question generation, transformer models can generate distracting answers in a multiple-choice quiz use case (Rodriguez-Torrealba et al., 2022). It is the case of EduQuiz[2], a quiz generation system for the reading comprehension use case based on GPT-3 that creates correct and distracting answers. However, ChatGPT and transformer models, in general, still require human intervention (Kitamura, 2023). Moreover, when it comes to sustainability, these models need to be energy-efficient in terms of maintenance and training costs (algorithms, data processing, representation, storage, etc.). The data mining must be managed ethically and following regulatory compliance (Kasneci et al., 2023).

Regarding the use of transformer models, Bhat et al. (2022) integrated GPT-3 in an assessment application. The questions generated by the system were manually evaluated in terms of their usefulness with positive results. Moreover, Bulut et al. (2022) used GPT-2 for reading passage generation in a reading comprehension assessment solution. Note that the authors pointed out that further research is still needed to adapt the outcome of the AI-based model for students. Conversely, MacNeil et al. (2022) used GPT-3 to generate descriptions about code programming, while Moore et al. (2022) presented another assessment system, in this case, for chemistry. A more sophisticated approach was followed by Topsakal et al. (2022), who designed a language

[2]Available at https://github.com/RamonDijkstra/EduQuiz, April 2023.

learning chatbot with ChatGPT, augmented reality, and gamification. Finally, Zong et al. (2022) used GPT-3 for math problems about two linear equations (word problem classification, equation extraction, and word problem generation) with competitive results.

Based on the aforementioned works, chatbots seem appropriate for cooperative and self-regulated learning since they are easy to use and promote engagement (Smutny & Schreiberova, 2020). Accordingly, Ethnobot (Tallyn et al., 2018) has question generation capabilities but is limited to the ethnographic field. Aligned with our work, Ng et al. (2020) developed WeChat, which became the most popular assistant for collaborative education among Chinese learners. More recently, Liu (2022) presented a language learning application combining collaborative approaches with gamification in a competitive learning environment.

Summing up, this work contributes with an innovative application combining ChatGPT with NLP for the teachers and researchers to keep control over the outcome of the model, avoiding inadequate pedagogical practices, named CAKE. Regarding the pedagogical basis of the proposed solution, it seeks to contribute to the following educational perspectives: cognitive (enhance reasoning during the learning process), emotional (promote positive feelings during training), behavioral (improve the results of the learning process), and contextual (encourage long-term training inside and outside the classroom). Note that the cognitive and behavioral perspectives are related to the application of NLP techniques over the content of the dialogue, while the emotional one is handled with emotion analysis. Finally, since the solution can be accessed anywhere and anytime, the users can benefit from the contextual perspective.

3. PROPOSED METHOD

The CAKE solution is shown in Figure 1. It has multi-language support and was designed bearing in mind students and teachers. Moreover, it provides automatic adjustment of the difficulty level of the proposed tasks.

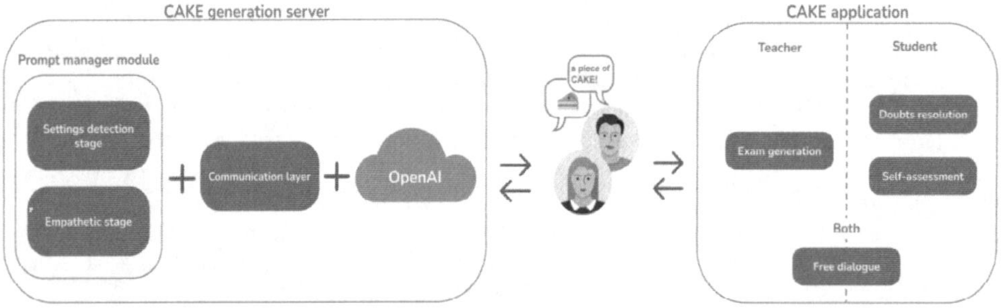

Figure 1. CAKE application for sustainable education and training

3.1 CAKE Generation Server

It is responsible for text generation. Note that NLP is exploited to both control the quality of the outcome and to ensure that an appropriate dialogue is established between students and teachers with the proposed solution. In addition, the generation server is in charge of past utterances management since the transformer models, and also ChatGPT, lack memory. Note that a cache of the six last interactions was established motivated by the cost-efficiency trade-off (for the free dialogue functionality, see Section 3.2).

3.1.1 Prompt manager module

This is the main module of the system. It defines the prompts of the model used for dialogue contextualization about a certain educational topic. The latter prompts are created using natural language and serve to direct the functioning of Open AI's ChatGPT model. Note that CAKE automatically adapts to the language of the users.

In the end, this module enhances the outcome of the system with empathetic utterances through feedback and personalization processes. For each of the following stages, a detailed prompt is defined in Table 1.

- **Empathetic stage**. There exist two prompts (7 and 8 in Table 1) related to the empathetic nature of the CAKE system. In order to promote engagement, the user utterances are analyzed to evaluate the student's level of knowledge on the subject. The difficulty level is automatically handled based on the number of correct/incorrect answers provided by the students to prevent frustration. Moreover, for each question correctly answered and depending on the degree of difficulty, a score is assigned (low: 10 points, medium: 20 points, high: 30 points). A badge will be granted when a user reaches the 100 points threshold. The latter rewards will be stored in their Hall of Fame.

- **Settings detection stage.** It consists of detecting: (i) the language, (ii) the role of the user (student or teacher), (iii) the functionality (see Section 3.2), and (iv) the area and topic. For the specific case of the exam generation functionality, the number of questions and difficulty level are extracted from the teacher's utterance. Otherwise, a random number of medium-level questions is generated.

3.1.2 Communication layer

It acts as an abstraction layer between the CAKE generation server and Open AI's ChatGPT model. The communication layer defines a structure based on tags to identify the sender (CAKE or the end user) and the context. The tags used are (see Table 1):

- **"System" tag**. It sets the application context in terms of (i) language, (ii) the sender (CAKE or user) and role (student or user), (iii) functionality, (iv) the area and topic, and (v) the number of questions and difficulty level. The empathetic nature of CAKE (see prompts 7 and 8 in Table 1) is also configured with the "System" tag.
- **"CAKE" and "User" tags**. These tags identify the utterances of the CAKE system or the end user, respectively.

3.2 CAKE Application

The client application was developed following the popular instant messaging approach of applications like WhatsApp. Note that the chat can be ceased with the command "terminate". The following functionalities are available in the CAKE application:

- **Free dialogue (F)**. This functioning mode is intended for both students and teachers for them to search for information or even generate or solve more sophisticated context-dependent questions.
- **Exam generation (E)**. It consists in generating exam questions for the teacher role. There are a total of 3 difficulty levels: low, medium, and high.
- **Doubts resolution (D)**. This functionality is intended for students to solve their doubts. Questions about exercises and problems can be formulated to obtain the correct answer and a detailed explanation.
- **Self-assessment (S)**. It was designed for students to practice for an exam, with empathetic feedback.

Table 1. Prompts used

ID	Detection	Prompt	Temp.	Result
1	Language	SYSTEM: If I told you this phrase/word: "X", in what language is it? Return {"language":""}. If it is not a language, return "not_applicable".	0	LANG
2	Role	SYSTEM: If I told you "X", what would you say I am? A teacher or a student? Answer with one word.	0	ROLE = T/S
3	Functionality	SYSTEM: I am a [ROLE]. If I told you "X", what would you say I have chosen from the available options: [AVAILABLE OPTIONS]? Answer with only one of the options from the list.	0	MODE = F/E/D/S
4	Number of questions, area, topic, level	SYSTEM: If I told you "X", what number, area, topic, difficulty level am I referring to? Answer in English using only the next format: {"number":"", "area":"", "topic":"", "difficulty_level":""}.	0	NUM, AREA, TOPIC, LEVEL
5	Question generation	SYSTEM: I am a [ROLE] with advanced knowledge, let's talk in [LANG]. I want to generate [NUM] exam questions in [LANG] about [AREA] and [TOPIC] with [LEVEL] difficulty.	1	QUESTION
6	Answer generation (AG)	SYSTEM: I am a [ROLE] with advanced knowledge, let's talk in [LANG]. I want you to answer this question [QUESTION].	1	ANSWER

7	AG + empathy	SYSTEM: I am a [ROLE] with advanced knowledge, let's talk in [LANG]. For the question [QUESTION], I have answered [ANSWER]. Tell me in [LANG] if it is correct, partially correct or incorrect. Try to detect in my answer frustration [0.0 to 1.0]. Always explain to me why it is incorrect and the correct answer. Adapt the answer to a student with [LEVEL] knowledge (maximum 100 words). Use this format: {"eval":"", "explanation":"", "frustration":"", is_correct:"0.0/0.5/1.0"}.	1	EVAL, EXPLANATION FRUSTRATION IS_CORRECT
8	Free dialogue + empathy	SYSTEM: The assistant speaks only in [LANG]. The assistant is dynamic (it never repeats the same thing twice), creative, intelligent and very kind. The assistant always asks a question when it finishes speaking. USER: I am a [ROLE] with advanced knowledge. Let's talk in [LANG]. I wanted to talk about "X".	1	RESPONSE

4. EXPERIMENTAL RESULTS

In this section, the implementations used are detailed to ensure reproducibility. Moreover, the results obtained are presented and discussed.

4.1 CAKE Generation Server

It was programmed in Python (version 3.8) using a Flask server (version 2.2.2) with a Gunicorn traffic balancer (version 20.1.0). The ChatGPT model (version 3.5) provided by OpenAI was used.

4.1.1 Prompt manager module

For the ChatGPT model, the *temperature* configuration is shown in Table 1 for each stage in the CAKE scheme. Note that this parameter allows adjusting the randomness of the generated texts (1 stands for random, and 0 for deterministic). Thus, it contributes to creating human-like dialogues.

4.1.2 Communication layer

It makes requests from our Flask server to the ChatGPT model using the OpenAI Python Library[3]. This library allows us to interact with the model using a REST API.

4.2 CAKE Application

The Flutter programming language was used to make it multi-platform, *i.e.*, to be integrated with Android and iOS smartphones and tablets, and also as a web application with macOS and Windows operating systems. For the instant messaging design, we used the Flyer Chat library[4].

4.3 Case Study

In order to evaluate the effectiveness of the proposed solution, two evaluation scenarios were designed.

Scenario 1. It evaluates the exam generation and self-assessment functionalities with the research questions in Table 2. Particularly, CAKE is used to generate 150 questions: 5 questions about 5 topics with 3 levels of difficulty (low, medium, high) in 2 languages (English and Spanish). The selected areas and topics are natural sciences - the cell, social sciences - history, mathematics - derivatives, language - grammar, and arts - painting.

Scenario 2. Explanatory answers are generated for each of the questions in the first scenario. The research questions used for evaluation are also displayed in Table 2.

[3]Available at https://github.com/openai/openai-python, April 2023.
[4]Available at https://github.com/flyerhq/flutter_chat_ui, April 2023.

Table 2. Research questions used for evaluation

Number	Scenario	Research question
1	1	Is the language correct?
2		Are the questions related to the selected area and topic?
3		Are the questions appropriate to the expected level of difficulty?
4	2	Is the language correct?
5		Is the answer related to the selected area and topic?
6		Is the answer appropriate to the expected level of difficulty?

4.4 Evaluation Results

The evaluation scenarios were composed of 300 utterances that were manually annotated by a teacher and an NLP researcher. Table 3 details the evaluation results for the two aforementioned scenarios. The best results are obtained for English, as expected since this is the original language of the ChatGPT model. Regarding area and topic, and difficulty level, the results are similar between the two languages in the two evaluation scenarios.

Table 3. Percentage of affirmative answers of the two annotators for the English and Spanish languages

Language	Annotator	RQ1	RQ2	RQ3	RQ4	RQ5	RQ6
English	1	100.00	100.00	88.00	100.00	100.00	88.00
	2	100.00	100.00	93.33	100.00	100.00	90.67
	Avg.	**100.00**	**100.00**	**90.67**	**100.00**	**100.00**	**89.34**
Spanish	1	89.93	98.67	86.67	98.65	98.67	90.67
	2	88.00	98.67	88.00	98.65	98.67	88.00
	Avg.	**88.96**	**98.67**	**87.34**	**98.65**	**98.67**	**89.34**

5. CONCLUSION

Smart education has become a relevant field of research with its own entity driven by the latest AI advances. Among them, the most disruptive change is associated with the transformer language models like OpenAI's ChatGPT. These AI-based solutions exploit the practical perspective of the learning process to promote the sustainability of real-time and continuous, high-quality, and active learning. Even though smart educational systems provide adaptive and personalized support for students and are also able to detect emotions in their speech, human intervention is still required to take the most of them for the global education community.

Consequently, we take advantage of the willingness of the new generation of educators and learners towards technology adoption to propose CAKE, a functional and well-designed solution that integrates transformers in the chatbot paradigm, combined with NLP techniques through prompt engineering. Its sustainability lies in its multi-platform and language-independent features. Moreover, it is flexible regarding areas and topics covered and the education level of the users. CAKE has empathetic capabilities and promotes engagement due to the easy, self-regulated learning process and thanks to the gamification techniques used. Note that it has been designed for both students and teachers with several functionalities (free dialogue, exam generation, doubts resolution, self-assessment). The generation server is composed of the prompt manager with empathy handling, the communication layer, and the OpenAI's model. Figure 2 shows the proposed solution in the self-assessment use case. The results obtained were evaluated in two scenarios in two different languages (English and Spanish). The first analyzes the exam generation and self-assessment functionalities for the teacher and student roles, respectively; while the second focuses on the doubts resolution and self-assessment functionalities for the students. Both the questions (scenario 1) and answers (scenario 2) generated by CAKE were manually annotated, obtaining promising results regarding language, area and topic, and level of difficulty.

In future work, we plan to apply data mining techniques for the system to learn from the end users to create more personalized learning experiences. Moreover, prompt engineering will be exploited to enhance the management of CAKE regarding non-English and mathematical languages. Ultimately, sustainability issues of transformer models, like the carbon footprint impact and governance issues, will be discussed.

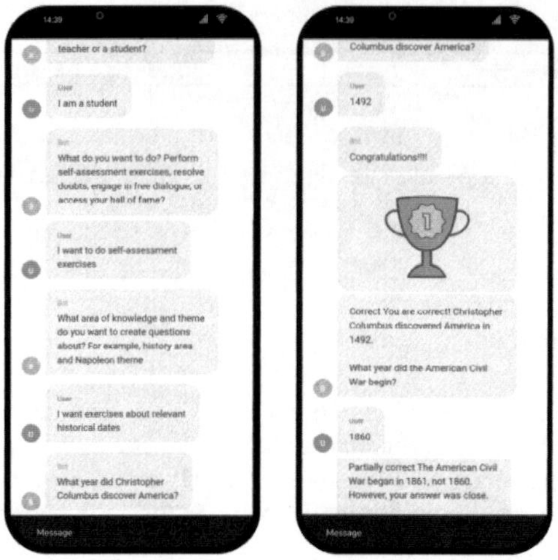

Figure 2. System application example

ACKNOWLEDGEMENT

This work was partially supported by Xunta de Galicia grants ED481B-2021-118, ED481B-2022-093, and ED431C 2022/04, Spain; and Ministerio de Ciencia e Innovación grant TED2021-130824B-C21, Spain.

REFERENCES

Baidoo-Anu, D., & Owusu Ansah, L. (2023). Education in the Era of Generative Artificial Intelligence (AI): Understanding the Potential Benefits of ChatGPT in Promoting Teaching and Learning. *SSRR*, 1-20. 10.2139/ssrn.4337484

Bhat, S., Nguyen, H. A., Moore, S., Stamper, J., Sakr, M., & Nyberg, E. (2022). Towards Automated Generation and Evaluation of Questions in Educational Domains. *Proceedings of the International Conference on Educational Data Mining*, 701-704. 10.5281/zenodo.6853085

Brata, K. C., Brata, A. H., & Lukman, E. P. (2019). Hanasu: Interactive Japanese Language M-Learning Application to Support Listening and Speaking Exercise. *Proceedings of the International Conference on Education and Multimedia Technology*, 311-315. 10.1145/3345120.3345155

Brown, T. B., Mann, B., Ryder, N., Subbiah, M., Kaplan, J., Dhariwal, P., Neelakantan, A., ... & Amodei, D. (2020). Language models are few-shot learners. *Advances in Neural Information Processing Systems*, 33, 1877-1901.

Bulut, O., & Yildirim-Erbasli, S. N. (2022) Automatic story and item generation for reading comprehension assessments with transformers. *International Journal of Assessment Tools in Education*, 9, 72-87. 10.21449/ijate.1124382

Cope, B., Kalantzis, M., & Searsmith, D. (2021). Artificial intelligence for education: Knowledge and its assessment in AI-enabled learning ecologies. *Educational Philosophy and Theory*, 53(12), 1229-1245. 10.1080/00131857.2020.1728732

Crompton, H., & Burke, D. (2018). The use of mobile learning in higher education: A systematic review. *Computers & Education*, 123, 53-64. 10.1016/j.compedu.2018.04.007

García-Gorrostieta, J. M., López-López, A., & González-López, S. (2018). Automatic argument assessment of final project reports of computer engineering students. *Computer Applications in Engineering Education*, 26(5), 1217-1226. 10.1002/cae.21996

Kasneci, E., Seßler, K., Küchemann, S., Bannert, M., Dementieva, D., Fischer, F., ... & Kasneci, G. (2023). ChatGPT for good? On opportunities and challenges of large language models for education. *Learning and Individual Differences*, 103, 1-9. 10.1016/j.lindif.2023.102274

Kitamura, F. C. (2023). ChatGPT is shaping the future of medical writing but still requires human judgment. *Radiology*, 230171. 10.1148/radiol.230171

Kumar Basak, S., Wotto, M., & Bélanger, P. (2018). E-learning, M-learning and D-learning: Conceptual definition and comparative analysis. *E-Learning and Digital Media*, 15(4), 191–216. 10.1177/2042753018785180

Li, N., Shen, Q., Song, R., Chi, Y., & Xu, H. (2022). MEduKG: A Deep-Learning-Based Approach for Multi-Modal Educational Knowledge Graph Construction. *Information*, 13(2), 1-18. 10.3390/info13020091

Liu, I.-F. (2022). Gamified mobile learning: effects on English learning in technical college students. *Computer Assisted Language Learning*, 1-24. 10.1080/09588221.2022.2080717

MacNeil, S., Tran, A., Mogil, D., Bernstein, S., Ross, E., & Huang, Z. (2022). Generating diverse code explanations using the GPT-3 large language model. *Proceedings of the ACM Conference on International Computing Education Research*, 2, 37-39. 10.1145/3501709.3544280

Malik, S. I., Al-Emran, M., Mathew, R., Tawafak, R. M., & Alfarsi, G. (2020). Comparison of E-Learning, M-Learning and Game-based Learning in Programming Education - A Gendered Analysis. *International Journal of Emerging Technologies in Learning*, 15(15), 133-146. 10.3991/ijet.v15i15.14503

Moore, S., Nguyen, H. A., Bier, N., Domadia, T., & Stamper, J. (2022). Assessing the Quality of Student-Generated Short Answer Questions Using GPT-3. *Proceedings of European Conference on Technology Enhanced Learning*, 243-257. 10.1007/978-3-031-16290-9_18

Mota, J. M., Ruiz-Rube, I., Dodero, J. M., & Arnedillo-Sánchez, I. (2018). Augmented reality mobile app development for all. *Computers & Electrical Engineering*, 65, 250-260. 10.1016/j.compeleceng.2017.08.025

Ng, S. F., Azlan, M. A. K., Kamal, A. N. A., & Manion, A. (2020). A quasi-experiment on using guided mobile learning interventions in ESL classrooms: Time use and academic performance. *Education and Information Technologies*, 25, 4699-4719. 10.1007/s10639-020-10191-7

Oyelere, S. S., Suhonen, J., Wajiga, G. M., & Sutinen, E. (2018). Design, development, and evaluation of a mobile learning application for computing education. Education and Information *Technologies*, 23, 467-495. 10.1007/s10639-017-9613-2

Parsazadeh, N., Ali, R., & Rezaei, M. (2018). A framework for cooperative and interactive mobile learning to improve online information evaluation skills. Computers & Education, 120, 75-89. 10.1016/j.compedu.2018.01.010

Pedro, L. F. M. G., Barbosa, C. M. M. de O., & Santos, C. M. das N. (2018). A critical review of mobile learning integration in formal educational contexts. *International Journal of Educational Technology in Higher Education*, 15, 1-15. 10. 10.1186/s41239-018-0091-4

Peng, H., Ma, S., & Spector, J. M. (2019). Personalized adaptive learning: an emerging pedagogical approach enabled by a smart learning environment. *Smart Learning Environments*, 6(1), 1-14. 10.1186/s40561-019-0089-y

Rodrigues, C., Reis, A., Pereira, R., Martins, P., Sousa, J., & Pinto, T. (2022). A Review of Conversational Agents in Education. *Proceedings of the International Conference on Technology and Innovation in Learning, Teaching and Education*, 461-467. 10.1007/978-3-031-22918-3_37/COVER

Rodriguez-Torrealba, R., Garcia-Lopez, E., & Garcia-Cabot, A. (2022). End-to-End generation of Multiple-Choice questions using Text-to-Text transfer Transformer models. *Expert Systems with Applications*, 208, 1-12. 10.1016/j.eswa.2022.118258

Romero-Rodríguez, J.-M., Aznar-Díaz, I., Hinojo-Lucena, F.-J., & Cáceres-Reche, M.-P. (2020). Models of good teaching practices for mobile learning in higher education. *Palgrave Communications*, 6(1), 1-7. 10.1057/s41599-020-0468-6

Rudolph, J., Tan, S., & Tan, S. (2023). ChatGPT: Bullshit spewer or the end of traditional assessments in higher education?. *Journal of Applied Learning and Teaching*, 6(1), 1-22, 10.37074/jalt.2023.6.1.9

Singh, Y., & Suri, P. K. (2022). An empirical analysis of mobile learning app usage experience. *Technology in Society*, 68, 1-14. 10.1016/j.techsoc.2022.101929

Tallyn, E., Fried, H., Gianni, R., Isard, A., & Speed, C. (2018). The Ethnobot: Gathering ethnographies in the age of IoT. *Proceedings of the Conference on Human Factors in Computing Systems*, 1-13. 10.1145/3173574.3174178

Topsakal, O., & Topsakal, E. (2022). Framework for A Foreign Language Teaching Software for Children Utilizing AR, Voicebots and ChatGPT (Large Language Models). *The Journal of Cognitive Systems*, 7(2), 33-38. 10.52876/jcs.1227392

Troussas, C., Krouska, A., & Sgouropoulou, C. (2020). Collaboration and fuzzy-modeled personalization for mobile game-based learning in higher education. *Computers & Education*, 144, 1-18. 10.1016/j.compedu.2019.103698

Wang, Z., Lan, A., & Baraniuk, R. (2021). Math Word Problem Generation with Mathematical Consistency and Problem Context Constraints. *Proceedings of the Conference on Empirical Methods in Natural Language Processing*, 5986-5999. 10.18653/v1/2021.emnlp-main.484

Zawacki-Richter, O., Marín, V. I., Bond, M., & Gouverneur, F. (2019). Systematic review of research on artificial intelligence applications in higher education - where are the educators?. *International Journal of Educational Technology in Higher Education*, 16(1), 1-27. 10.1186/s41239-019-0171-0

Zong, M., & Krishnamachari, B. (2022). Solving math word problems concerning systems of equations with GPT-3. *Proceedings of the AAAI Symposium on Educational Advances in Artificial Intelligence*, 1-8.

CONSUMER ATTITUDES AND GREEN SCEPTICISM TOWARDS ENERGY-EFFICIENT HOUSEHOLD APPLIANCES: FINDINGS FROM A PRELIMINARY STUDY

Gabriella Francesca Amalia Pernice[1], Valeria Orso[2] and Luciano Gamberini[2]
[1]*Departments of General Psychology, University of Padova, Padova, Italy*
[2]*Departments of General Psychology,*
Human Inspired Technology Research Center, University of Padova, Padova, Italy

ABSTRACT

Customers are becoming increasingly conscious of the environmental implications of their behaviour, prompting businesses to incorporate information about the level of sustainability of the products they consume. Nevertheless, poor or inaccessible information might result in consumers' negative attitudes towards companies and concerns about the quality of their products, leading to mistrust, commonly referred to as green scepticism. Such belief can hinder the translation of consumers' sustainability awareness into actual behaviour, with potential consequences for the environment, society, and firms. Despite the rising awareness of green scepticism toward sustainable marketing, the literature in this field is still limited. This exploratory study sought to comprehend the attitudes of working-age adults towards high-energy efficient household appliances and their perceived green scepticism towards sustainability-related information provided by companies. Actual Italian consumers who had purchased a household appliance in the last six months were recruited for the study, as they could offer valuable insights into attitudes and beliefs towards energy efficiency. The study also investigated consumers' green scepticism towards information companies provide, such as energy consumption. Preliminary findings indicate that even highly educated consumers who are conscious of environmental issues and perceive themselves as knowledgeable about energy efficiency are distrustful of the sustainability-related information provided by companies and experts in the field. The lack of clear information conveyed to consumers could exacerbate their scepticism, negatively influencing their decisions to purchase household appliances considering their energy consumption. The study offers several implications for companies and scholars, providing insights into overcoming green scepticism towards sustainable information of household appliances and encouraging more energy-saving habits.

KEYWORDS

Energy consumption, Green scepticism, Sustainability, Household appliance, Green attitudes, Eco-anxiety

1. INTRODUCTION

As the urgency of environmental issues grows, promoting the consumption of sustainable goods has become an important objective for many institutions. "Green marketing" refers to marketing strategies that aim to reduce a product's negative environmental impact. These strategies include communication campaigns and the inclusion of information regarding a product's environmental impact, such as energy-savings or CO2 emissions (Bhardwaj et al., 2023; Schwartz et al., 2020). However, green marketing initiatives are hindered by the increasing scepticism of consumers regarding the environmental claims of companies ('green scepticism'), due to the not-always clear definition of sustainability and information shared (Do Paço & Reis, 2012; Vlachos et al., 2013; Bhardwaj, et al., 2023). This lack of trust may be also a consequence of companies' irresponsible environmental conduct, such as making statements about sustainability, without taking real action, known as 'greenwashing' (Leonidou & Skarmeas, 2015). In 2009, a Eurobarometer Survey (EU) revealed that 48% of participants showed mistrust towards green marketing campaigns.

To date, research has primarily focused on positive drivers of green purchasing behaviour (Gho, 2016; Carrete et al., 2012; Atkinson & Kim, 2015). However, attention should also be given to the factors that hinder the purchase of green products. In recent years, the majority of studies have focused on the food industry, and only recently has the focus extended to the household appliance sector (Butt et al., 2022; Asif et al., 2022).

Further research is needed on consumer scepticism towards the environmental claims of companies, particularly in the context of energy efficiency and environmental choices (Raska & Shaw, 2012; Matthes & Wonneberger, 2014), to help firms and scholars design successful strategies and incentives to enhance energy-saving habits.

1.2 Related Work

As the urgency of environmental issues grows, consumers are becoming more conscious of the environmental impact of their actions (Dagher & Itani, 2014), and companies started to provide information about the environmental impact of their products to address these concerns (Schuitema & De Groot, 2015). Consumers may develop negative attitudes toward companies if the provided information is unclear or inaccessible. Specifically, green scepticism is related to consumers' distrust and questioning of the quality of "green" products, leading to negative assumptions about a company's values and decreased willingness to purchase (Nyilasy et al., 2014; Akturan, 2018).

Moreover, the rise in greenwashing, where some companies falsely claim their products to be environmentally safe, exacerbates consumers' negative attitudes towards companies (Mitchell & Ramey, 2011; Polonsky et al., 2010). Although environmental regulations exist, they are not adequately enforced, contributing to consumers' scepticism towards firms' green initiatives. Providing accurate information can encourage consumers' pro-environmental behaviours (Schuitema & De Groot, 2015; Schwartz et al., 2020). Evidence suggests that people with a high self-concept, green identity, and belief in the impact of personal consumption on the environment are more likely to purchase green products (Schuitema & De Groot, 2015). Despite concerns about environmental issues, consumers' attitudes may not always translate into actual green behaviour and purchasing decisions (Dagher & Itani, 2014). This may be due to ambivalent attitudes towards these products and towards companies (Al Mamun et al., 2018).

Further exploration of these factors could facilitate sustainable consumption and promote environmental responsibility. The Theory of Planned Behaviour (TPB), a social-psychological paradigm used to comprehend green behavior (Ajzen, 1991), has been widely used to measure green consumerism levels in the European Union and inform policy development (Rahimah et al., 2018). TPB states that behavioural intentions can be predicted by three sociocognitive variables: attitudes, subjective norms, and perceived behaviour control. The concepts of eco-knowledge and self-efficacy affect attitudes towards green products. When environmental knowledge increases, consumers are more concerned about their purchase decisions, which positively influences their attitude towards eco-friendly purchasing, as reflected in their behaviour (Zhao & Zhong, 2015; Tseng & Hung, 2013). Subjective norms and perceived behavioural control can affect an individual's beliefs in their ability to conduct behaviour and intention to buy green products based on situational and internal factors (López-Mosquera et al., 2014; Chen & Tung, 2014).

Based on the above theoretical and empirical premises, in this preliminary study, we aimed to understand working-age adults' attitudes toward high-energy efficient (HEE) household appliances, focusing on how they perceived greenwashing and scepticism towards sustainability information. We were also interested in their levels of eco-anxiety to assess their general concern regarding climate change, which has received less attention in the field of green purchases (Yang, 2021). Additionally, we intended to investigate their actual understanding of energy efficiency and appliance usage, which has been found to be significantly different from perceived knowledge in previous research (Pernice et al., 2022). Furthermore, the purchase phase has been investigated to determine empirically which attributes are most important to consumers when purchasing an appliance.

2. MATERIALS AND METHODS

2.1 Materials

A total of two questionnaires were conducted online. A brief description of both questionnaires follows.

First questionnaire and related sections: *Green attitude and green scepticism*

The questionnaire comprised 69 items grouped into five sections: Demographics, Attitude towards HEE household appliances, Green scepticism toward sustainability information regarding those appliances, Concerns about climate change, and Energy-efficiency actual knowledge. The Demographic section investigated participants' backgrounds, namely gender, age, education, job position, monthly income, how many people they live with, their previous engagement in other household appliances purchases and finally, their willingness to purchase household appliances online (8 items).

The second section explored the respondents' attitudes and knowledge about HEE household appliances. Items were completed on a 7-point scale Likert ranging from 1 "totally disagree" to 7 "totally agree":

- **Attitude (ATT)**; The degree of consumer's favourable or unfavourable evaluation of the HEE household appliance (6 items adapted from Hossain et al., 2022).
- **Subjective Norms (SN);** Social pressure exerted on consumers from the surrounding environment to perform, or not, a certain behaviour related to an HEE household appliance (4 items adapted from Issock et al., 2018).
- **Perceived Behavior Control (PEB);** Individual's degree of having the opportunity and ability to perform a "green" behaviour using HEE household appliances (5 items adapted from Bhutto et al., 2020).
- **Perceived Ecoknowledge (PEK);** Consumer knowledge of energy efficiency issues (6 items adapted from Waris & Hameed, 2020).

The third section explored respondents' perceptions of sustainability information, their scepticism toward the sustainability information of HEE appliances and the social actors who can drive this information. On a 7-point scale ranging from 1 "completely disagree" to 7 "absolutely agree," respondents rated the following items:

- **Greenwashing (GW);** The perception that the information regarding the sustainability of HEE household appliances is a tactic to manipulate consumers (6 items adapted from Nguyen et al. 2019).
- **Green scepticism toward Shop assistants, Companies, and Experts (GS)**; Untrusting the sustainability information that a shop assistant (4 items), a company (2 items), an expert (2 items) shares about the household appliances (adapted from Akturan, 2018).

The fourth section explored respondents' climate change worries using a standardised questionnaire (ECOANX). Items were scored using a 7-point scale Likert ranging from 1 "never" to 7 "always" (10 items from Innocenti et al., 2021).

In the final section, we aimed to explore participants' actual knowledge of energy efficiency by designing 8 *ad-hoc* true or false items. The items covered general information about energy efficiency, such as "*Energy efficiency is the use of less energy to perform the same task or produce the same result"* (4 items, of which 2 were true). We also included more specific information about how to use household appliances efficiently. One example statement was "*Laundry is only completely clean at temperatures above 30°C"* (4 items, of which 2 were true). These statements were based on information provided by the European Union in 2022.

<u>Second questionnaire</u>: *Users' preferences during the purchase phase*

The study utilised a pairwise comparison approach to investigate consumer choices concerning household appliances (Figure 1). A pairwise comparison is one of the most widely used methods for investigating consumer preferences in the field of sustainability (Kinoshita, 2020). This method consists in asking the respondent to compare two options at a time, and to choose their preferred one based on the predictor variables, called attributes and their attributes, the levels. The aim of the pairwise comparison is to determine the contributions of the attributes and their levels in the purchase.

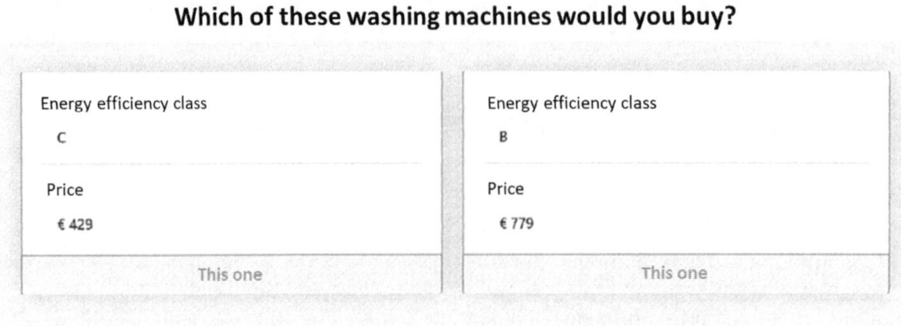

Figure 1. Example of pairwise comparison used in the study

In this study, respondents were asked to choose between two alternative washing machines (the most common appliance in Italy (according to Statista, 2023) based on several attributes, such as price, capacity, energy class, CO2, kWh, brand and their levels (Table 1).

The pertinent attributes and levels based on those attributes were identified based on the existing literature (i.e., Sonnenberg, Erasmus & Donoghue, 2011; Pernice et al., 2022). To ensure consistency with the current market demands, RStudio (R Core Team, 2021) was used to conduct a systematic web analysis of the three Italian main e-commerce and unrealistic options were eliminated. An average of 30 alternative pairs were shown to each participant.

Table 1. Pairwise comparison attributes and levels

Attribute	Levels
Energy class	A; B; C
Price	€ 829; € 779; € 429; € 379
kWh	175; 160; 145; 130
CO2	1500; 1000
Brand	Unknown; Known
Capacity	9 kg; 7 kg

2.2 Participants

The inclusion criteria to be part of the study were having purchased at least one large household appliance (e.g., dishwasher, refrigerator, washing machine) in the past six months and being of working age. The sample comprised 16 respondents (10 female, 6 male) with an average age of 39.31 years (SD = 12.55).

Concerning education, 25% of the respondents had a high school degree, 18.75% had a bachelor's degree, and 56.25% had a master's degree. With regard to income, 25% of the respondents had a monthly income below 1000€, 50% had a monthly income between 1000€ and 1499€, and 25% had a monthly income above 1500€.

The majority of employed respondents worked in the private sector (31.25%) or public sector (37.5%), while 25% were freelancers and 6.25% were small entrepreneurs. In terms of living arrangements, 37.5% of the respondents lived alone, 25% lived with one other person, and 37.5% lived with multiple people. Participants reported having bought at least one large household appliance in the last six months, with 56.25% purchasing a refrigerator, 37.5% purchasing a dishwasher, 31.25% purchasing a washing machine, and 18.75% purchasing an oven. Regarding their intention to buy household appliances online, 37.5% of the respondents preferred to buy online, 31.25% had no clear preference, and 31.25% preferred not to buy online.

2.3 Procedure

The study was conducted between December 2022 and January 2023. Questionnaires were pre-tested to minimise errors and potential misunderstandings. For participant recruitment, the researchers initially contacted acquaintances who had purchased large household appliances in the past six months. Once eligibility was confirmed, participants were sent a link to an online survey to complete both questionnaires. Subsequently, each participant was given the option to refer other eligible participants to the study. Participation was voluntary, and no compensation was provided. The study was approved by the local Ethical Committee.

2.4 Analysis and Results

All analyses were conducted using the software Rstudio (R Core Team, 2021). Non-parametric analyses were performed and the Benjamini & Hochberg (Benjamini and Hochberg, 1995) correction was applied.

To analyse participants' attitudes toward the energy-saving of household appliances and the perceived scepticism toward household appliance sustainable information, a series of one-sample Wilcoxon tests were run. In particular, we compared the scores that participants assigned to each dimension with the median value of the scale (Mdn= 4), which indicates a neutral attitude (Table 2).

Generally, the participants believed in their ability to make energy-saving choices (PEB), reporting to prefer buying HEE household appliances (ATT). They believed that they influence and were influenced by others (SN). They also reported to search for information about HEE household appliances and other sustainability-related information, such as CO2 and water consumption (PEK). They did not perceive the sustainability-information regarding the HHE household appliances as a marketing strategy to manipulate the consumers (GW). However, participants reported a lack of confidence in both appliance companies and experts (GS: companies and experts), while their beliefs toward shop assistants seemed neutral (GS shop assistants). They believe that the information they release has the ultimate purpose of the sale. Finally, they reported high worry about climate change (ECOANX).

Table 2.Values of one-sample Wilcoxon tests comparing score with median (Mdn=4); values means, standard deviations and medians

Dimension	V	p (BH)	M	SD	Mdn
PEB	120	<0.01	5.93	0.99	6.1
ATT	136	<0.01	5.83	0.81	5.91
SN	128	<0.01	4.87	0.71	4.87
PEK	130	<0.01	5.12	0.94	5.33
GW	88	0.059	4.45	0.96	4.58
GS Shop assistants	53	0.665	3.96	1.46	4.12
GS Companies	69	<0.01	5.03	1.37	5.25
GS Experts	115	<0.01	4.55	0.99	4.45
ECOANX	92	<0.05	4.55	0.99	4.45

Exploring participants scoring based on their sociodemographic characteristics, we performed a between-subject analysis based on income, and whether or not they were actively engaged in other household appliances purchase. For the sake of brevity, we only reported the analyses that showed significant differences. The Kruskal-Wallis test was performed between the levels of monthly income (low VS medium VS high). The test identified that the PEB was significantly higher in those with high income compared to those with medium income ($\chi2$ (2) = 6.84, p = .03). Surprisingly, GS towards experts was significantly higher in those with low income compared to those with high income, and medium income compared to those with high income ($\chi2$ (2) = 6.88, p = .03). The Mann-Whitney tests was performed considering whether or not they were actively engaged in other household appliances purchase (yes VS no). Those who were not actively engaged had significantly higher scores of ECOANX (W=41.00, p= .04). GW and GS towards the brand was higher in those who were engaged (W= 3.5, p= .01; W=4.5, p= .02) compared to those who were not. GS toward experts was significantly higher in those who were not engaged (W=42.50, p= .03).

Frequencies and proportions were first calculated to describe the respondents' actual knowledge. We set the sufficient threshold to 50%, meaning that participants had to answer correctly to at least 50% of the items. Results show that 53% of participants did not know how to manage household appliances to the highest level of efficiency. Additionally, 77% of them were not aware about the negative environmental impacts associated with the waste of energy.

Finally, we analysed respondents' preferences concerning the characteristics of household appliances. An overall importance score was calculated to determine what the most important attributes were (Table 3).

The price was found as the most important within the attributes, while kWh was the less important attribute.

Table 3. Pairwise comparison. Attributes, Levels and Relative importance

Attributes	Levels	Relative importance (equal 100% in total)
Price	429 €	25,93
Capacity	9 kgs	17,97
Energy Class	B	17,78
CO2	1000gr	12,98
Brand	unknown	12,77
kWh	145	12,54

3. DISCUSSION AND CONCLUSION

This study is a first attempt to provide insights into the attitudes and perceptions of real consumers towards HEE household appliances and the sustainable information provided by manufacturers and experts. Overall, participants expressed concern about environmental issues and showed a high level of awareness regarding the energy efficiency of their household appliances. They also perceived themselves as capable of implementing sustainable and energy-saving behaviours. However, they found the sustainable information provided by manufacturers and experts to be ambiguous and potentially driven by sales purposes. In contrast, participants had a neutral perception of the role of shop assistants in providing information on energy-efficient products.

According to the literature, purchasing green household appliances may not directly influence future green purchases if consumers' attitudes towards the sustainable market are generally negative (Rahimah et al., 2018). As our results show, being engaged in the purchase seems to be not enough to reduce the scepticism toward sustainable information Additionally, consumers reported that external social influence might play a key role in gathering information and shaping their energy-efficient habits. However, this information may not be accurate or helpful in promoting energy-saving behaviours. In fact, this could even reinforce incorrect habits and hinder sustainability efforts. Even though respondents were well-educated, their actual knowledge about sustainability and energy efficiency is often limited. A clear example could be their choice of the kWh as the least preferred characteristic in the purchase phase of a household appliance.

Fully comprehending the complexity of the sustainable information about the household appliances, such as energy consumption, could promote coherent purchase choices and enhance consumers' energy-saving behaviours in managing their household appliances. Further research in this field could be an additional step towards bridging the household energy efficiency gap (Kallbekken & Hermansen, 2013; Gamberini et al., 2012). Moreover, in line with the literature (Do Paço & Reis, 2012), our findings suggested that those attitudes are sensitive to sociodemographic characteristics, even if the sample was small. Companies and marketers need to improve their messaging and communication efforts around the energy efficiency benefits of their products, specifically addressing consumers' needs. Consumers would have more information regarding their appliances, but these have to be clear and reliable. Providing technical information to back up their statements could decrease consumers' negative attitudes towards companies and experts in the HEE household appliances market. Future research should focus on identifying effective communication strategies to promote sustainable and energy-saving-oriented behaviours among consumers.

REFERENCES

Ajzen, I. (1991). The theory of planned behavior. *Organizational behavior and human decision processes, 50*(2), 179-211.

Akturan, U. (2018). How does greenwashing affect green branding equity and purchase intention? An empirical research. *Marketing Intelligence & Planning.*

Al Mamun, A., Mohamad, M. R., Yaacob, M. R. B., & Mohiuddin, M. (2018). Intention and behavior towards green consumption among low-income households. *Journal of environmental management, 227*, 73-86.

Asif, M. H., Zhongfu, T., Irfan, M., & Işık, C. (2022). Do environmental knowledge and green trust matter for purchase intention of eco-friendly home appliances? An application of extended theory of planned behavior. *Environmental Science and Pollution Research, 1*-13.

Atkinson, L., & Kim, Y. (2015). "I drink it anyway and I know I shouldn't": Understanding green consumers' positive evaluations of norm-violating non-green products and misleading green advertising. *Environmental Communication, 9*(1), 37-57.

Benjamini, Y., & Hochberg, Y. (1995). Controlling the false discovery rate: a practical and powerful approach to multiple testing. Journal of the Royal statistical society: series B *(Methodological), 57*(1), 289-300.

Bhardwaj, S., Nair, K., Tariq, M. U., Ahmad, A., & Chitnis, A. (2023). The State of Research in Green Marketing: A Bibliometric Review from 2005 to 2022. *Sustainability, 15*(4), 2988.

Bhutto, M. Y., Liu, X., Soomro, Y. A., Ertz, M., & Baeshen, Y. (2020). Adoption of energy-efficient home appliances: Extending the theory of planned behavior. *Sustainability, 13*(1), 250.

Butt, A., Naeem, M. S., Ali, P. I., & Hameed, S. (2022). Impact of firms' Greenwashing practices on customer green trust and green brand attachment: Pakistans home appliances consumers' perspective. *Pakistan Business Review.*

Carrete, L., Castaño, R., Felix, R., Centeno, E., & González, E. (2012). Green consumer behavior in an emerging economy: confusion, credibility, and compatibility. *Journal of consumer marketing, 29*(7), 470-481.

Chen, M. F., & Tung, P. J. (2014). Developing an extended theory of planned behavior model to predict consumers' intention to visit green hotels. *International journal of hospitality management, 36*, 221-230.

Dagher, G. K., & Itani, O. (2014). Factors influencing green purchasing behaviour: Empirical evidence from the Lebanese consumers.*Journal of Consumer Behaviour,13*(3), 188–195.

Do Paço, A. M. F., & Reis, R. (2012). Factors affecting skepticism toward green advertising. *Journal of advertising, 41*(4), 147-155.

European Commission. (2009). Special Eurobarometer 220: barometric questions. https://data.europa.eu/data/datasets/s1489_62_2_ebs220?locale=en

Gamberini, L., Spagnolli, A., Corradi, N., Jacucci, G., Tusa, G., Mikkola, T., ... & Hoggan, E. (2012, June). Tailoring feedback to users' actions in a persuasive game for household electricity conservation. In *International conference on persuasive technology* (pp. 100-111). Springer, Berlin, Heidelberg.

Goh, S. K., & Balaji, M. S. (2016). Linking green skepticism to green purchase behavior. *Journal of Cleaner Production, 131*, 629-638.

Hossain, I., Nekmahmud, M., & Fekete-Farkas, M. (2022). How Do Environmental Knowledge, Eco-Label Knowledge, and Green Trust Impact Consumers' Pro-Environmental Behaviour for Energy-Efficient Household Appliances?. *Sustainability, 14*(11), 6513.

Innocenti, M. (2022). *Ecoansia: i cambiamenti climatici tra attivismo e paura*. Edizioni Centro Studi Erickson.

Issock, P. B. I., Mpinganjira, M., & Roberts-Lombard, M. (2018). Drivers of consumer attention to mandatory energy-efficiency labels affixed to home appliances: An emerging market perspective. *Journal of cleaner production, 204*, 672-684.

Kallbekken, S., Sælen, H., & Hermansen, E. A. (2013). Bridging the energy efficiency gap: A field experiment on lifetime energy costs and household appliances. *Journal of Consumer Policy, 36*, 1-16.

Kinoshita, S. (2020). Conjoint analysis of purchasing behavior for energy-saving appliances. International Journal of *Energy Sector Management, 14*(6), 1255-1274.

Leonidou, C. N., & Skarmeas, D. (2017). Gray shades of green: Causes and consequences of green skepticism. *Journal of business ethics, 144*, 401-415.

López-Mosquera, N., García, T., & Barrena, R. (2014). An extension of the Theory of Planned Behavior to predict willingness to pay for the conservation of an urban park. *Journal of environmental management, 135*, 91-99.

Matthes, J., & Wonneberger, A. (2014). The skeptical green consumer revisited: Testing the relationship between green consumerism and skepticism toward advertising. *Journal of advertising, 43*(2), 115-127.

Mitchell, L., & Ramey, W. (2011). Look how green I am! An individual-level explanation for greenwashing. *Journal of Applied Business and Economics, 12*(6), 40-45.

Nguyen, T. N., Lobo, A., & Greenland, S. (2016). Pro-environmental purchase behaviour: The role of consumers' biospheric values. *Journal of Retailing and Consumer Services, 33*, 98-108.

Nyilasy, G., Gangadharbatla, H., & Paladino, A. (2014). Perceived greenwashing: The interactive effects of green advertising and corporate environmental performance on consumer reactions. *Journal of business ethics, 125*, 693-707.

Pernice, G.F.A., Orso, V., Gamberini, L., (2022). Exploring consumer attitude toward sustainable energy-efficient appliance: Preliminary findings for augmented reality application. MCCSIS 2022 - Multi Conference on Computer Science and Information Systems; *Proceedings of the 15th International Conference on ICT, Society and Human Beings*, 2022

Polonsky, M. J., Grau, S. L., & Garma, R. (2010). The new greenwash?: Potential marketing problems with carbon offsets. *International Journal of Business Studies: A Publication of the Faculty of Business Administration, Edith Cowan University, 18*(1), 49-54.

Rahimah, A., Khalil, S., Cheng, J. M. S., Tran, M. D., & Panwar, V. (2018). Understanding green purchase behavior through death anxiety and individual social responsibility: Mastery as a moderator. *Journal of Consumer Behaviour, 17*(5), 477-490.

Raska, D., & Shaw, D. (2012). When is going green good for company image?. *Management Research Review, 35*(3/4), 326-347.

R Core Team (2021). R: A language and environment for statistical computing. R Foundation for Statistical Computing, Vienna, Austria. URL https://www.R-project.org/.

Schuitema, G., & De Groot, J. (2015). Green consumerism: The influence of product attributes and values on purchasing intentions. *Journal of Consumer Behaviour,14*(1), 57–69.

Schwartz, D., Loewenstein, G., & Agüero-Gaete, L. (2020). Encouraging pro-environmental behaviour through green identity labelling. *Nature Sustainability, 3*(9), 746-752.

Sonnenberg, N. C., Erasmus, A. C., & Donoghue, S. (2011). Significance of environmental sustainability issues in consumers' choice of major household appliances in South Africa. *International Journal of Consumer Studies, 35*(2), 153-163

Statista (2023). Household ownership rate of major appliances in selected countries worldwide in 2023. Retrieved from https://www.statista.com/statistics/1117972/major-appliances-ownership-selected-countries/

Tseng, S. C., & Hung, S. W. (2013). A framework identifying the gaps between customers' expectations and their perceptions in green products. *Journal of cleaner production, 59*, 174-184.

Vlachos, P. A., Koritos, C. D., Krepapa, A., Tasoulis, K., & Theodorakis, I. G. (2016). Containing cause-related marketing skepticism: A comparison across donation frame types. *Corporate Reputation Review, 19*, 4-21.

Waris, I., & Hameed, I. (2020). Promoting environmentally sustainable consumption behavior: an empirical evaluation of purchase intention of energy-efficient appliances. *Energy Efficiency, 13*(8), 1653-1664.

Yang, X., Tseng, Y., & Lee, B. (2021). Merging the Social Influence Theory and the Goal-Framing Theory to Understand Consumers' Green Purchasing Behavior: Does the Level of Sensitivity to Climate Change Really Matter?. *Frontiers in Psychology, 12*, 766754.

Zhao, R., & Zhong, S. (2015). Carbon labelling influences on consumers' behaviour: A system dynamics approach. *Ecological indicators, 51*, 98-106.

YOUNG PROFESSIONALS' CDW RECYCLING OBSERVATIONS

Anne-Marie Tuomala
LAB University of Applied sciences, Finland

ABSTRACT

Finland was the first country in the world to prepare a national road map to the circular economy (CE) in 2016. Since that many industrial fields have been planning and implementing various business applications and the authorities have been developing new regulation.
Demolition industry in Finland is unified under Finnish Demolition Association (FDA). It mainly consists of small and medium sized companies (SMEs), which have managed to grow and create new jobs. At the same time there is a growing need for qualified employees and engineers. The FDA noticed that it is a remarkable player as an implementer of the CE goals, as the construction and demolition waste (CDW) is the second biggest waste producer after the mining industry. The demolition industry has done and is able to do new business by innovating, but it needs more qualified forces.
The applied university level virtual course Demolition Waste Recycling has been running twice in 2021-2022. The interviewees were 19 recently graduated from universities of applied sciences, who worked in demolition sites during the past years and the interview gathered their observations and innovative ideas.
The result has important implications for national policymaking when designing new incentives to speed up the transformation to the CE. Moreover, demolition SMEs lack the resources to promote reuse of side streams, especially where essential actors are missing from the value chain.

KEYWORDS

Circulation Economy, Recycling, Construction and Demolition Waste, CDW, Sustainable Engineering

1. INTRODUCTION

1.1 Rationale and Background

The circular economy (CE) is a megatrend, where circulation of materials has a significant role. Finland was the first country in the world to prepare a national road map to the circular economy in 2016, under the leadership of Finnish Innovation Fund Sitra. The 2nd version published in March 2019 to update Finland's plans to reform its economic model to ensure successful sustainability by the circular economy.

Construction and demolition waste (CDW) accounts for more than a third of all waste generated in the European Union (European Commission 2023). According to the EU Waste Framework Directive (2008/98/EC), Finland was committed to reuse 70% of the construction and demolition waste (CDW) by 2020. However, the present utilization rate is still less than 60% (Ministry of Environment 2022). In Finland, about 84 % of the CDW is generated from repair projects and demolition of buildings and about 16 % from new buildings. The specialty in Finland is the wood-based waste, which is 41 % of the total CDW, whereas in the Central and South-Europe it is ca. 5 % (Peuranen & Hakaste 2014, 11). The new target year for 70 % of CDW recycling is 2027 (Finnish Environment Institute 2022, 23). To note, the current CSW utilization percentage is mainly generated in downcycling processes like energy production. The upcycling, such as recycling in new construction materials, keeps at much lower percentage level.

CDW circulation can be divided in two groups: backfilling and recycling. Backfilling is in practice downcycling because the material application is of low value. The upcycling is that the CDW is transformed into new materials and products that increase its value compared the material just after the demolition. It is always more beneficial to apply upcycling not only for the market itself but also to increase the number of products produced with this waste and therefore increasing the demand of it as source mate (Ferriz-Papi &

Thomas 2020, 2). This requires that there are enough remanufacturers for the CDW utilization.

According to the EU's Waste Framework Directive (WFD), preventing waste is always the preferred option and sending waste to landfill should be the last resort. Each industrial field, like the construction, renovation and demolition industry among others, should understand the CE applications and take into use so, that materials and wastes can be recycled and upcycled. The figure below, Waste hierarchy to construction and demolition activities (Dodd et al. 2020, 19) introduces separately applications of EU's waste hierarchy and in demolition and in construction activities.

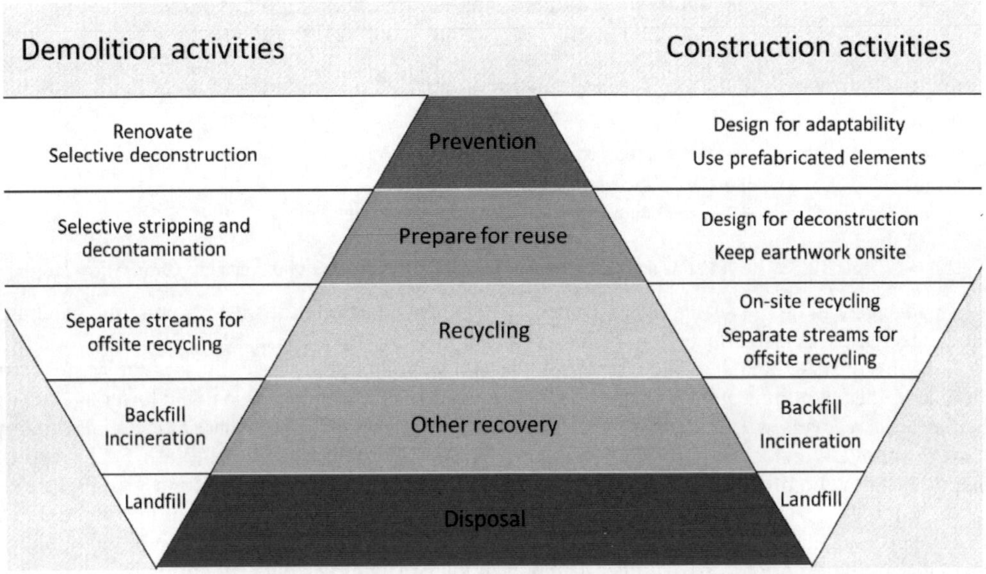

Figure 1. Waste hierarchy to construction and demolition activities based on Waste Framework Directive (Dodd et al. 2020, 19)

In Finland, the Government Decree on Waste from 2012 was the starting point. That time the country was far away from the recycling targets for CDW and below the average European level of 47%, being only 26% (Kojo & Lilja, 2011, 73). Another study (Høibye & Sand 2018, 45) states that in 2012, ca. 16 million tons of construction and demolition waste were generated in Finland; ca. 14 million tons of soil waste and 1.5–2.0 million tons from buildings. It was estimated that utilization of construction and demolition waste is now ca. 60%. However, due to the state of construction and demolition statistics, this figure is surrounded by a high degree of uncertainty.

The decree defined that by the end of year 2020 the recycling rate of CDW in Finland must be 70%. Dahlbo et al. (2015, 338-340) estimated that achieving the 70% recycling rate of CDW by 2020 would have required the overall recovery rate to be at least 80 % in practice, since some waste will inevitably be utilized as energy. This research group was of the opinion, that overall system produced environmental benefits and was economically profitable, but it was far behind the target of a 70%. Major changes will be needed for sorting, separation, and recovery within the system. Wood was the critical component for increasing recycling, because it is a large fraction currently recovered as energy.

The amount of plastic waste keeps growing in all over the world. Häkkinen et al. (2019, 5) state that the building and construction sector is the second largest consumption sector of plastics, accounting for 19% of all non-fiber plastics in the EU. PVC, PE, PP, EPS, and PU are typical plastics used in buildings. Different kinds of recycling techniques have been investigated and proposed. However, in Finland the current recycling rate for mechanical recycling is low (<20%), and most of the plastics in construction and demolition waste goes for energy recovery and disposal.

The Figure 2 illustrates the waste amount and treatment situation development from 2010 until 2019, when there was one year left until the target year 2020, when 70 % recycling rate should have been achieved. In 2019 total waste amount was declining as well as material recovery.

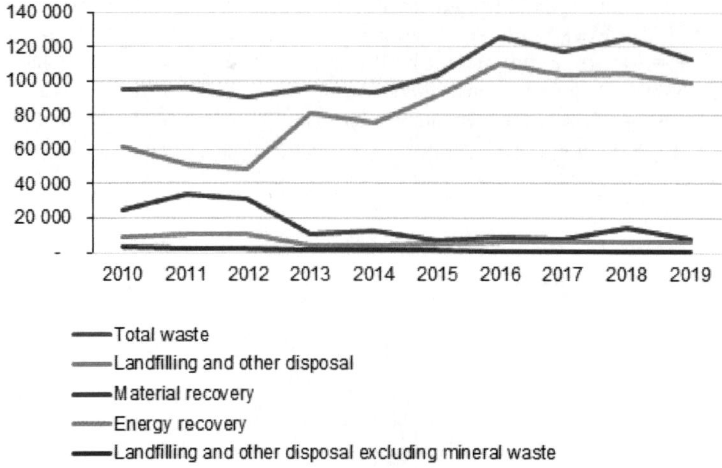

Figure 2. Waste amounts and treatments in Finland 2010-2019, 1,000 tons per year (OSF 2021)

The decrease in 2019 was mainly explained by the changed amount of waste from mining and quarrying, but the decrease was also affected by the fallen amount of waste from construction and demolition in general. The amounts of waste treatment were the result of the lower amount of waste. Most of the waste was mineral waste landfilled in mining areas and landfilled soil masses of construction. Due to the lower amount of waste, their material recovery was lower than in previous years, which decreased the share of material recovery, being 7.2 per cent of all treated waste (OSF 2021.). The Finnish Environment Institute (2022, 5, 23) reports for the year 2019 as total amount of waste 116 million tons of which share of construction was 13,7 million tons (ca. 12 %). CDW re-use as materials was 720 000 tons, which equals with the 48% of the total CDW amount. Researchers estimate that Finland presents a steep decline after 2010, which is likely to be the result of intense demolition activities prior to 2010, rather than prevention strategies (Zhang et al. 2022, 7).

The statistics of the year 2020 in Table1 below show growth in total waste amount. The mineral waste, the biggest among in tons, consists of concrete, brick, gypsum, asphalt, asbestos, and soil.

Table 1. Waste generation by industry by NACE, Information, Waste class and Year (OSF 2021).

	Year 2020
NACE F **Construction** (41-43)	1 000 tons
Amount of waste 1000 tons, TOTAL	13 689
Chemical waste	0
Metallic waste	217
Paper and cardboard waste	0
Wood waste	273
Animal and vegetal waste	1
Household and mixed waste	9
Sludges	721
Mineral waste	12 453
Other waste	16
Of which hazardous waste	314

Finally, the recycling rate of 48% was achieved by 2020, which was maybe embarrassing for the Ministry of Environment, as in the previous year the recycling rate was a bit higher exceeding 50 % clearly. Some media channels give for the year 2020 over 50% recycling rate. The official percentage in the Ministry of Environment (2022) site states that the present utilization rate is less than 60%, but the more exact percentage is not available.

CDW is one of the focus areas of the National Waste Plan in Finland. The national strategy also aims at increasing the safe utilization of CDW as well as improve the availability of accurate data on CDW (Karppinen 2021, 2.). The Figure 3 serves as an example of the new and more updated data dissemination.

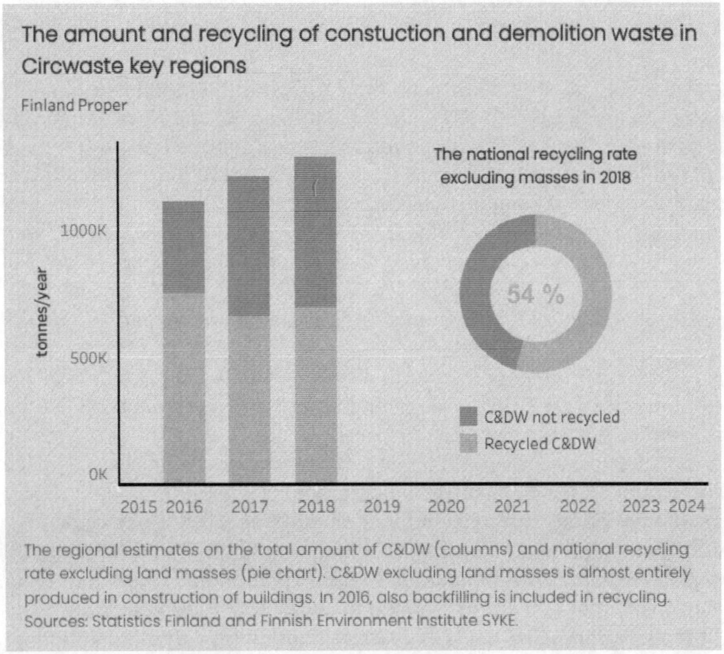

Figure 3. The national recycling amounts of CDW (Finnish Environment Institute SYKE 2021)

In Finland the CDW markets have been developed quite slowly, but new businesses are established like Ecoup and Conenor among others and supply chains are developing. There are also several tools and measures launched by the Ministry of Environment to speed up the progress. It has been published three guidebooks for the acting bodies to understand sustainability in building demolition. Furthermore, Materials Marketplace is a free of charge internet service and contact point for the producers and users, sellers, and buyers, of waste and side streams. In parallel, also private companies have established sharing platforms. To add, the reform of the Land Use and Building Act is ongoing and to be finished, hopefully, during this year. Important measures in operational level are e.g. the agreement on the collection of plastic films and use of recycled material and the GreenDeal on Sustainable Building between Ministry of the Environment and Finnish Property Owners RAKLI.

1.2 Research Question

The development tendency of CDW recycling has not been in Finland as expected in the national level, coordinated by the Ministry of Environment. It is good to shed light what kind of issues – positive or negative - we can find from the root level, that contribute to the overall situation.

This study explores the practices of CDW management in Finnish demolition field work. The primary research question is what kind of information a self-initiated employee can bring from the CDW field work into the common CDW management development activities.

1.3 Study Aims and Objectives

This study aims to explore employees' experiences and development ideas of CDW management on demolition sites. To add, through a review of above introduced literature, which is mainly based on statistical information of the country, focuses on the factors, barriers, motivations that influence the generation and management of CDW in the actual field work.

The following objectives facilitate the achievement of this aim (1) examine the current recycling activities in the demolition sites that contribute to keep materials in loops, (2) explore the behavioral implications of the employees in the context of environmental sustainability, and (3) propose the development ideas for the recycling or upcycling activities.

2. METHODS

Very little research has been made on construction employees with respect to the CDW recycling development. Employees, workers on the demolition sites, are the actual implementors of the society level of regulations, business contracts, rules etc. and, at the same time, they should proceed well with the goals. Therefore, their roles as agents of the CE transformers in practice is valuable as well as feedback on development ideas and because of that their contribution should not be overlooked.

This study employs two qualitative methods for data collection and analysis, which include two parts, thematic interviews, and employee-to-employee observations. The methods were chosen to collect data to formulate a systemic exploration of major recycling activities on demolition sites and to find needs for the further development. It is supposed that empirical data could be used to add value and deepen the theoretical background.

Interviewing relies on an interactive method in which mutual learning occurs between those involved in the interview process. Interviews are flexible, allowing in-depth analysis from a relatively small sample size and place the focus of research on the views of participants. (Young et al. 2018, 11). In relation to the chosen topic, the semi-structured thematic interview was chosen that interviewees could be encouraged to speak freely about their opinions and experiences and not be guided too much.

Field observations are usually extremely difficult as observers need to overcome several barriers, including employers' resistance to perceived additional liabilities, employees' fear that their level of activity will be reported to managers and many other practical and technical difficulties.

The research of Yun & Shen (2011, 1) conclude that that survey and case study are major methods for data collection in CDW management research, and the data are mostly processed through descriptive analysis. In Finland case studies have been used in calculations of CDW amounts. In this research we ended up using something else than a case study and decided to use two complementary methods. And, as the multimethod approach enables data triangulation and can limit the bias associated with any one single method (Young et al. 2018, 17), the combination of two methods was assessed to be more reliable for the results and bring more new information.

2.1 Thematic Interviews

Data gathering begins with the identification of interviewees (Young et al. 2018, 11). The used sampling strategy was a key informant sampling. This follows from the fact that the interviews were targeted to key people that are knowledgeable about the issue as the course participant of the topic and as employees in the demolition fields. More precisely, the interviewees were 19 persons recently graduated from the universities of applied sciences and they worked during the interview time or maximum in past two years on demolition sites during the last. Their study backgrounds were from construction engineering (10), environmental engineering (7), and business administration (2). They all had participated during their studies Bachelor level course Demolition Waste Recycling (5 ECTS) before their graduation.

The timing of the interviews was April – May 2021 and June 2022. The interviews were recorded by the Teams virtual meeting program by the permission of each interviewee. The cameras were open during the session, which enabled good communication. The interviewees were from 6 different Finnish regions from the south and west part of the country. The length of interviews varied mainly between 30 to 45 minutes, 2 exceeded this time as the start was delayed because of the technical problems in use of Teams software or internet connection.

When defining and identifying the interview themes, the following ones were picked up in the interview strategy analysis and taken in the use:

1. General practical experience in demolitions industry and participated projects.
2. Description how the material collection and waste recycling is organized. Highlights and development needs.
3. Reasons behind the current situation, personal assessment.
4. Future perspectives of the CDW management development.

It was taken into the consideration that the purpose of the study was explorative, and the thematic questions should be issues that are familiar to the interviewees and essential for to the study subject.

In the following we give quotations as samples for each theme. The interviewees are numbered randomly as persons 1, 2, 3 etc.

General practical experience in demolitions industry and participated projects

P10. "If we could get rid of incineration of the wood waste, that would be good and for the nature and for the circular economy".

P11: "In demolition it is easy to see, that we should now remove all harmful construction materials in new buildings – less problems in the future".

Description how the material collection and waste recycling is organized. Highlights and development needs.

P5: "It is about the willingness to do thing in a different way. There is so much wood material among the mixed waste. I think we could easily reuse the wood e.g., in manufacturing smaller furniture like shelves in parts, wood boxes etc. And, of course, to reuse as repair materials."

Reasons behind the current situation, personal assessments

P 1: "The waste collection points, bins and containers have been placed here and there on many sites…that confuses and the right one is not necessarily found in a hurry. If the bins and containers are too full, that causes disorder as well. Sometimes they are not discharged and left for waiting too long time."

Future perspectives of the CDW profession

P4: "If the pre-demolition audit becomes compulsory, I think that the persons must be accredited specialists of asbestos and hazardous components…such a requirement.".

P 7: "5G is coming, which speed up the data management. When users can manage the data by the web-based applications, the data is always available. This development, for sure, brings buzz for the field".

Remarkable amount of interview data included suggestions to tightening up regulation and rules and in general to have consequences of the possible irresponsibility. One interviewee did a calculation, which indicates the remarkable savings when demolitions materials are carefully sorted, and the amount of mixed waste is minimized. Another consideration is that the majority of the interviewees talked about the possibilities of digitalization, but that is not so much in focus in top administration of the employer companies – or not disseminated to the citizens and companies very efficiently from the administration.

As demolition project itself is run by the companies, the general development is much about the cost-effectiveness and new business opportunities The private companies do not like any more to take the materials and waste to the regional municipal waste management companies, which they see as competitors. They rather want to develop new products for own business, which is a good sign for the circular economy development. Crushed concrete, one of the mineral waste sorts, is a good example, as each of the leading companies have a product name for their demolished and crushed concrete. Companies have also started to demonstrate their recyclability skills in their own office and warehouse buildings. In February 2022 it was opened a new waste treatment plant in Vantaa. The value of investment is 35 M€ and the capacity fulfills 1/3 of the waste handling need of the capital region. The handling capacity of the CDW will be 70% in the capital area, which is remarkable for the progress in the whole country. As interviewees were totally aware of the importance of the practices on-site, 7 of them saw some help in sorting by robots and AI-utilization.

Like mentioned, the demolition companies are actively developing new recyclable products. Still, the interviewees told, that it is not always about the money (it is done what is written in the contract), at least in some extent willingness to do in a new way and behind this unwillingness is lack of knowledge. This concerns many activities, like organizing waste collection points and their logistics. Orientation and familiarizing for new workers are not always systematic. As there is not yet enough knowledge, on every site there is not a recycling/sustainability specialist, who could guide the other properly and organize activities.

2.2 On-Site Observations

The observers of this study were 11 from the group of 19 interviewees. Their selection was done first by their own willingness, but also based on their possibility to do the observation itself. They wrote observation diaries in May-July 2021 and 20222, depending on the work occupation schedule. The diaries were returned into the electronic platform's drop box, where they were accessible only for the researcher and the observer personally. Before the observation actions, the diary form was ready made and sent for the observers. It is recommended to use in observation research a diary, which can be submitted as form (Kananen 2014, 66-69), like shown in Table 2 below.

Table 2. Observation diary model (Kananen 2014, 69).

Time	Place	Situation	Persons and their roles	Activity	Description

In most of the cases, and the superiors and the employee colleagues were aware of the observation project on-sites, but in all cases the superior(s) were aware of that and authorized the activity. The notes were written as fresh as possible, but many times the observed issue required re-checking or asking from other the reasons behind or is the noticed issue new, continuing, once-off etc.

In data analysis it was possible to distinguished five idea types, which are listed in the Table 3 and their characteristics are described by examples. For each idea type categories were found 2-5 ideas.

Table 3. Characteristics of interviewees ideas based on-site observations

Idea type	Example	Amount by idea type
Need for legislation improvement	End of Waste (EoW) definition	3
Staff training	How to organize training especially for often changing employees and subcontractors	5
Connections with stakeholders	Active attitude, practices vary in different regions	2
New circular product	Horizontal utilization of materials between industries incl. composites	4
Best practice	How to organize work more efficiently in the process or in the demolition site	5

Sometimes the End-of-Waste (EoW) procedures and its definitions and interpretations might even cause the situation to be on the verge of **bankruptcy**. The cases are known among demolition professionals and that is why very rarely a company wants to be a forerunner. Demolition companies rather follow, which company takes the first step and then apply the similar idea by themselves, when the "product case is closed and working".

The need for staff training is also well known, but the characteristics of the industry has not yet enabled systematic approach in on-site project level. There are much sub-contracts, temporary employee agencies are in active use and every operator takes care of own training activities. There is also lack of workers in general and emerging lack of qualified workers.

Connections with stakeholders are developing and becoming diversified. This concerns especially public-private partnerships in cases when the municipalities and cities are main contractors. This have led to e.g. organizing seminars and training together. The demolition companies' contacts with the educational institutions have also become more intensive.

Like stated earlier, the new circular products are in the focus of the demolition industry. For the recycling and upcycling it is also to search horizontal utilization of materials between industries. Wood waste material needs urgent recycling solutions. Fortunately, some innovations have appearing like wood stone and products of wood fiber plastic composites.

There were notified many best practices, which might bring efficiency in the CDW on-site management. One observation was that if the temporary employees on sites would be trained to recognize waste fractions, the recycling rate 70 % would easily be achieved. Storing materials properly, new, or demolished ones, would prevent from entering in the waste stream and keeping at least the possibility for recycling.

3. RESULTS AND DISCUSSION

This section presents the key results of the study, how young professionals proceeded with the information and innovative ideas and how that could contribute to the development of CDW recycling. First, when it was examined the current recycling activities in the demolition sites that contribute to keep materials in loops, the gathered information indicates that managers seek more actively company partnerships, which work in reuse or recycling business. We also found that field workers had an impression, that recycling is in general steadily developing, but Finland looks like to be behind e.g., Sweden. This is based on the observations of two employees, who have worked and in Sweden and in Finland. In Sweden they did not see any bins or containers for mixed waste, but in Finland that is still a standard waste stream.

Second, the results are discussed in the context of the demolition company in the supply chain as it gives the framework for the sustainability implementation in the demolition project. The starting point of the demolition project is the procurement and the main contractor, who is in the top of the supply chain. The main contractor is either a private company or a public body, like city administration. The demolition project itself follows the accuracy in sorting how it is determined in the contract with the main contractor. Demolition companies, the next in order in the supply chain, have sub-contractors as well, which are consultants, laboratories, temporary employee agencies etc. Demolition companies must work according to the contract and budget. If there is a need to require more of their implementation, the main contractor must be ready for a bigger budget. However, there are simple things, what demolition company could do for the progress of CDW management. The most emerging on-site activity is the orientation training that every employee knows the CDW basics, starting from the recognition of waste fractions and person-in-charge of the site plans the material and waste logistics as carefully as possible with the notable markings on the bins and containers, also safeguarding enough often emptying of containers. And demolition companies can transfer these requirements for the own sub-contractors including **compliance with** the principle of sincere **cooperation**. Lastly, small and medium sized companies (SMEs) lack the resources to promote reuse of side streams, especially where essential actors are missing from the value chain. These are the most remarkable behavioral implications of the employees in the context of environmental sustainability, found in this research.

Third, even if employees on-site are not decision makers, and thus cannot implement own innovative ideas just like that, they also haven't seen – nearly at all – new initiatives or instructions coming from the top management. The most remarkable change has been the new Waste Act Regulation 2021, which added asphalt and mineral wool as well as bitumen and roofing felts among separately collectible waste fractions, which means that those streams have now own waste collection containers and spot, but not enable new ideas to flourish very much. The interviews or observations did not support the approach that employees would be proposed to suggest ideas for the recycling or upcycling activities.

The primary research question, what kind of information a self-initiated employee can bring from the CDW field work into the common CDW management development activities is found to be suitable from the research methods point of view. The observations and thematic interviews in the field are methods, brought the reality to the researcher's table and created new perspectives and generated more detailed information. To summarize, employees can bring information, which might look like minor or modest, like a suggestion that waste sorting instructions should also be visual and locate in coffee break premises, but if specialists in the field see this worth to try, would it be better to react from the management side as well? The costs would be minor, and it would be easy to put in practice. It looks like that objective, the need to increase CDW utilization rate consists of many projects specific practical barriers and lack of information.

4. CONCLUSIONS

The interviews and observations supported a high degree of common **situation awareness in the field work** and assessment that managers behave like let business as usual continue. A proposed development activities form a framework to evaluate and govern the overall sustainability potential of recycling activities.

In recent years, environmental problems caused by the CDW have become increasingly prominent. The share of the CDW in the total emissions is ca. 30%. That is why the Ministry of Environment has begun to pay attention to this industrial field and has launched tools and guidance and started to prepare new regulation. The new goal is to achieve 70 % utilization rate of the CDW by 2027.

The thematic interviews and demolition on-site observations provided an overview of various activities performed and what kind of obstacles and practices are currently in use. The interview analysis and observation reports delivered together insights into this fast-developing professional area of CDW recycling. To do more field work investigations, would also help in creation of new practices and policies.

ACKNOWLEDGEMENTS

I gratefully acknowledge the European Unit project Sustainable Horizons (GA 101071300), which has enabled the report writing. I also thank each interviewees-observers for their valuable time and contribution to the development of the circulation of demolished materials.

REFERENCES

Conenor. 2022. Company website. Available at http://www.conenor.com/. Last accessed 2022/10/27.

Dahlbo, H. & Bacher, J. & Lähtinen, K. & Jouttijärvi, T. & Suoheimo, P. & Mattila, T. & Sironen, S. & Myllymaa, T. & Saramäki, K. 2015. Construction and demolition waste management - A holistic evaluation of environmental performance. Journal of Cleaner Production. March 2015. DOI:10.1016/j.jclepro.2015.02.073

Dodd, N., Donatello, S., & Cordella. M. 2020. Level(s) indicator 2.2: Construction and Demolition waste and materials User manual: overview, guidance and instructions (Publication version 1.) Available at https://environment.ec.europa.eu/topics/waste-and-recycling/construction-and-demolition-waste_en

Ecoup.2022. Company website. Available at https://ecoup.fi/en/. Last accessed 2022/10/18.

European Commission.2023. Construction and Demolition Waste. Available at https://environment.ec.europa.eu/topics/waste-and-recycling/construction-and-demolition-waste_en Last accessed 2023/04/07.

European Commission. 2022. Waste Framework Directive. https://ec.europa.eu/environment/topics/waste-and-recycling/waste-framework-directive_en. Last accessed 2022/10/18.

Ferriz-Papi, J. & Thomas, S. 2020. Aggregates from Construction and Demolition Waste in Production of Concrete Blocks. Journal of Construction Materials. October 2020.

Finnish Environment Institute SYKE. 2021. Construction and Demolition Waste. Available at https://www.materiaalitkiertoon.fi/en-US/Monitoring/Construction_and_demolition_waste Last accessed 2022/10/18.

Finnish Environment Institute SYKE. 2022. Finland's Waste Management in Figures. Available at (in Finnish) https://issuu.com/suomenymparistokeskus/docs/valtsu-4.3-d, Last accessed 2022/10/27.

Häkkinen, T.& Kuittinen, M. & Vares, S. 2019. Plastics in buildings – A study of Finnish apartment buildings and day-care centre. Ministry of the Environment.

Høibye, L. & & Sand, H. 2018. Circular economy in the Nordic construction sector Identification and assessment of potential policy instruments that can accelerate a transition toward a circular economy. Nordic Council of Ministers. http://dx.doi.org/10.6027/TN2018-517.

Kananen, J. 2014. Laadullinen tutkimus opinnäytetyönä. Miten kirjoitan kvalitatiiviseopinnäytetyön vaihe vaiheelta. Jyväskylän ammattikorkeakoulun julkaisuja. Suomen Yliopistopaino. Oy – Juvenes Print, 2014.

Karppinen, T. 2021. Finnish Environment Institute SYKE. Description on the Calculation Method. Construction and Demolition Waste (C&DW) – Regional Amounts and National Recycling Rate Excluding Land Masses.

Kojo, R., Lilja, R., 2011. Talonrakentamisen materiaalitehokkuuden edistäminen (Removing the barriers to material efficiency in housing construction). Reports of the Ministry of the Environment 21/2011. Available at https://helda.helsinki.fi/handle/10138/41495. Last accessed 2022/10/27.

Ministry of Environment of Finland. 2022. Circular economy in the construction sector. https://ym.fi/en/circular-economy-in-the-construction-sector. Last accessed 2022/10/27.

Official Statistics of Finland (OSF). 2019. Waste statistics. Helsinki. Statistics Finland. Available at http://www.stat.fi/til/jate/2019/jate_2019_2021-06-16_tie_001_en.html. Last accessed 2022/10/27.

Official Statistics of Finland. (OSF). 2021. Waste Statistics. Statistics Finland. Available https://pxweb2.stat.fi/PxWeb/pxweb/en/StatFin/StatFin__jate/statfin_jate_pxt_12qw.px/table/tableViewLayout1/ Last accessed 2022/10/16

Peuranen, E. & Hakaste, H. 2014. Rakentamisen materiaalitehokkuuden edistämisohjelma. Ramate-työryhmän loppuraportti. Reports of Ministry of Environment. 17/2014. Available at https://julkaisut.valtioneuvosto.fi/bitstream/handle/10138/135172/YMra_17_%202014.pdf?sequence=1&isAllowed=y Last accessed 2022/10/28.

Sitoumus2050. Green Deal on Sustainable Building in Finland. https://sitoumus2050.fi/fi_FI/kestavapurkaminen#/ Last accessed 2022/10/27.

Sitra. Finnish road map toa circular economy 2016-2025. https://www.sitra.fi/en/projects/leading-the-cycle-finnish-road-map-to-a-circular-economy-2016-2025 Last accessed 2022/10/16.

Tam, V. W-Y. & Lu, W.2016. Construction Waste Management Profiles, Practices, and Performance: A Cross-Jurisdictional Analysis in Four Countries. Sustainability; Volume 8, Issue 2. https://www.mdpi.com/2071-1050/8/2/190

Young, JC., Rose, DC., Mumby, HS. et al. 2018. A methodological guide to using and reporting on interviews in conservation science research. Methods Ecol Evol. 2018; 9: 10–19. https://doi.org/10.1111/2041-210X.12828

Yuan H. & Shen L. Trend of the research on construction and demolition waste management. Waste Management. 2011 April ;31(4):670-9. DOI: 10.1016/j.wasman.2010.10.030

Zhang C., Hu M., Di Maio F., Sprecher B., Yang X., Tukker A. 2022. An overview of the waste hierarchy framework for analyzing the circularity in construction and demolition waste management in Europe, Science of The Total Environment, Volume 803, 2022. DOI: https://doi.org/10.1016/j.scitotenv.2021.149892

SHORT

PAPERS

DESIGNING A PICTURE E-BOOK INTERFACE WITH A READ-ALOUD FUNCTION TO IMPROVE ELEMENTARY SCHOOL STUDENTS' READING INTEREST

Gloria Yi-Ming Kao and Pei-Wen Lin
Graduated Institute of Digital Learning and Education, National Taiwan University of Science and Technology
Taipei 106, Taiwan, ROC

ABSTRACT

This study used the online software system, Publuu, to design an e-book reading interface, with the aim of enhancing students' reading interest and motivation through applying the reading aloud strategy in the process of reading picture e-books. We used a pilot study to test our e-book interface design for this purpose. The participants were four elementary school fourth-grade students who experienced using the e-book interface to read the text and the picture of the assigned e-book material. Meanwhile, they were instructed to use the read-aloud strategy in order to enhance their interest in reading. The preliminary results showed that this e-book design helped enhance students' reading performance and interest in reading. In the future, we plan to further improve the e-book interface design for elementary school teachers to teach reading activities in the classroom.

KEYWORDS

Picture e-book, Digital reading, Read-aloud, Reading motivation

1. INTRODUCTION

With the advent of the digital era, new forms of books are constantly being introduced. Different forms of e-books such as picture e-book applications on mobile devices or web-based e-books are flourishing, and these texts and pictures are usually displayed with multimedia and interactive features (Kucirkova, 2019). Common functions of e-books include graphics, hotspots, narration, interactive buttons, non-linear reading, giving feedback, animations and multimedia media (Lim, 2021; Rubegni et al., 2021; Yang, 2013; Yokota et al., 2014; Zhang et al., 2020). The use of e-books in the teaching field can effectively promote reading motivation and reading effectiveness (Danaei et al., 2020; Lin et al., 2018; Takacs et al., 2015; Wu, 2016).

On the e-book reading interface, e-books usually have three major functions, namely zooming, hotspot, and page control buttons. In the study of reading interactive e-books, 80% of students think the page control button is very important, and the high autonomy of page turning also enhances the number of times they read back and forth (Lim, 2021). Hotspot clicks are the most important feature of e-books as they provide students with supplementary reading information. Through teachers designing hotspots in the reading story and limiting the number of hotspots, students are free to explore the hotspots after reading without distracting their attention (De Jong & Bus, 2004; Korat, 2009; Shamir et al., 2008). Johnston (2020) surveyed university students' habits of using e-books, and found that 9% of students thought that the function of enlarging the page was helpful for reading.

However, when using e-books in reading activities, teachers or researchers usually focus on reading or writing skills. For example, when Lim (2021) investigated how the use of interactive e-books by 30 secondary school students affected secondary school students' reading performance, 66% of the students agreed that the picture dictionary feature in the e-book helped with reading activities in terms of its effectiveness in aiding reading comprehension. In early reading experiences, if readers encounter difficulties with unfamiliar word recognition, it can be very challenging to read, which may result in lower comprehension of the story and lower willingness to read (Danaei et al., 2020), so adding hyperlinks to hotspots helps students understand the words in the text and reduces the reading difficulties of students who have low levels of reading literacy. Wu (2016) developed an English reading e-book system that includes the functions of translation, pronunciation, reading and annotation to effectively and progressively improve students' reading comprehension skills through a personalized reading instruction mechanism.

Paivio's dual coding theory (1986) suggests that teaching materials can not only be presented visually, but also through voice input or output to help students achieve better learning outcomes. Altun (2018) investigated the effectiveness of reading e-books for pre-school children, and used story re-telling to test the students' reading levels. The relationship between practicing oral comprehension and text comprehension is relevant in the context of reading aloud, where comprehension of written text is stimulated through reading aloud, and the higher the level of engagement in the reading aloud activity, the more beneficial it is for language and reading development (Acosta-Tello, 2019; Batini et al., 2021). Therefore, this study will focus on the read-aloud strategy when designing and applying the interface features of picture e-books in reading activities.

In addition, Mayer (2003) proposed the multimedia learning theory, which provides a methodology for the production of effective teaching materials and guides instructors to design effective multimedia materials to reduce cognitive load. Therefore, this study drew on the dual coding theory and multimedia design theory to design the interface function of the electronic picture book used for this study.

This study used the online software system, Publuu, as the basis for the design and development of e-book materials, in the hope of helping students' reading scores and interest in reading. A prototype of the e-book interface was designed, and a pilot experiment was conducted to investigate the ease of use and applicability of the e-book interface, and to improve students' reading scores and interest in reading.

2. METHODS

The participants of this pilot study were four 10-year-old students (two girls and two boys) in fourth-grade of an elementary school in Taiwan. Since this pilot study was conducted to investigate the students' operational fluency of the e-book interface and their preliminary experience of the reading activity, no control group was used. To prepare the required materials and tools, we spent one and a half months using the Publuu software to redesign paper-based picture books into e-books, and designed multimedia functions according to our teaching objectives. The teaching objective was to provide a picture e-book interface with a read-aloud function to improve elementary school students' reading interest.

In the design process of the e-book, we firstly set up the interface with the software, and incorporated learning support functions into the reading interface, including free page flipping, hotspot, and page zooming, in the hope of enhancing students' comfort and ease of reading e-books and of reducing students' cognitive load. Secondly, in our didactical setting, we required the students to read aloud the text to engage them in the reading process. In order to do that, besides using the above functions of Publuu, we also used an "add links" function to link to external website to extend the functions of Publuu. The "add links" function could allow Publuu to connect to external websites. In our setting, the students use "add links" to connect to a teacher-assigned Google Drive folder. In this Google Drive folder, there are teachers' demonstration audio files. Students could listen to the demo first, then record their own versions of read-aloud videos with iPads. Listening to these demos might enhance their reading aloud fluency.

We spent four lessons on the pilot study, with each lesson lasting 40 minutes. To collect students' experience of using our e-book interface, interviews were conducted to evaluate the effectiveness of the reading activity.

Figure 1. e-book Interface Design

Figure 2. Reading Interface

3. RESULTS

The following observations were made after the pilot study. In terms of multimedia interface design, the reading interface design of the electronic picture book is highly relevant to students' learning. Based on pedagogical considerations, it is necessary to delete buttons that are not relevant to teaching activities so as to simplify the learning pages. In addition, after observing students' reading patterns, the page display was changed from single-page to double-page to help students' reading fluency or continuity, and to avoid discontinuous reading because the display of a single page was cut in half. All students responded that the page magnification function is very different from the previous paper reading experience, as the text can be zoomed in on and the content can be seen more clearly. Hotspots need to be paired with story materials to help students solve word difficulties.

Through this research, the read-aloud strategy helped students' reading performance and facilitated the teacher's instruction. The results showed that the reading aloud function was new and interesting to the students. The teacher could evaluate whether the students were engaged in the reading activity by evaluating their reading aloud audio files. The result indicated that all students' read-aloud performance was better on the posttest. From the teacher`s observations, the read-aloud strategy helped the students' reading fluency. The teacher found that the students' reading performance in normal classroom reading activities was also improved after participating in this pilot study. In the interview, students reflected that telling a story in a dramatic way is an interesting experience. In the past, seldom had students' oral reading performance been recorded, meaning that students had no chance to listen to their own or to others' voices.

4. CONCLUSION

In this study, based on our pedagogical considerations, we proposed a new way of using Publuu. We designed a picture e-book interface combined with a read-aloud function for elementary school children. The teacher observed that the reading performance of the students was improved and their interest in reading was enhanced, especially for students with lower levels of reading literacy. In the future, we will adjust the design of this current picture e-book interface based on our observations, and perform a formal experiment to further test our hypotheses. It is expected that the e-book interface can be used for students of different ages and backgrounds, to provide interesting experiences for readers. In particular, using an e-book interface with a read-aloud function has the potential to further boost elementary school children's reading motivation and performance.

REFERENCES

Altun, D., 2018. The Efficacy of Multimedia Stories in Preschoolers' Explicit and Implicit Story Comprehension. *Early Childhood Education Journal*, Vol. 46, No.6, pp. 629-642.

Acosta-Tello, E., 2019. Reading Aloud: Engaging Young Children during a Read Aloud Experience. *Research in Higher Education Journal*, Vol. 37.

Batini, F. et al, 2021. Reading and Linguistic Comprehension: A Systematic Review. *Ricerche di Pedagogia e Didattica. Journal of Theories and Research in Education*, Vol. 16, No. 1, pp. 79-86.

Danaei, D. et al, 2020. Comparing Reading Comprehension between Children Reading Augmented Reality and Print Storybooks. *Computers & Education*, Vol. 153, pp. 103900.

De Jong, M. T., and Bus, A. G., 2004. The Efficacy of Electronic Books in Fostering Kindergarten Children's Emergent Story Understanding. *Reading Research Quarterly*, Vol. 39, No. 4, pp. 378-393.

Johnston, N., and Ferguson, N., 2020. University Students' Engagement with Textbooks in Print and E-book Formats. *Technical Services Quarterly*, Vol. 37, No. 1, pp. 24-43.

Korat, O., 2009. The Effects of CD-ROM Storybook Reading on Israeli Children's Early Literacy as a Function of Age Group and Repeated Reading. *Education and Information Technologies*, Vol. 14, No. 1, pp. 39-53.

Kucirkova, N., 2019. Children's Reading with Digital Books: Past Moving Quickly to the Future. *Child Development Perspectives*, Vol. 13, No. 4, pp.208-214.

Lim, J. et al., 2021. Interactive E-book Reading vs. Paper-based Reading: Comparing the Effects of Different Mediums on Middle School Students' Reading Comprehension. *System*, Vol. 97, pp. 102434.

Lin, P.H. et al, 2018. Exploring Imaginative Capability and Learning Motivation Difference Through Picture E-Book. *IEEE Access*, Vol. 6, pp. 63416-63425.

Mayer, R. E., & Moreno, R., 2003. Nine Ways to Reduce Cognitive Load in Multimedia Learning. *Educational Psychologist*, Vol. 38, No. 1, pp.43-52.

Mayer, R. E., 2009. The Promise of Multimedia Learning. *Multimedia Learning*, pp. 3-27.

Rubegni, E. et al, 2021. "The girl who wants to fly": Exploring the Role of Digital Technology in Enhancing Dialogic reading. *International Journal of Child-Computer Interaction*, Vol. 30, pp. 100239.

Shamir, A., Korat, O., & Barbi, N., 2008. The Effects of CD-ROM Storybook Reading on Low SES Kindergarteners' Emergent Literacy as a Function of Learning Context. *Computers & Education*, Vol. 51, No. 1, pp. 354-367.

Takacs, Z.K. et al, 2015.Benefits and Pitfalls of Multimedia and Interactive Features in Technology-Enhanced Storybooks: A Meta-Analysis. *Review of Educational Research*, Vol. 85, pp. 698–739.

Wu, T. T., 2016. A Learning Log Analysis of an English-reading E-book System Combined with a Guidance Mechanism. *Interactive Learning Environments*, Vol. 24, No. 8, pp.1938–1956.

Yang, C-H., 2013. The Effects of Different Levels of Interactive Electronic Picture Books for The Effects of Different Levels of Interactive Electronic Picture Books in Influencing the Reading Motivation and Reading Effect [Unpublished master's thesis]. *National Taiwan University of Science and Technology.* Taipei, Taiwan.

Yokota, J. and Teale, W.H., 2014. Picture Books and the Digital World: Educators Making Informed Choices. *The Reading Teacher*, Vol. 67, pp. 577–585.

Zhang, R. et al, 2020. A Systematic Review of Research on E-book-based Language Learning. *Knowledge Management and E-Learning*, Vol. 12, No.1, pp. 106-128.

CHALLENGES IN DESIGNING AN ONLINE ACADEMIC WRITING TUTORIAL FOR INTERNATIONAL EFL/ESL SPEAKING GRADUATE STUDENTS

Zhengxiang Wang[1], Zhi Li[2] and Veronika Makarova[2]
[1]*Stony Brook University, Department of Linguistics, Social and Behavioral Sciences, Stony Brook, NY, USA*
[2]*University of Saskatchewan, Department of Linguistics, 9 Campus Drive, Saskatoon, SK, Canada*

ABSTRACT

International graduate students (IGSs) whose mother tongues are other than English are known to experience difficulties with academic writing in graduate programs offered in English. The authors designed a set of extracurricular online tutorials aimed at helping IGSs with academic writing development and tracking their writing progress. The article summarizes challenges experienced by the developers in building and running this e-learning tutorials set. The results of the study indicate that while some outlined minor technical challenges in the tutorial design can be addressed, running the tutorials as an extracurricular non-credited activity is not sustainable long-term due to the learners' low motivation resulting with high dropout rates. Revamping the tutorial as a credited online Academic Writing elective course is suggested as a potential solution.

KEYWORDS

Online Tutorials Design, E-Learning, Academic English Writing, International Graduate Students

1. INTRODUCTION

EFL/ESL speaking International graduate students (IGSs) are a disadvantaged group in English-medium universities. As compared to their peers who are native speakers of English, IGSs face much greater challenges in university environments (Devedžić et al., 2022) and specifically in Academic Writing (Chen, 2022). Many universities in Canada are slow to respond to the needs of the IGSs in providing writing support (Okuda & Anderson, 2018). E-learning has the potential to provide those "missing links" that "schools usually omit in the teaching they provide" (Devedžić et al, 2022, p. 3). To address the "missing links" in the teaching of academic writing to IGSs, we developed an extracurricular writing tutorial set with a free open access on MoodleCloud. In this article, we address the challenges experienced by the team in developing and running the tutorials. Our article can be helpful to instructors who are not e-learning specialists but would like to create an online course or tutorial. The research questions we address in this article are: 1. What are the challenges in developing and running an online extracurricular academic writing tutorial? 2. How do three raters (with expertise in ESL) and the learners themselves evaluate the learners' progress with Academic writing in the tutorials?

2. ACADEMIC ESL/EFL WRITING

Academic writing in a foreign language is extremely difficult, since it requires a mastery of the formal grammar and academic vocabulary, of the logic, structure and cohesion of a narrative as well as of academic conventions (such as citation practice, paragraph organization, connectors, etc.) (Chen, 2022). Issues with Academic writing negatively impact IGSs' grades in their programs of study (e.g., Singh, 2015). However, most research of academic writing has been conducted in the context of standard tests of English proficiency (e.g., Riazi, 2016), or of undergraduate level students (Dafouz, 2020). Multiple versions of online courses for academic writing have been proposed (e.g., Shirvani & Porkar, 2022). Relatively few available studies of online writing by graduate students have been conducted and they mostly addressed structured Academic Writing courses (e.g., Wang & Qin, 2022). The novelty of our study is its focus on an extracurricular e-learning of Academic English writing by IGSs.

The tutorial set focuses on the genre of Literature Review (LR) since it is one of the most frequent assignments among different Canadian graduate programs (Shi & Dong, 2015) and is also a required component of any graduate thesis or an academic paper. While developing the tutorials set, we adopted "the linguistic and genre" approach accommodating for acquisition of linguistic structures and writing within a specific genre (Xu, 2019).

3. ONLINE TUTORIAL SET AND ITS TOOLS

The purpose of the tutorial is to provide a cost-effective and user-friendly tool for IGSs to improve their academic writing online working individually or in groups with a minimal involvement of the instructor. The tutorials are split into 5 units addressing the LR genre requirements, logic and structure, grammar, academic vocabulary, and reported speech. Each unit consists of an H5P interactive e-book with embedded exercises, a writing task, and an end-of-unit survey, a final quiz, plus an evaluation tool by the raters. The tutorial set is delivered via MoodleCloud over the course of 3.5 months. Each tutorial unit lasts for about two weeks, and has fixed dates of writing task submissions. Tutorials 1, 3, 5 contained individual, and tutorials 2 and 4 – collaborative writing tasks, since L2 online writing courses often include collaborative writing to encourage student interactions (Li & Kim, 2016). In total, 30 participants from 4 different universities completed the tutorials.

The major tools in the tutorials were selected with the goal of making tutorials easily accessible and simple enough to be used by learners and instructors with little e-learning experience or computer skills. Moodle is a learning management system (LMS) utilized to host all the materials (e.g., surveys, e-books, quizzes etc.) and functional modules (e.g., announcement, workshop, discussion forum etc.) of the tutorials. Moodle was selected as a platform because it is fully open to anyone for registration. In addition, Moodle also allows the developers/instructors to manage the participants, e.g., split them into groups, send announcements, etc. It enables onsite communication among the instructor and the participants. Survey Monkey was chosen as a support tool, since it is commonly used in Canadian universities and is also supported by the authors' home university. Survey Monkey is employed for the participants to sign up for the study, report their language and academic background, as well as their perceived progress during the study. Survey Monkey is also used by raters assessors to provide ratings of the participants' writings following a set of rubrics.

4. CHALLENGES OF THE ONLINE TUTORIAL DELIVERY

Moodle-related challenges. First, participants are required to create a Moodle account. To do this, they have to plough through endless policies pages. After they finally create an account, they must first log out, and then log back in. However, they are not automatically directed to the course (tutorials in our case). Instead, the participants need to click the "Site Home" on the sidebar to see the tutorial and get themselves enrolled in it. All these steps are rather counterintuitive. To help the participants navigate through creating the account and finding the tutorials, we provided detailed instructions on how to complete the enrolment through multiple channels (i.e., project website, automatic email after registration of a Moodle account). The manuals took a considerable extra time and effort of the developers. Even despite the manuals, more than half of the participants failed to register and had to be manually enrolled by a research assistant (RA), and dozens of potential participants dropped out prior to their enrollment due to the above difficulties.

Second, Moodle lacks several functionalities that are important for running the tutorials and conducting research based on the outcomes of the tutorials. For example, although Moodle provides a Module called "Assignment" for the participants to submit their writings, this Module does not support more than one rater assessing a submitted writing. For this reason, we had to transfer this task to Survey Monkey, which in turn, had its issues as outlined below. Likewise, although there is a Module called "Workshop" available in Moodle for the participants to collaborate in a group, it does not allow real-time online document editing, which is essential for group writing tasks. To address that, the team had to create multiple editable documents via Google

Docs, one Google Doc per each group in each tutorial, and post the links to them in the groups' working spaces on Moodle. This did not only multiply the numbers of tools to use, but also required the participants to learn how to edit Google Docs. In addition, Moodle has no support for automated extraction of participants' posts or communication records on the Discussion Forum and the Workshop, which added to extra time and costs to have these types of data extracted manually.

Survey Monkey related challenges. Survey Monkey does not have a built-in function for extracting and aggregating the response data automatically in an individualized way. Automated data extraction is pertinent for collecting information or feedback from the participants and for tabulating raters' assessments of the participants' writings. For example, the authors needed to trace participants' progress self-evaluation over each tutorial, but there was no automatic way supported by Survey Monkey to aggregate the individualized response data for every participant across different tutorial units. Survey Monkey does provide aggregated response data for all submissions within a specific survey with basic summary statistics and charts, but this aggregation is done at population-level and does not allow individualization (by each participant). Individualized survey data aggregation is also important for informing the participants of the respective raters' assessments in a comprehensive and timely manner. Due to the lack of individualized aggregation in Survey Monkey automatic support, the aggregation has to be done manually, which is extremely time-consuming. For instance, it takes about 20-30 minutes to aggregate three raters' assessments for every submitted writing. Moreover, manual aggregation creates a concern for human error.

H5P-related challenges. An inherent challenge of the tutorial sets was making the content interactive while maintaining a depth of instructional content. The initial content was developed by the course instructors in text format. To convert this content into interactive e-books in H5P additional elements had to be developed (e.g., figures, small quizzes) and the initial text had to be rearranged to make the content interactive and more easily readable online (Ref. Fig.1). This increased the time spent on communication within the team related to clarifications of the content, figures, and quizzes to maximize the tutorial quality. In addition, although H5P provides various options and modules to design and make interactive e-books, it does not support open-ended quizzes where an expected response should be provided in text format. As a trade-off, quizzes in the study were spilt into two parts, with the open-ended quizzes placed in Moodle, and non-open-ended quizzes - in H5P e-book, which could be confusing for the participants.

Challenges related to participant dropout. From over 300 participants who signed up for the study, only 30 participants (10%) completed the whole tutorials set. As mentioned above, dozens of participants

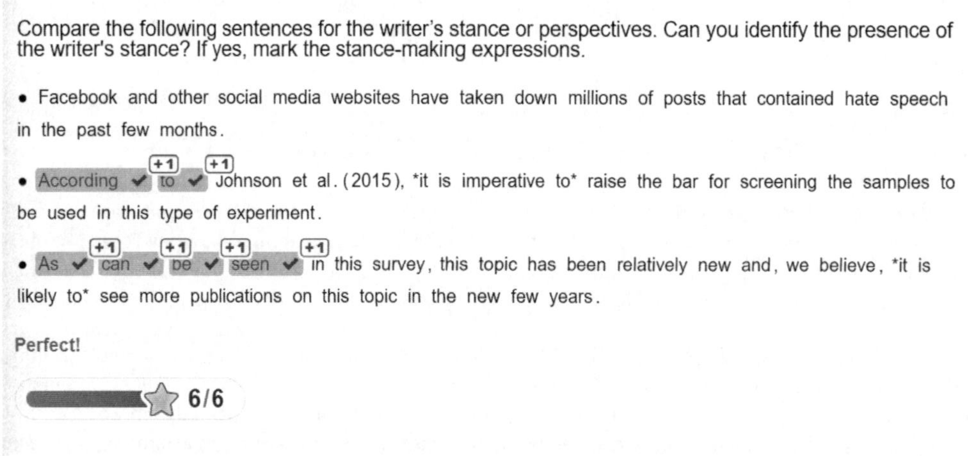

Figure 1. An example of an interactive exercise in Tutorial 4 on the Moodle Platform (in h5p)

dropped out even before starting their work on tutorials. Others dropped out in subsequent weeks of the tutorial commenting on not being able to handle the extra work on writing along with their coursework and research. We offered remuneration of $100 to the participants, but it was not a sufficient motivation to continue non-credited work.

5. THE ASSESSMENT OF THE PROGRESS WITH ACADEMIC WRITING IN THE ONLINE TUTORIALS SET

Three raters' aggregated assessment demonstrates an increase in the scores for all the writing assessment rubrics (Ref. Table 1). The participants who completed the tutorials ranked their experience and progress with different aspects of Academic writing and LR genre awareness - highly (Ref. Table 2).

Table 1. Raters' aggregated assessment (10-point scale) of the participants' writing by 9 rubrics in Tutorials 1 and 5

	rubrics	T1	T5
1	incorporating sources	6.8	8.1
2	referencing	6.3	7.8
3	structure	6.9	7.8
4	logic	7.1	7.9
5	clarity	6.8	8
6	cohesion	6.8	8
7	connectors	6.8	8.1
8	grammar	6.9	7.9
9	academic vocabulary	7.1	8.2

Table 2. Participants' assessment of their progress upon the completion of the tutorials set (4-point)

rubric	Average score (1-4)
overall progress	3.4
genre knowledge	3.4
incorporating sources	3.3
referencing	3.3
structure, logic	3.3
cohesion	3.3
academic vocabulary	3.1
grammar	3.3
overall experience	3.5

6. DISCUSSION

Developing academic writing skills by IGSs has been shown to be an extremely difficult task in earlier research (e.g., Chen, 2022). Various types of virtualization and e-learning forms for improving IGSs' writing have been suggested (e.g., Barbour El Rassi, 2022). However, to the best of our knowledge, online tutorials for teaching Academic Writing to IGSs have not been proposed so far, and our research is innovative in this respect.

Our study shows that first, completing an online writing tutorial has a positive impact on the participants' academic writing as per raters' and self-assessment scores, which need to be further verified in future with measuring objective writing parameters. Second, as noted in earlier literature, the success of online learning largely depends on the students' "readiness and intentions to continue" (Barbour El Rassi, 2022, p. 18), which was not the case in our study. While we, of course, had no way to identify the actual reasons for dropouts, many learners sent apologetic e-mails to the instructors stating that they are too busy with their regular programs of studies to continue. Due to university policies, we could not pay the participants more than $100 for taking part in the study, and it was simply not enough to motivate them to continue with the tutorial set when their program or research priorities were at stake. It appears that taking a non-credited tutorials set was too overwhelming for most participants to handle in combination with work on their regular programs of studies and theses. High dropouts in online MOOC courses have also been reported earlier (Kinash, 2013). A solution is seen in the conversion of our tutorials into an elective credited course on the home university platform familiar to the learners.

The study was limited by the number of participants that persevered to the end of the tutorials (30), and by types of assessment (expert and self-assessment). Text analysis techniques for writing assessment should be added in future.

7. CONCLUSION

In conclusion, the online tutorials are an effective way to improve Academic writing according to the expert scores and participants' progress self-assessment. We identified a range of challenges and solutions related to navigating different tools to create an online e-learning tool for Academic Writing to IGSs. We also found that teaching non-credited tutorials leads to a high dropout. Offering such tutorials for partial or full credit as an elective would yield more consistent enrollments. Our future research will also involve a detailed analysis of texts produced by the participants following parameters identified in earlier studies (e.g., Kyle & Crossley, 2015; Wang, & Qin, 2022).

ACKNOWLEDGEMENT

This project is funded by the Insight Development Grants (430-2020-00179) from Social Sciences and Humanities Research Council of Canada (SSHRC).
All the authors made equal contributions to this article.

REFERENCES

Barbour El Rassi, M. A., 2022. Virtualization Process of Learning in a Developing Country. *Proceedings of EL 2022*. Lisbon, Portugal, pp. 18-26.

Chen, M., 2022. Research on Multi-modal Teaching of Graduate Students Academic English Writing Based on Information Technology. *MATEC Web Conference*. 359 01002.

Dafouz, E., 2020. Undergraduate Student Academic Writing in English-medium Higher Education: Explorations through the ROAD-MAPPING Lens. *In Journal of English for Academic Purposes*, Vol. 46, 100888.

Devedžić, V. et al., 2022. *Proceedings of EL 2022*. Lisbon, Portugal, pp 3-9.

Kinash, S., 2013. MOOCing about MOOCs. *In Education Technology Solutions*, Vol. 57, pp. 56–58.

Kyle, K., & Crossley, S. A., 2015. Automatically Assessing Lexical Sophistication: Indices, Tools, Findings, and Application. *In TESOL Quarterly*, Vol. 49, No 4, pp. 757–786.

Li, M., & Kim, D., 2016. One Wiki, Two Groups: Dynamic Interactions Across ESL Collaborative Writing Tasks. *In Journal of Second Language Writing*, Vol. 31, pp. 25–42.

Okuda, T., & Anderson, T., 2018. Second Language Graduate Students' Experiences at the Writing Center: A Language Socialization Perspective. *In TESOL Quarterly,* Vol. 52, No 2, pp. 391-413.

Riazi, A. M., 2016. Comparing Writing Performance in TOEFL-iBT and academic assignments: An exploration of textual features. *In Assessing Writing*, Vol. 28, pp. 15–27.

Shi, L. & Dong, Y., 2015, Graduate Writing Assignments Across Faculties in a Canadian University. *In Canadian Journal of Higher Educatio*n, Vol. 45, No 4, pp. 123-142.

Shirvani, M., & Porkar, R., 2022. Online-Based L2 Writing Courses and Practicing Metacognitive Strategies: Teacher-Regulated or Individualized? *In Theory and Practice in Language Studies*, Vol. 12, No 7, pp. 1419–1427.

Singh, M. K. M., 2015. International Graduate Students' Academic Writing Practices in Malaysia: Challenges and Solutions. *In Journal of International Students*, Vol 5, N1, pp. 12-22.

Wang, & Qin, L. T., 2022. Facilitating and Assessing Academic Writing to Graduate Students in a Pilot English for Academic Purposes Course: An Activity Theoretical Perspective. *In International Journal of Educational Reform,* Vol 31, N- 3, pp. 300–316.

Xu, F.,2019. Learning the Language to Write for Publication: The Nexus Between the Linguistic Approach and the Genre Approach. In Habibie P., Hyland K. (Eds.), *Novice writers and scholarly publication: Authors, mentors, gatekeepers* (pp. 117–134). Palgrave Macmillan.

AN ANALYSIS OF PATHS TO RECOVER DEFICIENCIES IN THE LOGICAL-MATHEMATICAL FIELD DURING TWO YEARS OF THE PANDEMIC

Vittoria Bonanzinga[1] and David Barilla[2]
[1]*Dipartimento di Ingegneria dell'Informazione, delle Infrastrutture e dell'Energia Sostenibile, Università Mediterranea di Reggio Calabria, Via Graziella, (Feo di Vito), 89100 REGGIO CALABRIA, Italy*
[2]*Dipartimento di Economia, Università di Messina, Via Dei Verdi, 75, 98123 MESSINA, Italy*

ABSTRACT

Two short university courses of recovery of logical-mathematical deficiencies were completed by 210 students of a degree course in Primary Education Sciences, for future elementary school teachers. The courses were held during the years of the pandemic and were therefore online. Students had completed an entrance test with pen and paper, in person, when they first applied for university. The test was organized by an external organization given that there were more than 1000 candidates. Many deficiencies in logic and mathematics were revealed, but without specifics. We therefore transformed some of the test questions into electronic form, with automatic assessment so that students could verify their level themselves and lecturers could plan teaching activities. These questions were used in the courses of recovery. Having to verify the results of more than 1000 students at the original entrance tests, this method for identifying specific deficiencies of the successful candidates was of great support.

KEYWORDS

Automatic Assessment, Logical-Mathematical Deficiencies, E-Learning, Blended Learning

1. INTRODUCTION

There are three main international assessment systems used to evaluate students in primary and secondary education: the OECD's Programme for International Student Assessment (PISA), (OECD, 2019) that focuses on assessing student competencies in reading, mathematics, and science, with a particular emphasis on problem-solving skills, evaluating 15-year-old students, and is conducted every three years (Mullis et al. 2017); the Progress in International Reading Literacy Study (PIRLS) that focuses on assessing reading comprehension skills of 9-10 year-old students and is conducted every five years (Mullis et al. 2017); the Trends in International Mathematics and Science Study (TIMSS) that focuses on assessing mathematics and science competencies of 9 and 12 year-old students; it is conducted every four years and includes both content and cognitive domains (Mullis et al. 2020). The choice of which program educators wish to use depends on the specific needs of the users. In (Rindermann, 2007) the author conducted a comparative study that analysed data from the PISA, TIMSS, and PIRLS of many countries, and found that the three assessments were strictly correlated. He also found that the results of PISA and TIMSS were strictly correlated with a country's performance in scientific and technological disciplines, such as patent production, research and development, and employment in these fields. Likewise, these assessments can also highlight deficiencies at high-school level. However, the author has also raised some criticisms regarding the limitations of these assessment systems, including the possibility that some nations may improve their performance through intensive preparation for these specific tests, thus not fully representing students' overall educational experience (Rindermann, 2007). Any significant shortcomings of high school students will have repercussions on university performance, with the possibility of numerous dropouts. For instance, in Italy, PISA scores tend to reflect weaknesses in science subjects. Thus, to cope with the difficulties many students have, almost all universities around the world plan recovery or strengthening courses for first-year students.

2. PARTICIPANTS

In the Mediterranean University of Reggio Calabria (MURC), in the 2020/2021 academic year, the number of students with deficiencies in the mathematical and scientific field, detected by a selection test for admission to the degree course in Primary Education Sciences, was 158 out of 250 students admitted to the limited number course, while in the 2021/2022 academic year, 110 students out of 250 were found to be deficient. The number of participants in an offered logic-mathematical remedial course was 210 (instead of 268) in total in the two academic years.

3. METHODOLOGY

In AYs 2020-2021 and 2021-2022, MURC planned two courses during the covid-19 pandemic that took place remotely on the Microsoft Teams platform. The shortcomings of students entering the primary education degree course were highlighted in the results of an entrance test, which is used for admission selection. The entrance test consisted of 80 questions in three areas of essential knowledge, 40 questions are of linguistic competence and logical reasoning, 20 of literary, historical-social and geographical culture and 20 of mathematical-scientific culture. Students who correctly answered less than half of those foreseen in the various fields were obliged to attend a specific remedial course. In the 2020-2021 academic year, all the recovery paths in the three disciplinary areas included 4 hours on the learning methodology (Chiovaro, 2012) and 4 specific hours in the specific areas of competence, which took place in 4 different meetings from 19 April to 7 May 2021. MURC provided a tablet for each enrolled student in AY 2020/2021 and this facilitated the use of different e-learning platforms for exercises, ongoing checks, sharing of materials, and remedial activities. The use of one's own portable device, such as a mobile phone, a tablet or a PC proved to be very useful for activities in different digital learning environments, (Bonanzinga et al. 2020), (Bonanzinga et al. 2022). In the academic year 2021-2022, the courses took place from 23 to 30 June 2022 with 2 meetings of 3 hours each for each area: linguistic and logical, literary, historical-social and geographical, mathematical-scientific. The logical and mathematical deficiencies of incoming university students are analyzed and some strategies for their recovery are presented.

The proposed didactic model suggests a synergistic use of pen-paper and software, (Farihah, 2019), and combines the 5Es instructional model and that of Revised Bloom's Taxonomy (RBT, Anderson et al. 2001). The 5E instructional model is a student-centered model for teaching, which consists of five stages: engagement, exploration, explanation, elaboration and evaluation (Bybee, 1997). As regards the definition of learning outcomes we followed Bloom's taxonomy, reviewed by Anderson et al. in 2001, which provides a list of six categories of skills: remember, understand, apply, analyze, evaluate, create (from the simplest to the most complex), (Anderson et. al. 2001).

In the first phase of the project, the objectives to be achieved according to the model were formulated "S.M.A.R.T." conceived by G. T. Doran and published in the November 1981 issue of Management magazine Review, (Doran, 1981). With the acronym S.M.A.R.T. the characteristics of the objectives to be pursued are indicated:

- Specific: Results should not be too general
- Measurable: the results must be able to be measured, it is therefore necessary to devise methods for quantify progress
- Achievable: the results must be achievable
- Relevant: the desired outcomes must be relevant to the course
- Time-limited: It must be clear how long the goals should be achieved.

Here we present three examples of the questions given to the students, with automatic assessment on the Microsoft Teams platform.

1. A circle has a diameter of 20 cm. The area = (i) 7cm, (ii) 3.5 cm; (iii) 14 cm; (iv) 8 cm.

Twenty-five percent of the students did not know the correct answer to this simple question.

2. A car travels at 36 km/hr. If one wanted to express this speed in meters/seconds it would be = (i) 10m/s; (ii) 127.6 m/s; (iii) 360 m/s; (iv) 36 m/s. Thirty-one percent of the students did not answer correctly.

3. Given a triangle, ABC, if the square built on the side AB has an area = the sum of the squares built on the side BC and AC, the angle at C = (i) right angle; (ii) acute angle, (iii) obtuse angle, (iv) indeterminate angle, there is no data to determine it. Thirty-two percent of the students did not answer correctly.

Albeit these are relatively elementary questions, the percent of students not answering correctly was high.

Having to select some topics to be treated in the recovery paths, it was decided firstly to consider the length of a circumference and the area of the circle. The Teams platform was used. For example, to explain the circumference, pupils were invited to use three different sized objects such as a coin, a cylindrical glass, a cap or other cylindrical objects to draw circles. The circumference of the coin was called C_1; C_2 the circumference of the base of the glass and C_3 the circumference built starting from a cap. Using a piece of tape, the students were asked to measure the circumference and then draw three lines of the three different circumferences and call them L_1, L_2, and L_3. In this way, they obtained the so-called rectified circumferences and after measuring the diameter of the three circles, they then experimentally calculate the ratios between the different lengths of the circumferences L_1, L_2 and L_3 and their respective diameters. They obtained an approximation of π and the inverse formula was obtained, i.e. the length of a circumference is given by the diameter x π, i.e. $2 \pi r$, where r is the radius of the circumference, (Lima and Jordão, 2022). To calculate the area of a circle, we used a model of a pizza and divided it into 4 pieces, then into 8 slices and then into 16. With these 16 pieces the students were asked to lay out the pieces in alternating triangles (up and down). By making the number of slices tend to infinity, we can approximate the area of the circle to that of a rectangle having for base $\pi \times r$ and for height the radius of the circle; they thus obtained the well-known formula for calculating a circle $\pi \times r^2$, (Casselman, 2012).

Secondly, in the context of analytical geometry, straight lines in Cartesian form and the conditions of parallelism and perpendicularity were considered, along with the distance between two points on the plane. Using GeoGebra we showed how to represent points in the Cartesian plane and how to calculate the distance between two points. Furthermore, using the GeoGebra CAS Calculator we have shown how to enter the Cartesian equations of straight lines, how to represent and visualize them in the Cartesian plane and how to verify the conditions of parallelism and perpendicularity, (Hohenwarter, 2002), (Hohenwarter, and Preiner, 2007).

Thirdly, through the toss of a coin or a dice the concept of probability of an event is introduced. The probability of an event is defined for equally probable events and is given by the number of favourable cases divided by the number of possible cases: it was pointed out that this definition applies only when the number of possible cases is finite. Furthermore, statements of mathematical logic were introduced along with application of the negation connective and, in the numerical field, the ordering of rational numbers, expressions with powers and fractions, equations of I and II degree, inequalities of I degree, and decomposition of polynomials of II degree.

4. RESULTS

From the analysis of the entrance tests of the students of the degree course in Primary Education it is clear that the students have greater difficulties in the questions of mathematical logic, in the recognition of a set relationship, through Euler-Venn diagrams, in the decomposition of second-degree polynomials, but also in changing units of measurement, in expressions with powers of rational numbers and in finding a figure that completes a series. The tests administered to students for the recovery of logical-mathematical deficiencies, with an automatic assessment system, proved to be very useful for both students and teachers (Barana et al. 2015), (Bonanzinga et al. 2022). The choice of the April-May recovery activity period proved to be more effective than in June. The students participated with interest in the planned activities and in the questionnaires prepared via Microsoft Teams Forms.

5. CONCLUSION

Questionnaires with Microsoft Teams Forms prove to be a very convenient tool for surveys and for collecting information. The remote teaching activity presented critical issues determined by the difficulty of the internet connection, accentuated by the presence of numerous participants and by the difficulty of verifying the autonomous progress of the students. Moreover, there is software that allows students' activities to be monitored remotely and sometimes recordings are used to verify the regularity of the tests. Undoubtedly, blended teaching is more effective than distance learning, as it allows for greater student involvement. But in some cases, it is extremely useful for working students, or students with disabilities or who live in very peripheral areas and far from universities, as evidenced by the enormous growth of all online university courses.

ACKNOWLEDGEMENT

The author would like to thank the anonymous referees for useful and detailed comments on the previous version of the manuscript.

REFERENCES

Anderson, L.W., et al. (2001), *A taxonomy for learning, teaching, and assessing. A revision of Bloom's taxonomy of educational objectives*, New York, Addison Wesley Longman.

Barana A., et al. (2015) Automated Assessment in Mathematics, IEEE 39th Annual Computer Software and Applications Conference (COMPSAC), pp. 670-671.

Bonanzinga, V. et al (2020). Blended learning an experience with use of mobile devices for the course "Fundamentals of mathematics for Basic Education", *19th Conference on Applied Mathematics, Aplimat 2020, Proceedings*, pp. 119-125.

Bonanzinga, V. et al (2022). University training for future primary school teachers on automatic formative assessment, *Proceedings of the International Conference e-learning 2022*, IADIS press, pp. 119-126.

Bybee, R. W. (1997). *Achieving Scientific Literacy: From Purposes to Practices*. Heinemann, Portsmouth, NH.

Casselman, B. (2012). Archimedes on the Circumference and Area of a Circle. AMS Feature Column.

Chiovaro V. (2012). Apprendere lungo la via. Spunti per una vita serena, Attendiamoci.

Doran, J. T. (1981). There's a S.M.A.R.T. Way to Write Management Goals and Objectives, *Management Review* (AMA Forum), November 1981, pp. 35-36.

Farihah, U. (2019). The synergy of students' use of paper-pencil techniques and GeoGebra in solving analytical geometry problems. *J. Phys.: Conf. Ser.* 1211.

Hohenwarter, M. (2002). GeoGebra - Ein Software system für dynamische Geometrie und Algebra der Ebene. Master thesis, University of Salzburg.

Hohenwarter, M. and Preiner, J. (2007). Dynamic Mathematics with GeoGebra. *The Journal of Online Mathematics and Its Applications*, 7, Article ID: 1448.

Lima F. M. S., Jordão P. G. F. (2022). Why is it that the Ratio of Any Circle's Circumference to its Diameter is a Constant? The College Mathematics Journal, 53:3, 171-182.

Mullis, I.V.S., et al. (2017). PIRLS 2016 International Results in Reading. Chestnut Hill, MA: TIMSS & PIRLS International Study Center, Boston College.

Mullis, I.V.S., et al. (2020). TIMSS 2019 International Results in Mathematics. Chestnut Hill, MA: TIMSS & PIRLS International Study Center, Boston College.

OECD (2019), PISA (2018) Results (Volume I): What Students Know and Can Do, OECD Publishing, Paris, https://doi.org/10.1787/5f07c754-en.

Rindermann H., (2007). The g-factor of international cognitive ability comparisons: the homogeneity of results in PISA, TIMSS, PIRLS and IQ-tests across nations, in European Journal of Personality, n. 21, pp. 667-706, DOI:10.1002/per.634, ISSN 08902070 (WC · ACNP).

ROLE OF METAVERSE IN E-LEARNING AND ITS CONTRIBUTION TO IMPROVING THE QUALITY OF EDUCATION

Kyriakos Kouveliotis[1] and Maryam Mansuri[2]
[1]*Prof. Dr.- FRSA - Provost & Chief Academic Officer, Berlin School of Business and Innovation, Germany*
[2]*Dr. - Head of Academic Operations, Berlin School of Business and Innovation, Germany*

ABSTRACT

Metaverse is a virtual world where an immersive experience through the use of an avatar takes place within it, which is in many cases accessible by a VR headset. This research is about the use and benefits of metaverse for e-learning. Metaverse helps e-learning by being an attractive medium where 3d models and virtual presentations with the help of AI are possible to improve the current state of education. In this research first, we introduced the metaverse and stated that there are two approaches for its role in the metaverse, first the facilitating role where it exactly mirrors the real-world classrooms, and second the accommodating role where there are extra functionalities compared to the real world. Then we explored how metaverse improves the quality of education, under two major categories. First, one being attractiveness, discussed those characteristics that are originating from the external world, like how entertaining the metaverse is to an external agent. The second one was about assistive approaches, which discussed those characteristics that originated from within the metaverse. Next, there were challenges, and finally the conclusion. The conclusion contains a brief mention of all the characteristics of the metaverse that result in improving the quality of education.

KEYWORDS

Metaverse, Education, E-learning, Improving, Learning

1. INTRODUCTION

The term "metaverse" was coined in 1992 for science fiction purposes (Benrimoh, Chheda, & Margolese, 2022), however, it emerged in the early 2000s as a part of experiences in the gaming industry (Hwang & Lee, 2022). Metaverse can be defined generically in its present usage as "a three-dimensional virtual shared space (sometimes called a virtual world), where users can have immersive experiences by entering it as digital avatars and interacting with other users, while its emphasis is on artificial reality and augmented reality, and where creatively pushes the boundaries of technology" (Tesorero, 2022; Jacob, 2022; Cannes lions, 2022; Hwang & Lee, 2022). As of today, there are many "metaverse platforms" available to access virtual worlds which are provided by different companies. Among them, Horizon Worlds (by Meta), Altspace VR (by Microsoft), BlueJeans (by Verizon), Decentraland, Roblox, Sandbox, and Second Life are some examples (Shein, 2022).

Concerning education-specific metaverses, currently, there are some specialised metaverse platforms available for education, which streamline VR and AR educational content based on the respective syllabus of the university ("Metaverse in a," 2022). Even Meta has a dedicated team for the creation of education applications within their metaverse (Horn, 2022), and it can be seen that as the metaverse unfolds rapidly, the arena for learning, training and upskilling gets wide open ("Metaverse meets," 2022).

Before covid, it was much more difficult to imagine that one-day e-learning technologies like metaverse can play their role in the field of education, but today it is much more feasible ("Metaverse meets," 2022). Just as the learning tools like laptops in the past resulted in the creation of online learning, today also metaverse is poised to be a dominant medium for education in the future (Horn, 2022). Metaverse can improve learning increasingly; Based on a PwC report, those who were trained using VR were up to 275% more confident to behave in real tasks than what they learned in virtual reality. They showed a 40% improvement compared to what they learned in the face-to-face classroom, and a 35% improvement compared to e-learning (Vikram, 2022). This article covers how, and by which approaches metaverse results in the improvement of education.

2. METAVERSE FOR E-LEARNING

There are two choices while implementing metaverse for e-learning: first, we should expect that metaverse adapts to e-learning (where it "Facilitates"), and second, we should expect that e-learning should adapt to metaverse (where it "Accommodates").

2.1. Facilitate

Here metaverse adapts to traditional learning and acts just as a medium. There are no extra tools of metaverse to be used in the process of education, only the teacher and students just as traditional learning have the basic interacting actions in a virtual classroom. No 3d models, no AI tools, and no such things that provide an additional characteristic to the metaverse, but it's absolute tutor-based learning. There are only lectures through virtual 3D classrooms, educating seminars, and virtual campus activities (Horn, 2022; "Metaverse meets," 2022).

2.2. Accommodate

In this case, the e-learning gets adapted to the metaverse and its capabilities. Here in addition to tutor-based learning, there are additional tools like 3d models for better understanding of the topic, AI is in use to assist the learning process, or there are numerous varied landscapes for students to be present as a part of the learning process.

In other words, experts are expecting the metaverse to not only mirror the real world in a three-dimensional medium but also to extend it to allow us to do what previously could only be imagined. This makes it by far, more engaging and robust than the real world. (Needleman, 2021) There are co-working spaces, virtual reality environments, and learning applications. (Horn, 2022) The Brookings Institution states that "when education lags the digital leaps, the technology rather than educators defines what counts as an educational opportunity." (Horn, 2022).

3. IMPROVING THE QUALITY OF EDUCATION

There are two main directions where the metaverse helps in improving the quality of education, first those aspects that act from outside of the metaverse and encourage students to be involved in the metaverse, like its "attractiveness", and second are those aspects that act from within the metaverse, like its unique tools and applications, which are "assistive".

3.1. Attractiveness

Metaverse is attractive to many students when they see it from the outside world. That is because it is the same area where their entertainment including their games, virtual concerts and hangouts with friends have been conducted. So, the correlation between the fun time they had in the metaverse, and the e-learning that will be done through the metaverse, results in it being attractive. What matters here, is the perceptions that students have about their intention to use a metaverse learning environment (Yang & Ryu, 2022). We see that today; Generation Z is more willing to accept online education. For them, the digital world has the same importance as the real one. Based on these facts, the metaverse has to be an area of choice for them (Lin, Wan, Gan, Chen, & Han-Chieh Chao, 2022).

3.1.1 Being Entertaining

The current popular metaverse platforms, which are mainly in the gaming sector, have dominant populations of children and young adults. One reason that universities are using metaverse, is to attract children and young adults, due to the already proven attraction of them towards it (Kshetri, 2022a). As the children can create an avatar to represent themselves in a virtual classroom, they can have more fun compared to traditional education (Puncreobutr, Dhamacharoen, & Tapaneeyakorn, 2022). Metaverse makes education more entertaining and interactive ("Metaverse meets," 2022). Some companies even have called their education metaverse, "edutainment", which combines the two words education and entertainment (Schorsch, 2023). Therefore, attracting students more towards education can increase the overall quality of education.

3.1.2 Gameful Experience

The gaming industry with great experience in the development of games might adopt "serious games", where gaming technologies, experiences, and storytelling will be used for training and simulation ("Six trends," 2022). As in the metaverse, the education gets gamified, and the skills learned by students can stay for the lifelong with them (Agarwal, 2022). This is especially valid for those students with eidetic memory (Kumbhat, 2022). In general, gameful and practical ways of training have a greater effect on learning efficiency. For example, physics experiments, building automobile prototypes, chemistry experiments, running simulations, and visiting places of historical significance, are all examples of learning through a gameful experience and are more efficient than their non-practical and non-gameful counterparts (Agarwal, 2022). As an example, a pilot by Atlanta's Morehouse College showed that there was a 10% increase in students' GPAs for the world history class through virtual reality, compared to traditional e-learning via Zoom and face-to-face learning of its past year (Kshetri, 2022a). Also, immersion can increase engagement in learning (Kshetri, 2022). In addition, there is a meaningful relationship between a gameful experience in the metaverse, and the motivation for learning (Park & Kim, 2022).

3.1.3. Promoting Communication

Compared to the online classes at the present, metaverse will provide more effective communication. Both tutors and learners can create their respective rooms for meetings, collaborations, study, or even just for socializing. They can see each other, share files easily, and play their favourite games. Under these circumstances, the friendship of classmates will be enhanced (Lin et al., 2022).

3.1.4. Lower Costs

In general, metaverse causes lower costs for many aspects of learning, including instructor, travel, and in some cases facility costs (Gronstedt, 2022). For an institution as a whole, there might be an initial high expense there to create the metaverse software and implement it by using devices like headsets, but after the first installation, the upcoming costs will be reduced. For example, there will be no need to buy expensive laboratory instruments to do experiments, as all experiments can be done virtually in the metaverse.

3.1.5. No Bias and Zero Risk

Biases either unconscious or conscious, whether about being disabled, dark-skinned or being a woman, all can be solved by the use of avatars under a metaverse (Cook, 2021). In addition, the use of avatars (instead of actual presence) in different sorts of virtual situations, results in zero risks if things go wrong. This advantage has already been tested in case of the flight simulations (Freifeld, 2022).

3.2. Assistive

If we view the metaverse absolutely from within, we can discover its assistive capabilities, like 3d models, AI tools, and virtual applications.

3.2.1 Avatars and Population

An avatar of a deceased individual or a historical figure can be represented by its respective avatar, and with the help of AI, it can talk, such that it feels it was brought back to life (Needleman, 2021). For example, as a part of the history class, bringing Herodotus to talk to students, can result in a better educational experience. It is even possible to swap out characters, their speaking languages, and other variables, to customize the respective training to match the demographics where they are being used (Freifeld, 2022).

Another positive aspect of the metaverse is that it is capable of handling an infinite number of users at the same time (Needleman, 2021). Even a normal-sized virtual class can be made a multi-story class to get more capacity, or the tutor can be mirrored in another class or on a large screen.

3.2.2 Space and Environment

Teachers can arrange their lesson plans over a landscape, where for each topic students will travel to the next area, which is specific to that topic, and as such due to a more interactive and immersive manner, it will have better results in overall learning of students (Agarwal, 2022). Also, every day they may attend the class in a different-looking class. Another possibility is that when viewing any event, instead of having to view it from a specific perspective, we have the freedom to be able to change the angle of view based on our choice (Needleman, 2021). Students can even be suspended or on the fly while attending the class.

One advantage of travelling in spaces in the metaverse is that, not only the space can be collapsed by the implementation of higher speeds in VR, but also to be more precise, time is also collapsed. As such, it reduces training time by 5 or 10 times (Gronstedt, 2022). It can be understood better when considering the time, we spend going to university or to pass through the corridors of our real universities. It is worth knowing that currently there are metaverse universities (For example the Invact Metaversity in India), that function like fully equipped real universities but under virtual reality. They can have a virtual reality library, cafeteria, hallways, and playground, and offer different courses to be accomplished using VR headsets (Vikram, 2022).

3.2.3 Hard/Impossible-To-Create Phenomena

Metaverse can make education more affordable. For example, instead of purchasing cadavers which cost a lot, they can use a 3d model of it in the metaverse, at a lower cost (a, 2022). Sometimes some experiments in the real world are risky, like those related to chemical reactions or flying aeroplanes. In such cases, virtual reality headsets and gloves can come into play to assist with such tasks (Kshetri, 2022a). Another example is about how to administer anaesthesia to a virtual patient a medical student in a simulated operating room (Vikram, 2022).

In addition to above mentioned hard-to-create phenomena, there are some impossible-to-create phenomena too. For example, historical monuments that are destroyed can be visited in their rebuilt forms, or the arrangement of atoms in a protein can be visualized (Kshetri, 2022a).

3.2.4 Hardware

Conventional metaverses can only simulate two of the human senses, visual and auditory. But if in addition to that, they could synthesize touch (for example by use of haptic gears), taste, and smell, and could integrate all of these into a single model, that could be wonderful ("Tech I'd," 2022). It is even expected that shortly, the mainstream headsets provide possibilities like facial expression, eye tracking, biometrics, etc., which results in more realistic responses by the users (Freifeld, 2022).

3.2.5 Software

Some researchers regard the most recent version of the internet to be the metaverse (Duan et al., 2021). Therefore it is expected that all well-known applications, especially in the field of education, create updated versions of their applications which operate on a metaverse basis and well-known metaverse platforms. Due to the evolving nature of metaverse platforms, these applications should regularly be checked and updated to cope with the changes and technological trends of these metaverses (Dohan et al., 2022).

In the metaverse, any virtual content can be mixed with a real-world video and after that, any sort of virtual interaction is possible (Needleman, 2021). Also, it is expected that the prevalence of metaverse results in the replacement of traditional 2D presentations like PowerPoint, with 3D and interactive presentations within metaverse. Sometimes metaverses for education, explain concepts on various levels, from a more difficult level to a lower difficulty one, and the user can choose between them. It helps in case the user wants to get a deeper understanding of that concept ("Metaverse in a," 2022). Another advantage of the metaverse is that in some cases, it is accessible in variable timings, as such providing on-demand learning (Gronstedt, 2022). Another advantage of the metaverse is that the educational content in the metaverse can be monetized. Content creators can use tokenized assets in the metaverse, as a new business model that is provided to them. Under the metaverse, most of the tokenized assets in use are non-fungible token technologies (NFTs). By the use of NFTs, content creators can retain most of the revenue resulting from the sale of their works. In addition to virtual products and services, acquiring real-world goods is also possible ("Six trends," 2022).

The use of Artificial Intelligence in the metaverse is such that to access the AI for our respective questions, any avatar of choice can be assigned to be powered by AI, and as the questions will be asked, it will reply to them based on its AI knowledge (Needleman, 2021). The same can be done for access to the web, while the use of a virtual laptop is also a possibility.

4. CHALLENGES

However, for specialized purposes and in long-term use, the costs of education through the metaverse are lower for institutions, but one of the challenges that are faced at present for implementation of the metaverse is its high costs. Other issues are data privacy and protection, adaptability to new technology, and lack of rural access to advanced infrastructure (Kshetri, 2022b). The availability of computing infrastructure for a full-fledged worldwide metaverse also needs to be assessed. Its regulatory compliance issues, besides the possibility of unethical behaviour and even addiction, are among the other facing issues the implementation of the metaverse (Singal, 2022).

One major drawback of the metaverse is that by long-term use of it, we get used to it, and any return to traditional forms of learning will be with some burden because the way users were engaged with the virtual world is different from the way they get engaged with the real world (Needleman, 2021).

One other challenge that is faced for use of metaverse in education, is the absence of relevant use cases as of now because as the metaverse gets developed more and more, more possibilities for its usage emerge, so the future gets more adequate for the use of metaverse in education, but not necessarily the present (Murray, 2022). At least once in the past, there had been a failed attempt for educators to cope with technology, and that was when their obsession and excitement with the Second Life metaverse platform, faded fast and it was set aside (Horn, 2022). Such failures should be deeply assessed and prevented for future attempts.

5. CONCLUSION

In this paper we have discussed how metaverse can help in improving the quality of education. First of all, as it is entertaining and provides a gameful experience it is attractive to an outside viewer. Besides that, the metaverse encourages socializing and more communication, while it lowers costs, and avoids social biases (like for gender or ethnicity) by use of an avatar. By use of virtual replicas and simulations, it also avoids the regular risks that are present in real-life experiences, for example, the possible accidents in a test flight. One example of the usefulness of metaverse in education is that the deceased individuals can appear as their avatars and by use of AI can talk for audiences.

An infinite number of users can be present in the metaverse at the same time, and a lesson can be taught by travelling through a landscape. A virtual university with all the accommodations of a fully functioning university can be there in the metaverse. One advantage of the metaverse is that hard-to-create phenomena and impossible-to-create phenomena, both can be performed in the metaverse. For example, flying an aero plane, administering anesthesia, or even visiting historical monuments that no longer exist. In addition to the implementation of visual and auditory senses at present, other features like touch, taste, smell, facial expression, eye tracking, biometrics, etc., all can be implemented. Software applications that are based in the metaverse can emerge, PowerPoint presentations can be replaced with 3D presentations within the metaverse, the level of difficulty of a concept can be chosen, on-demand learning at variable timings will be possible, the created educational content can be monetized, and questions from AI and the web can be asked from an avatar who was assigned solely for this purpose.

In general, the metaverse improves the quality of education substantially, through the approaches mentioned above. A report by KPMG suggests that around 92% of users expressed their belief that metaverse is capable of enhancing their learning opportunities, including all aspects of job training, school, and higher education (Colvin, 2022).

REFERENCES

Agarwal, A. (2022, Sep 07). Metaverse Can Play a Crucial Role in Learning Skills. *Business World,* https://www.proquest.com/magazines/metaverse-can-play-crucial-role-learning-skills/docview/2710549277/se-2

Benrimoh, D., Chheda, F. D., & Margolese, H. C. (2022). The Best Predictor of the Future—the Metaverse, Mental Health, and Lessons Learned from Current Technologies. JMIR Mental Health, 9(10). https://doi.org/10.2196/40410

Cannes lions introduces new gaming award (2022). Cape Town: SyndiGate Media Inc. Retrieved from https://www.proquest.com/other-sources/cannes-lions-introduces-new-gaming-award/docview/2735377763/se-2

Colvin, C. (2022). Into the metaverse: VR catches workers 'interest. *Healthcare Dive,* https://www.proquest.com/trade-journals/into-metaverse-vr-catches-workers-interest/docview/2714986740/se-2

Cook, M. (2021, Dec 11). Leading, learning in the metaverse: We can learn to fluently navigate the rapidly shifting technological landscape with thinkers and doers embedded in industry. *The Business Times* https://www.proquest.com/newspapers/leading-learning-metaverse/docview/2608842068/se-2

Dahan, N. A., Al-Razgan, M., Al-Laith, A., Alsoufi, M. A., Al-Asaly, M., & Alfakih, T. (2022). Metaverse Framework: A Case Study on E-Learning Environment (ELEM). *Electronics, 11*(10), 1616. https://doi.org/10.3390/electronics11101616

Duan, H., Li, J., Fan, S., Lin, Z., Wu, X., & Cai, W. (2021). *Metaverse for Social Good: A University Campus Prototype.* Cornell University Library, arXiv.org. https://doi.org/10.1145/3474085.3479238

Freifeld, L. (2022). Masters of the Metaverse. *Training, 59*(3), 20-25. https://www.proquest.com/trade-journals/masters-metaverse/docview/2716599566/se-2

Gronstedt, A. (2022). THE DAWN OF METAVERSE LEARNING. *Talent Development, 76*(6), 30-35. https://www.proquest.com/trade-journals/dawn-metaverse-learning/docview/2672394658/se-2

Horn, M. B. (2022). Meet the Metaverse. *Education Next, 22*(3) https://www.proquest.com/scholarly-journals/meet-metaverse/docview/2733270801/se-2

Hwang, R., & Lee, M. (2022). The Influence of Music Content Marketing on User Satisfaction and Intention to Use in the Metaverse: A Focus on the SPICE Model. Businesses, 2(2), 141. https://doi.org/10.3390/businesses2020010

Jacob, T. (2022). AMBCrypto - Hardfork Media Private Limited: Ultraverse City to revolutionize advertising with Digital Billboards in the Metaverse. Newstex.

Kshetri, N. (2022, Aug 23). Six benefits that the metaverse offers to colleges and universities. *The Conversation: Education* https://www.proquest.com/newspapers/six-benefits-that-metaverse-offers-colleges/docview/2705502791/se-2

Kshetri, N. (2022, Sep 13). 5 challenges of doing college in the metaverse. *The Conversation: Education* https://www.proquest.com/newspapers/5-challenges-doing-college-metaverse/docview/2713785147/se-2

Kumbhat, N. (2022, Jul 10). 5 Education trends anticipated in the world of Metaverse. *Financial Express* https://www.proquest.com/newspapers/5-education-trends-anticipated-world-metaverse/docview/2687530412/se-2

Lin, H., Wan, S., Gan, W., Chen, J., & Han-Chieh Chao. (2022). *Metaverse in Education: Vision, Opportunities, and Challenges.* Cornell University Library, arXiv.org.

Metaverse in a Nutshell - All You Need to Know. (2022, Nov 28). *DQ Channels,* https://www.proquest.com/magazines/metaverse-nutshell-all-you-need-know/docview/2741102453/se-2

Murray, S. (2022). Business schools explore teaching in the metaverse. *FT.Com,* https://www.proquest.com/trade-journals/business-schools-explore-teaching-metaverse/docview/2758967172/se-2

Needleman, S. E. (2021, Oct 18). Tech Visionaries See Brave New Metaverse. *Wall Street Journal* https://www.proquest.com/newspapers/tech-visionaries-see-brave-new-metaverse/docview/2582697630/se-2

Network, d. (2022, 10). Metaverse meets Education: Transformation beyond traditional trajectory. *Digital Learning,* https://www.proquest.com/magazines/metaverse-meets-education-transformation-beyond/docview/2723772317/se-2

Park, S., & Kim, S. (2022). Identifying World Types to Deliver Gameful Experiences for Sustainable Learning in the Metaverse. *Sustainability, 14*(3), 1361. https://doi.org/10.3390/su14031361

Puncreobutr, V., Dhamacharoen, A., & Tapaneeyakorn, W. (2022). Factors Affecting the Readiness of Thai Universities to Organize Learning Activities in the Metaverse Era. *Webology, 19*(2), 9285-9296. https://www.proquest.com/scholarly-journals/factors-affecting-readiness-thai-universities/docview/2695105136/se-2

Schorsch, P. (2023). Florida Politics: ROYBI to debut first intelligent, immersive edutainment metaverse at CES 2023. Newstex.

Shein, Esther (2022, November 18). *Top metaverse platforms to know about in 2023.* TechTarget. https://www.techtarget.com/searchcio/tip/Top-metaverse-platforms-to-know-about

Singal, P. (2022). Improved Learning and Evolving Opportunities: Get Ready for An Immersive Digital Workspace in the Metaverse. Newstex.

Six trends driving near-term adoption of metaverse technologies. (2022). *Dataquest,* https://www.proquest.com/trade-journals/six-trends-driving-near-term-adoption-metaverse/docview/2713466642/se-2

Tech I'd like to see: Metaverse for P2P learning [Startups]. (2022, Sep 08). *The Times of India* https://www.proquest.com/newspapers/tech-id-like-see-metaverse-p2p-learning-startups/docview/2711040179/se-2

Tesorero, A. (2022, Oct 05). Arab Media Forum: Will the metaverse turn journalism into a two-way process? *Gulf News* https://www.proquest.com/newspapers/arab-media-forum-will-metaverse-turn-journalism/docview/2721349501/se-2

Vikram, J. G. (2022, Sep 01). For U & Me – In Sight: Make the Metaverse Work for You. *Open Source for You,* https://www.proquest.com/magazines/u-amp-me-sight-make-metaverse-work-you/docview/2716460622/se-2

Yang, E., & Ryu, J. (2022). Work-in-progress—The Effect of Students 'Perceptions on Intention to use Metaverse Learning Environment in Higher Education. The Institute of Electrical and Electronics Engineers, Inc. (IEEE). https://doi.org/10.23919/iLRN55037.2022.9815996

ENHANCING ACCESSIBILITY IN HIGHER EDUCATION

Paula Escudeiro and Márcia Campos Gouveia
Games, Interaction & Learning Technologies, ISEP/IPP,
Rua Dr. António Bernardino Almeida, 431, Porto, Portugal

ABSTRACT

As technology continues to advance and more institutions adopt assistive technologies strategies, it is becoming increasingly important to understand the most effective ways to implement assistive technologies strategies in higher education. The purpose of this paper is to explore how assistive technologies strategies can be effectively integrated into higher education for disabled communities. Orientation and mobility are often underestimated. People with disabilities, such as deaf, blind, and physically impaired people, are often challenged with navigation problems. Signage is the most common mechanism of wayfinding information, but orientation is much more complex. To address the lack of inclusion and accessibility that overwhelms indoor environments, such as schools and public services, an Indoor Mapping application was designed. The target is the ISEP community – its students, staff, and visitors. The aim is to create an assistive technology, that can guide disabled students, but also incoming and Erasmus students that struggle in their early days at the University.

KEYWORDS

Assistive Technologies, Higher Education Strategies, Indoor Mapping, Accessibility, Geolocation, Orientation & Mobility

1. INTRODUCTION

As of this moment, there are only a few options for navigational assistance in indoor environments for people with disabilities, especially those who are blind or visually impaired. Indoor environments can be geographically complex, and signage remains the most common mechanism for providing and receiving wayfinding information in such spaces.

People with disabilities are a significant share of the global population. According to the World Health Organization, at least 2.2. billion people live with some kind of visual impairment (World Health Organization, 2019) and over 1.5 billion people have hearing impairments (World Health Organization, 2023). Also, more than 1 billion people need at least one assistive product, i.e., a wheelchair, prostheses, or communication or memory aids, for example (World Health Organization, 2018). This means that assistive technology is of great importance for people that live with different disabilities, such as deafness, blindness, or reduced mobility.

To address challenges of inclusion, communication, and accessibility (Escudeiro, et al., 2017), ISEP Indoor Mapping was developed. The application intends to raise awareness and enhance equality, by creating an assistive technology that provides self-sufficiency in indoor environments. By using beacons, a small device that works based on Bluetooth Low Energy, the application can identify the geolocation of the user and provide him or her the information necessary to move through buildings of the *Instituto Superior de Engenharia do Porto* (ISEP) campus.

Serving as an assistant guide and indoor navigation system, the idea is to develop an assistive technology, a navigation electronic aid, that allows students to easily move through the facilities and quickly become familiar with them, while also being aware of easy access points, carefully adapted to their needs, such as wheelchair or elevator access.

2. THEORETICAL FRAMEWORK

Nowadays, there is an overwhelming amount of visual and text information in every street, building, public space, and all else. However, despite this available visual information, there is a general lack of location information in indoor environments, that are suitable for everyone. This also involves the lack of navigation aids that are made available in indoor environments.

With the advent of greater mobility, a greater need for localization and indoor positioning has emerged (Samama, 2019). As such, there is a growing relevance in the development of user-friendly indoor navigation systems (Hirtle & Bahm, 2015), systems that enhance inclusion and accessibility, but also systems that complement the outdoor based-positioning systems (Samama, 2019). Both of them relate to the ability to travel effectively, regardless of whether the travelers are blind, deaf, or physically impaired (Jacobson, 2013). However, information challenges often appear for people with disabilities (Escudeiro, et al., 2017).

The challenges they face are due to the lack of accessible information related to orientation and mobility. Orientation is seen as "the process of knowing where one is in space" and mobility is "the capacity or facility of movement" (Jacobson, 2013). This is important when navigating through complex environments, since access to information is often underestimated and mistreated in the case of blind individuals, jeopardizing both orientation and mobility. A person with vision constantly receives information when looking around. In contrast to the vast amount of wayfinding information that sighted people have access to, the blind often struggles when traveling (May & Casey, 2013). Thus, "way-finding challenges for individuals who are blind or visually impaired can be considered problems of information access" (May & Casey, 2013).

To overcome these problems of information access, there is a need to develop assistive technologies, not only for blind people but also for deaf and foreign visitors, that often struggle to communicate and understand (Escudeiro, et al., 2017).

3. METHODOLOGY

The research for this paper was conducted using a mixed methods approach, including a literature review and case studies. The literature review was conducted using a variety of academic databases, to gather relevant articles, books, and reports on e-learning in higher education and academic environment. The Indoor Mapping application aims to enhance the inclusion and accessibility of all students, by supporting new orientation and mobility methods. The application methodology and the outcomes of the project will enhance and value inclusiveness and accessibility while improving the orientation, mobility, and navigation of all students. In the future, the aim of the project is to be extended to other indoor environments. The ISEP campus is just the starting point of a new venture in assistive technologies.

3.1 Concept

The overall goal of the project was to create an application that implements an Indoor Mapping method with different access points distributed by the ISEP campus. The signal is then captured by a geolocation technology – in this case, beacons – that activates a mobile application to provide information.

Geolocation is not a new concept and different companies use it in different ways. In this case, geolocation will be used to provide an easy and accessible virtual space for students, workers, members of the institution, and occasional visitors, to navigate the surroundings of ISEP. As such, the work developed focuses essentially on a navigational component – the ISEP map.

The map defines a specific set of directions and information that guide the user within a certain location or set of locations. Each building has a designation, geo-location, and internal data – classrooms, bathrooms, elevator locations, and all else. Combining all the data of each building, the user can experience accurate guidance between buildings, which even specifies alternatives for users with reduced mobility that require an elevator or wheelchair access.

As mentioned, there are numerous people out there that live with some kind of disability. This project pretends to raise awareness of the importance of orientation and mobility, which is often underestimated. Thus, this application, with geolocation features, has two different targets. On the one hand, the main target is deaf, blind, and physically impaired people, but also Erasmus students and incoming students. All of them often struggle to find assistance and communicate.

The objective is to provide all people with audio, text, and visual guide assistance, that allows everyone to move from one point to another inside ISEP. Similar to the GPS, the application provides a numerous of information, indicating the more suitable path.

The technology used to geolocate the users were beacons. A Bluetooth beacon is a small wireless device that works based on Bluetooth Low Energy, that repeatedly transmits a constant signal that other equipment can find. It broadcasts a radio signal, that a Bluetooth-equipped device like a smartphone can identify once it is in its range. Based on precise positioning technology, it allows to geolocate the user in an indoor environment.

3.2 Objectives

The ISEP Indoor Mapping application was developed to offer assistance and self-sufficiency to all its users. Whether they are incoming students, Erasmus students, visitors, or students with disabilities, the application provides several wayfinding information similar to signage. However, in this case, the information is presented according to the user's needs: it shows wheelchair and elevator access, the more suitable path and the shortest one too, and it does so by using an audio, text, or signage guide to ensure everyone is included in the use of this technology. Thus, the objectives of the project are:

- The creation of an assistive tool that provides orientation and mobility assistance;
- The inclusion of a map of the ISEP campus, that offers information about every building;
- The development of a guide in the form of audio, text, or signage;
- Raising awareness for orientation and mobility challenges;
- Raising awareness for the lack of inclusion and accessibility of disabled students;
- To create an accurate and detailed map of the ISEP campus, including all buildings and floors;
- To provide a user-friendly interface that allows users to easily navigate the ISEP campus, search for specific locations and points of interest, and view real-time information such as room availability and event schedules;
- To integrate the indoor mapping application with other campus systems, such as the room reservation system and the events calendar, to provide real-time updates and improve efficiency;
- To provide accessibility features, such as voice navigation and alternative text descriptions, to make the indoor mapping application used for all users, including those with disabilities;
- To gather usage data and feedback from users to continuously improve the indoor mapping application and ensure it meets the needs of the ISEP community;
- To explore the potential for the application to be extended to other campuses and institutions, to provide a scalable solution for indoor mapping;
- To provide a wayfinding solution for visitors, students, and staff to easily find their way around the campus and reach their desired location.

It's important to note that the objectives are specific, measurable, achievable, relevant, and time-bound (SMART), and they are aligned with the overall goal of the project. Also, the objectives match the project's main goal and the available resources.

4. IMPLEMENTATION

The ISEP InDoor Mapping application is an innovative solution that helps students, faculty, and visitors navigate the complex indoor spaces of the institute. This application is to be implemented to provide real-time indoor location and navigation services to the users, by using advanced technology such as indoor positioning systems, Wi-Fi, and Bluetooth beacons to determine the user's location accurately.

The implementation of the ISEP Indoor mapping application requires significant planning and resources. As such, assessing current accessibility and mobility needs of students, faculty and staff is the first step in implementing and developing accessibility and mobility solutions in higher education. This can be done through surveys, focus groups, and interviews.

Developing a comprehensive plan that addresses the identified needs, including specific goals, objectives, and strategies for achieving them, is the next step. This plan should include specific goals and objectives, as well as strategies for achieving them.

Providing training and education for all members of the higher education community on the importance of accessibility and mobility and how to create an inclusive environment is also of extreme importance for the success of the implementation of this application.

Implementing technology solutions such as assistive technology, captioning and transcription services, and mobile apps can also help increase accessibility and mobility for students, faculty, and staff with disabilities, as technology plays an important role in accessibility and mobility.

Continuous evaluation and improvement of the effectiveness of the accessibility and mobility solutions in place, and adjusting as needed, is important and can be done through ongoing surveys and focus groups.

Collaborating with external organizations such as local disability organizations and government agencies can provide valuable resources and expertise to support the implementation and development of accessibility and mobility solutions in higher education.

Ensuring that accessibility and mobility solutions comply with legal requirements is important as higher education institutions must comply with laws and regulations.

Budgeting for and securing funding for these initiatives is important as developing and implementing accessibility and mobility solutions can require significant financial resources.

5. CONCLUSION

The ISEP Indoor Mapping application is a powerful tool that is designed to improve the mobility and accessibility of the ISEP campus. The application provides detailed and accurate maps of all buildings and floors, including important accessibility features such as ramps, elevators, and accessible restrooms. The user-friendly interface makes it easy for users to navigate the campus, search for specific locations and points of interest, and view real-time information such as room availability and event schedules. Additionally, the integration with other campus systems, such as the room reservation system and events calendar, allows for real-time updates and improved efficiency.

Accessibility features, such as voice navigation, alternative text descriptions, and the ability to adjust font size and contrast, make the indoor mapping application used for all users, including those with disabilities. The ISEP Indoor Mapping application also gathers usage data and feedback from users with different abilities and disabilities to continuously improve the application and ensure that it meets the needs of the ISEP community.

The ISEP Indoor Mapping application has the potential to be extended to other campuses and institutions, providing a scalable solution for indoor mapping and accessibility. Additionally, the wayfinding solution provided by the application allows visitors, students, and staff to easily find their way around the campus and reach their desired location, regardless of their mobility needs.

Overall, the ISEP Indoor Mapping application is a valuable tool that can improve the mobility and accessibility of the ISEP campus, making it more inclusive and accessible for all members of the community.

Within the scope of orientation and mobility, it is possible to say that the ISEP Indoor Mapping application has been well succeeded in the accomplishment of its purpose. The application was designed to ensure inclusion and accessibility for every individual, namely students, staff, and visitors. The intention was to create such an application that allowed everyone to move through the surroundings of the ISEP campus with easiness and autonomy.

As such, the ISEP Indoor Mapping is a technology of great value for people with reduced mobility, hearing, or visual impairments, and even for Erasmus students who do not speak Portuguese and often struggle to communicate with the staff. Accomplishing the overall goal of the project is to raise awareness for orientation and mobility challenges and to create a more inclusive and accessible school environment.

6. FUTURE WORK

The ISEP Indoor Mapping application has the potential for further development and improvement in several areas. Some possible future work for this application includes, one, the incorporation of augmented reality technology – the ISEP Indoor Mapping application could be enhanced by incorporating augmented reality technology to provide users with an interactive and immersive experience. This could include features such as 3D building models, virtual tours, and real-time navigation. Second, the integration with personal devices, such as smartphones, smartwatches, and fitness trackers to provide users with real-time location tracking and personalized navigation. Third, the integration of the application with public transportation schedules and routes to provide users with information on the best ways to reach the ISEP campus. Fourth, the development of an indoor navigation system that uses Bluetooth or Wi-Fi to help users navigate through the buildings and corridors of the ISEP campus. Fifth, the ISEP Indoor Mapping application could be customized to allow users to select their accessibility needs, such as wheelchair navigation or visual impairments, and then the application will provide personalized suggestions and recommendations. Sixth, the ISEP Indoor Mapping application could collect data on user behavior and preferences, such as frequently visited locations, to improve the overall user experience and make the application more efficient. Seventh, the application could be made available in multiple languages to accommodate international students and visitors. Last, the application could be continuously evaluated and tested with different user groups, including people with disabilities, to ensure that it meets the needs of all users and to identify areas for improvement.

By expanding this assistive technology to other markets, it is possible to expand its benefits to every individual that somehow suffers from a lack of accessible location and wayfinding information, and inclusion, due to orientation and mobility challenges.

ACKNOWLEDGEMENT

This work is being developed at the research group GILT – Games, Interaction and Learning Technologies from the Polytechnic Institute of Porto, with the support of FCT/MCTES - *Fundação para a Ciência e a Tecnologia I.P.* through national funds, within the framework of the Unit's Base Multi-year Financing (UIDB/05627/2020).

REFERENCES

Escudeiro, P. et al., (2017). Digital Assisted Communication. Porto, SCITEPRESS.

Hirtle, S. C. & Bahm, C. R., (2015). Cognition for the Navigation of Complex Indoor Environments. In: Indoor Wayfinding and Navigation. Boca Raton: CRC Press, pp. 1-12.

Jacobson, W. H., (2013). Orientation and Mobility. In: Assistive Technology for Blindness and Low Vision. Boca Raton: CRC Press, pp. 29-58.

May, M. & Casey, K., (2013). Accessible Global Positioning Systems. In: Assistive Technology for Blindness and Low Vision. Boca Raton: CRC Press, pp. 81-104.

Samama, N., (2019). Indoor Positioning: Technologies and Performance. New Jersey: JohnWiley & Sons, Inc.

World Health Organization, (2018). Assistive Technology. [Online] Available at: https://www.who.int/news-room/fact-sheets/detail/assistive-technology

World Health Organization, (2019). World report on vision. Geneva: WHO.

World Health Organization, (2023). Deafness and hearing loss. [Online] Available at: http://www.who.int/health-topics/hearing-loss

Yan, J. & Zlatanova, S., (2023). Seamless 3D Navigation in Indoor and Outdoor Spaces. Boca Raton: CRC Press.

PARTICIPANTS' NET PROMOTER SCORE ON THE STUDENTS MEET INTERNATIONALLY THROUGH LANGUAGE EDUCATION PROJECT

Jun Iio[1], Shigenori Wakabayashi[2], Junji Sakurai[3] and Yuya Nakagawa[4]
[1]Faculty of Global-Informatics, Chuo University
1-18 Ichigaya-tamachi, Shinjuku-ku, Tokyo, 162-8478, Japan
[2]Faculty of Letters, Chuo University
742-1 Higashi-nakano, Hachioji-shi, Tokyo, 192-0393, Japan
[3]Sekisaibo, LLC.
3-7-1 Akebono, Kashiwa-shi, Chiba, 277-0841, Japan
[4]Faculty of Education, Mie University
1577 Kurimamachiya-cho, Tsu-shi, Mie, 514-8507, Japan

ABSTRACT

Since 2020, we have been conducting the Students Meet Internationally through Language Education project, which offers students opportunities to participate in intercultural communication using online tools. In the academic year 2022, seven universities, 15 senior high schools, and two elementary schools from six countries will join the project. To evaluate the significance of the project, the organizers asked participants to answer a simple questionnaire. This paper reports the results of the investigation and demonstrates the efficiency of the project.

KEYWORDS

Intercultural education, Online meeting tools, Net promoter score, Participants' evaluation

1. INTRODUCTION

Over the past decade, online communication via the Internet has become popular. In addition, owing to the COVID-19 pandemic, online meeting tools such as Zoom, Webex, Google Meet, and Microsoft Teams have become common tools among not only businesspeople but also teachers and students. Although globalization makes us teach our students a sense of globalism, we cannot move internationally during the pandemic. Therefore, utilizing such online meeting tools will have a significant impact on realizing intercultural communication education.

The workshop initiative for language learning (WILL) started the Students Meet Internationally through Language Education (SMILE) project in 2020. The project focuses on intercultural education classes between overseas and Japanese schools and provides a typical curriculum from preparation through reflection (see Figure 1). It also offers a SMILE coordinator who supports teachers in matching schools, arranging meetings, and performing each series of activities using teaching materials. Furthermore, the project helps participants by offering a web application titled Dialogbook, which records students' activities (Iio and Wakabayashi, 2020; Wakabayashi et al., 2022).

Introduction	Preparation			Internet Face-to-Face Communication			Wrap-up
Foundation	Element 1	Element 2	Element 3	1st Collaboration Class	2nd Collaboration Class	3rd Collaboration Class	Review the achievement
Set up and share a clear image of the goal of this course	Prepare the contents and skills needed for your talk	Think of the questions you will ask	Confirm that you can use Webex/Zoom	Topic "Introduction of yourself and your school"	Topic "What is popular among Students"	Topic Varies	Check the contents of Dialogbook

Figure 1. Overview of the prototypical course

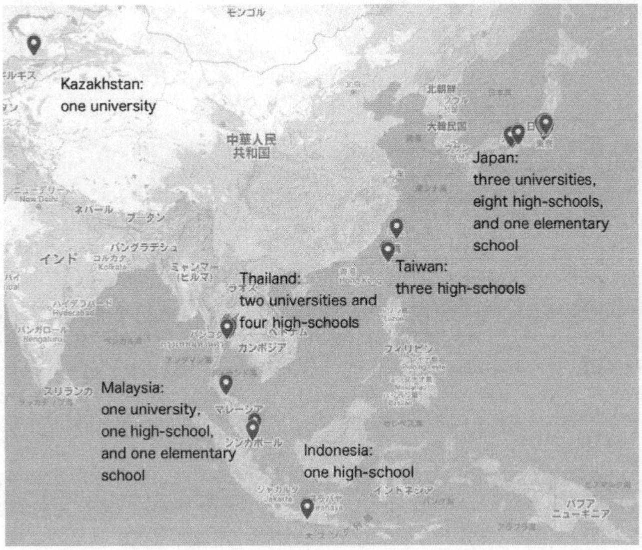

Universities	Faculty of Letters, Chuo University	Japan
	iTL, Chuo University	Japan
	Soka University	Japan
	King Mongkut's Institute of Technology Ladkrabang	Thailand
	King Mongkut's University of Technology Thonburi	Thailand
	Universiti Teknologi Malaysia	Malaysia
	Suleyman Demirel University (SDU)	Kazakhstan
High schools	Dokkyo Senior High School	Japan
	Toshimagaoka-joshigakuen Senior High School	Japan
	Seishin Gakuen Senior High School	Japan
	Tsukuihama Senior High School	Japan
	Ichihara Chuo Seninor High School	Japan
	Chuo University Senior High School	Japan
	Sakainishi Senior High School	Japan
	Taweethapisek school in Bangkok	Thailand
	Rattaphumwittaya School, Songkhla	Thailand
	Rajprachasamasai Phaimathayom Rachadabhisek	Thailand
	Watsuthiwararam School	Thailand
	Fu Cheng Senior High School	Taiwan
	Taipei Municipal Yong Chun Senior High School	Taiwan
	SMK Sri Mersing	Malaysia
	SMA Negeri 1 Yogyakarta (SMAN1)	Indonesia
ES	Kurima Elementary School	Japan
	SJK(T) Taman Tun Aminah	Malaysia

Figure 2. The map (left) and the list (right) of the schools participating in the SMILE project in AY 2022 (Iio, 2023)

In the academic year (AY)[1] 2020, the four senior high schools participated in the project. Two pairs of schools engaged in intercultural communication. In the AY 2021, the following year, the project included eight schools, including four universities and four senior high schools. This year, seven universities, fifteen senior high schools, and two elementary schools from six countries joined the project (Iio et al., 2023). More than 500 students participated in the project this year (see Figure 2); therefore, the WILL asked them to answer a simple questionnaire to confirm the efficiency of the project.

After collecting the answers, we calculated the net promoter score (NPS) (Reichheld, 2003) based on the answers to Question 3. The NPS is a measure of customer mindset and a predictor of sales growth. Although the SMILE project has not been a commercial activity thus far, it can be used as a key indicator of acceptability by the participants (students and teachers) in the project.

Section 1 briefly introduced the SMILE project. The remainder of this paper is organized as follows: Section 2 describes the literature review. Section 3 provides a brief description of the investigation and the results of the questionnaire. Finally, Section 4 concludes the paper.

2. LITERATURE REVIEW

Before providing an overview of the project evaluation, this paper describes related work, particularly from the perspective of international communication education and evaluation case studies using the NPS.

[1]The Japanese academic year starts from April and ends in March of the following year.

2.1 Importance of intercultural education

As internet communication technologies have improved, the significance of cross-cultural education, which enables students to engage in intercultural communication, has steadily increased. Several pedagogical methods have been proposed to provide such education. One well-known method is content- and language-integrated learning (CLIL), which is expected to foster intercultural understanding. For instance, Yang (2021) showed that content-driven CLIL learners obtained slightly higher cultural quotient scores than their language-driven CLIL English major peers.

In addition, collaborative online international language (COIL; Appiah-Kubi, 2020) is a typical method of learning international communication in English. However, COIL programs are led mainly by Western communities (Rubin, 2017). Because several Western languages, such as European languages, are similar to English, learning English is relatively easy. However, Eastern languages, including Japanese, are significantly different from Western ones. Therefore, learning English may be more difficult for Eastern students than for Western students. That is the reason why WILL started the SMILE project.

2.2 Evaluations using NPS

Originally, the NPS was proposed as a tool for commercial businesses to evaluate products and services. However, as previously mentioned, it can be used to assess educational services.

Kara et al. (2021) calculated the NPS for students' likelihood of recommending academic programs. Because they considered the application of the NPS in higher education to be limited, they calculated the score and performed deeper investigations. Lucero (2022) proposed using the NPS to assess accredited continuing medical education or continuing education. She also concluded that NPS is a valid indicator of the intention to act.

Wassen (2019) reported an interesting case in which the NPS was used to assess parental satisfaction with the education of students with autism. Furthermore, the NPS has been used in several cases to assess student satisfaction with higher education, including at a state university (Kara and Zeren, 2022), among engineering students (Cruz et al., 2019), and in virtual learning environments (Gamarra-Moreno et al., 2021).

3. INVESTIGATION

This section briefly describes the investigation and results of the questionnaire.

3.1 Overview of the Investigation

As described previously, to confirm the efficiency of the SMILE project, the WILL asked participants to answer a simple questionnaire to evaluate the project. The questions used in the questionnaire are listed in Table 1. The questionnaire was prepared using Google Forms as an online questionnaire.

Table 1. Questions in the questionnaire

Questions	Options
Q1. What is your affiliation?	each affiliation
Q2. Are you a teacher or a student?	teacher or student
Q3. Do you recommend this program to your friends and colleagues?	10: willing to recommend ~ 0: not at all
Q4. Let us know the reason for your evaluation in the previous question. (It is okay to write your opinion in your local language.)	open question

The answers to the online questionnaire were collected from 24th November 2022 to 31st January 2023, with 404 answers.

3.2 Investigation results

This section explains the investigation results. Figure 3 shows the results for Questions 1 and 2. Figure 3 (left) shows a bar chart illustrating the number of answers from each school, and Figure 3 (right) shows the 392 students and 12 teachers who answered the questions.

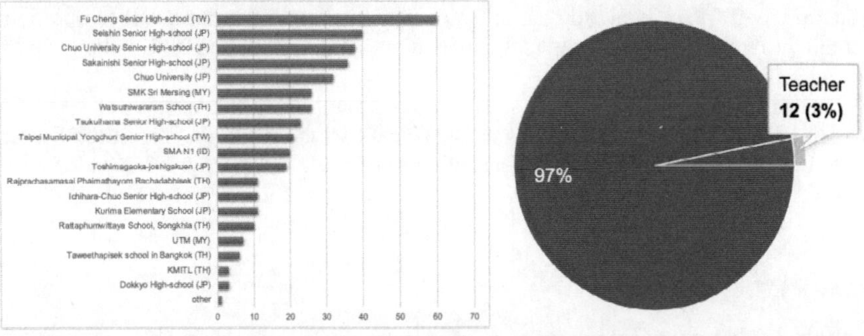

Figure 3. Results of Questions 1 and 2

Figure 3 (left) shows that several answers were collected from Japanese schools and schools in their counterpart countries. Although more than half of the participants were from Japan, several participants in the project from overseas schools answered the questionnaire (see Table 2).

Table 2. Number of answers from each country

Country	Number of answers	% of answers
Japan	213	52.7
Taiwan	81	20.1
Thailand	53	13.9
Malaysia	33	8.2
Indonesia	20	5.0
other	1	0.3

The results of Question 3 are shown in Figure 4. Based on these results, we calculated the NPS of the project. The NPS was calculated by subtracting the percentage of detractors from that of promoters. Promoters were defined as those who answered Question 9 or 10. In contrast, detractors are defined as customers who answered Question 6 or lower.

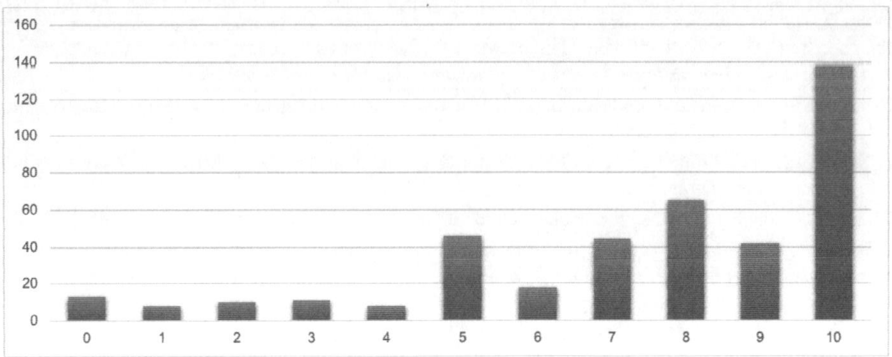

Figure 4. Results of Question 3

In this case, the number of promoters and detractors was 181 and 114, respectively. As previously described, the total number of responses was 404. Therefore, the NPS was calculated to be 0.166. NPS ranges from -1.0 to 1.0. In general, the positive NPS could be interpreted to imply that the customers who evaluated the products or services were satisfied, and the products were accepted by the customers.

4. CONCLUSIONS AND FUTURE WORK

This study introduces the SMILE project, an online intercultural education program, and reports the results of an assessment using the NPS as a case study. In the AY 2022, more than 500 students from seven universities, 15 high schools, and two elementary schools participated in the project. The participants were asked to complete a questionnaire, and 404 responses were collected. The NPS was calculated from the results of 11 responses on a scale of 0 (not likely to recommend) to 10 (willing to recommend) to question "Q3. Do you recommend this program to your friends and colleagues?" We obtained an NPS value of 0.166, which is a relatively better score.

A deeper analysis of the assessment data, such as aggregating values by country, will be performed in future studies. In addition, we obtained 291 answers to Question 4, which asked the reason for selecting the answer to Question 3, even though it was not a mandatory question. It is an open-ended question and is not easy to analyze; however, it can provide attractive findings through text-mining approaches.

ACKNOWLEDGEMENT

This project was supported by Grant for Establishing Interdisciplinary Research Cluster, Chuo University, and JSPS, KAKENHI 22K00689. We thank all the students and teachers for participating in the project and providing valuable data for research on their learning activities.

REFERENCES

Appiah-Kubi, P. and Annan, E., 2020. A review of a collaborative online international learning. *International Journal of Engineering Pedagogy*, 10(1), 109-124.

Cruz, J. O., Moreno, E. E., and Silupu, W. C., 2019. Effect of the Implementation of University Accreditation on the satisfaction of engineering students using the Net Promoter Score. *In 2019 IEEE Sciences and Humanities International Research Conference (SHIRCON)* (pp. 1-4). IEEE.

Gamarra-Moreno, A., Gamarra-Moreno, D., Gamarra-Moreno, A., and Gamarra-Moreno, J., 2021. Assessing Problem-Based Learning satisfaction using Net Promoter Score in a virtual learning environment. *In 2021 IEEE World Conference on Engineering Education (EDUNINE)* (pp. 1-5). IEEE.

Iio, J. and Wakabayashi, S., 2020. Dialogbook: A Proposal for Simple e-Portfolio System for International Communication Learning, *International Journal of Web Information Systems,* 16(5), 611-622.

Iio, J., Wakabayashi, S., Sakurai, J., and Nakagawa, Y., 2023, Cross-cultural Education in Asia using Online Communication Tools and the Dialogbook, *The 21st International Conference on e-Society (ES2023),* pp. 428-431, Lisbon, Portugal.

Kara, A., and Zeren, D., 2022. The relationship between the Net Promoter Score (NPS) and students' college experiences at a state university. *International Review on Public and Nonprofit Marketing*, 1-17.

Kara, A., Mintu-Wimsatt, A., and Spillan, J. E., 2021. An application of the net promoter score in higher education. *Journal of Marketing for Higher Education,* 1-24.

Lucero, K. S., 2022. Net Promoter Score (NPS): What Does Net Promoter Score Offer in the Evaluation of Continuing Medical Education?. *Journal of European CME,* 11(1), 2152941.

Reichheld, F. F., 2003. The One Number You Need to Grow. *Harvard business review,* 81(12), 46-55.

Rubin, J., 2017. Embedding collaborative online international learning (COIL) at higher education institutions. *Internationalisation of Higher Education*, 2, 27-44.

Wakabayashi, S., Iio, J., Kumaraguru, R., Komoto, R., and Sakurai, J., 2022. How ICT Tools Support a Course Centered on International Collaboration Classes, *IFIP WCCE 2022: World Conference on Computers in Education,* 2C-3, Hiroshima, Japan.

Wazzan, W. A., 2019. Higher education for students with autism, using net promoter score to assess parents satisfaction. *Creative Education,* 10(11), 2495-2503.

Yang, W., 2021. Developing intercultural competence: A comparison of CLIL and language majors before and after industrial placements. In *International Perspectives on CLIL,* pp. 129-148. Palgrave Macmillan, Cham.

INTERNATIONAL COOPERATION IN THE NORDIC AND BALTIC COUNTRIES. THROUGH SEMI-ONLINE TEACHER TRAINING AND RESEARCH

Kristina Mullamaa
Lecturer, PhD, University of Tartu, Estonia
Project leader for Nordplus cooperation projects with Nordic and Baltic universities from 2019
Lossi 3, 51003 Tartu, Estonia

ABSTRACT

The paper describes and analyses a semi-online format of international cooperation and further education of university teachers. Thanks to the Nordplus scholarship, seven universities in the Nordic and Baltic countries have been able to meet at seminars and discuss their work in the framework of Nordic-Baltic cooperation. Instead of traditional conference settings, the new semi-online format enables participants to create learning materials together and develop joint research. The format also allows participants to delve deeper into analyzing the specifics of their work, compare the modus operandi at their university with that of the others, experience and see work at different universities. Joint learning materials are created and published, the cooperation is analyzed at different levels and research is published together. We sum up the principles of cooperation and its format. The method is ethnographic research: descriptive analysis approach.

KEYWORDS

Teacher training through semi-online cooperation, International cooperation and material preparation, Motivation

1. INTRODUCTION

The positive co-factors of a self-leading professional definitely include astute choices of further education. In our article below, we introduce an enticing way we have found for carrying this through: the Nordplus-supported visit-your-colleagues format with study-visits, meetings and further education trainings in different Nordic and Baltic countries.

We developed this cooperation on the basis of concepts that are similar to blended learning (work individually; share online; meet for live meetings & discussions). Our team includes six universities from different Baltic and Nordic universities in Sweden, Finland, Estonia, Latvia, Lithuania and Iceland. The cooperation has kindly been supported by the Nordplus scholarship scheme supported by the Nordic countries, with their main office in Reykjavik, on Iceland.

By now, we have successfully finished two joint projects, and embarked on two consecutive cooperation projects. In the paper below we describe the projects´ format, results and opportunities for teacher development as *"semi-online (or blended) teacher training through international cooperation"*.

While blended learning in itself is nothing new, and can be used for any field, including methodology and further education, we suggest that having an international scope, where study visits and in-depth personal analyses of one´s role and work environment form an integral part of the project, is a case to be highlighted for positive inspiration for others. The meta-analyses combined with personal introspection, 1:1 tutoring and mentoring, and colleagues´ advice on different layers of professional life, at different stages of it, is an incredibly valuable tool of our model. This is why we have decided to share this experience on the conference on the possibilities of e-learning in Porto.

The following brief paper has two main foci: practical and theoretical. The form and results will be analyzed:

1. The format of the meetings and our products: the commonly created learning materials and joint publications;

2. The benefits of such cooperation – as the theoretical-methodological model of one possible set-up forms ("blended teacher education for international cooperation") for modern teacher development,

The methodology is ethnographic research: describing and analyzing an ongoing process ("composed of work in progress reports or fresh developments" as suggested for "short papers" at this conference (https://www.elearning-conf.org/call-for-papers/).

2. "BLENDED TEACHER TRAINING" FOR INTERNATIONAL COOPERATION

The quick changes in our modern lives and lifestyles also bring along the necessity to teach somewhat differently than some time ago. Even though the cornerstones of good teaching – friendliness, support, clear structure, and quality teaching – have remained the same, the teacher´s role is oftentimes even more of that of an advisor. Also, life has become more international. We have international students from many different countries of the world, we attend numerous scientific conferences in different countries, and concepts and approaches in methodology, values and principles seem to evolve, develop, and change sometimes in months.

At the same time, deeper introspection and capacity to analyze one´s own values, role, goal and methods are needed more than ever. It is exactly during many changes and "crosswinds" that staying true to your course and having clear and balanced views of your role, foci, possibilities - can support one to function efficiently and purposefully, at the same time developing. In our earlier research we have shared some thoughts both on student e-learning (Mullamaa 2009, 2011, 2015) and emotional balance and learning (Mullamaa 2016, 2018) as well as the teachers´ (Mullamaa 2020, 2022).

Teacher personal balance has been pointed out also by both internationally acclaimed research in language studies (e.g. Brophy 2004, Dörnyei 2001) as well as practitioners´ journals (e.g. IATEFL journal *Voices*). A good balance both in personal and professional lives is also a recipe for greater motivation (Schuz 2007, Reeve 2009) and achievement in popular psychology (cf. e.g. Goleman 2013).

This background suggests that the continuous introspection, self-analysis and further education, as well as discussions with one´s colleagues, and international experience can be supportive for professional and personal development and balance. Thus, in addition to the formal teacher training events, and different international conferences – which are all very good, inspiring and supportive for the teacher´s self-development – we felt that an additional format might be needed.

Inspired by this, we decided to apply for a Nordplus grant and sketch a vision for cooperation that would include a small number of colleagues for intensive, frank and deeply analytical get togethers, for analysing the specifics, difficulties and bonuses in our everyday teaching. We presume that cornerstones of a well-balanced professional in teaching include good personal and professional life constituents (positive outlook on life, capability to solve problems, ability to see the big picture, knowing and applying different methodologies, good communication and mentoring, as well as counselling skills, empathy, stamina, etc.). If the personal and professional skills and values match, this is beneficial for teaching, and the general atmosphere for learning in class, which is of paramount importance for good learning. In the implicit way, regular meetings with colleagues to discuss one´s role and teaching, may improve the above.

On the theoretical, more analytical level, this form of cooperation also enables participants to analyze and negotiate their role as university teachers. It helps one support the work process, but also conceptualize and reconceptualize the profession in the light of the new developments, find balance for one´s professional self in the discussions with colleagues, and reach the conceptual, academic, and emotional balance that are necessary constituents of motivation in their professional lives. The Nordic-Baltic context with the similar, yet somewhat different organization cultures offer additional perspectives to participants who also are experts in the field, and extra, meta-level research is conducted on our cooperation, Nordic-Baltic communication styles and principles.

More specifically, we have developed something, that could broadly be called "blended teacher training for international cooperation". In addition to the obvious benefits such a model may have in different situations where it is difficult to meet IRL (as we remember from the global situation a few years ago), it also gives teachers a. the possibility to experience different work cultures and environments in different countries and universities, b. exchange views with colleagues in these contexts and c. grow both academically and personally through Socratic conversations, introspection, and analyses of one´s own context and others´ "case studies". The project was granted, and the work could begin.

2.1 Nordplus International Meetings and Workshops: Format and Contents

Already from earlier contacts we had a good cooperation experience with some universities in Scandinavia and the Baltics. Now, we invited them into our two new projects – one of them: creating a textbook and Idioms word list, the other: on Methodology and teaching at a higher educational institution.

The universities gladly joined in, and we built up a format for meetings. This entails going round twice a year, visiting a different Nordic of Baltic university each time. The format we follow is the same:

1. The host is responsible for arranging the *further education training session* that takes place each time. The topics thus far have included: language learning theories: history and the modern dilemmas; The European Language Framework: new focus on language competencies; e-learning and the principles of safeguarding against digital perils. We have also discussed the principles of creating good textbooks, and will have seminars on student engagement and motivation.

2. In addition to the further education training session, each meeting includes a *visiting round of the hosting university*. We will see how teachers and students work, see the lecture rooms and exhibition areas, teachers´ joint working or resting areas, campus of the university and meet some colleagues of the hosting partner – thus getting a good overview of what everyday life looks at the respective university.

3. The third component is the *international meeting*. The meeting is for discussing our joint projects. Mostly, these have focused on creating joint learning materials. But an important part of the international meeting, beyond discussing our joint products, is the sharing on how one´s teaching is going, a briefing on the work environments and administrative arrangements.

3. a & b Especially during and after Corona times, when teaching was a rather solitary on-line project, the *sum up of some modus operandi at different universities*, as well as the creative solutions teachers might have had, provided good ideas and support for project participants. An additional topic is the increasing bureaucracy in some institutions, and here it may be comforting to see the parallels, and fruitful to suggest improvements. But also, the ventilation *and meta-level analysis of the administrative processes and hierarchies at different institutions* of higher education.

4. A necessary part of every person´s world view beyond the education, workplace, and academic and personal growth, is also the *geographical-cultural surroundings*, the spirit of the local university and university town, and how it influences us. Therefore, during our cooperation seminars we also have a glance of the recipient´s culture, some icons of their national lore and important cultural and societal phenomena for the country.

Based on these four tires, the projects have developed efficiently and with glee.

2.1.1 Products and outputs

Each project we undertake has a real, "hands on" product, which is not only usable for project members but open for the wider audience. For example, we have created a good country-study material on Sweden: based on the well-known children´s book Nils Holgersson´s travels. Also, in our book (and digital material) the main character travels through the same areas of Sweden, but this time in modern Sweden.

In the framework of the "Idioms" project we are creating a list of specific Swedish idiomatic expressions with Estonian counterparts. Methodology project II creates a joint Icelandic-Estonian-Swedish-Latvian homepage with links to different articles, tips and further education possibilities for university teachers of Swedish. Right now, it is meant for internal use only, but as it grows and is tested by our members, we also plan to open it to the wider audience. Also, a joint collection of that relate to learning and methodology will be published.

2.1.2 A Well Balanced Professional

Already from our earlier articles (see e.g. Mullamaa 2018; 2020), we have pointed to the importance of the psychologically and mentally agile and well balanced professional also as concerns teachers. This is in tune with other research. According to Brophy 2004: 391-393), for example, "the efficiency as a teacher", includes the strength and ability to have envision a broad picture of the purpose of one´s goals. Dörnyei (2001: 31) points to an important aspect in a student assessment study where "the participants considered the teacher´s own behavior to be the single most important motivational tool".

My personal belief is here that a person, who has a good balance in the professional life and is also positive, achieving and glad in the personal one, can serve as a reliable inspiration and role model also for students.

In this, the semi-official discussions around one´s professional role, goals, possibilities for developing have been especially valuable in our cooperation. In a small round of like-minded professionals, trust is built and as our colleagues also represent the different stages at the academic career and age, the experience of the longer career is in dialogue with the enthusiasm of the young university teachers at the beginning of their teaching profession – a valuable focus to analyze any profession *per se*. All this enables one to be more confident and happy in one´s job and role. This in its turn, hopefully, wears off to the students, their ways of handling *their* tasks and roles with confidence and pro-actively.

2.1.3 Teacher development: benefits and methodological implications

In addition to the feedback we get from the immediate work environment, formal seminars and conferences we attend and present at, semi-formal discussions with like-minded colleagues in different cultural contexts can be beneficial and illuminating. This is exactly what our project cooperation has enabled us to do. We express our sincerest thanks to the Nordplus project which has approved our project ideas and allowed us to use, develop and modify this format.

The outcome is not directly measurable (yet). We could devise charts, numbers and formulae on how "satisfied" one is with their work due to the current affiliation with our project. Yet, the processes are still too fresh to chart them numerically. Would an arbitrarily denoted, say, "number 7", e.g. – grant us mor knowledge – than just a happy and balanced professional saying they *"feel supported and good"*? A matter of schools of thought, probably. Currently, at this stage of our project development, we follow the "self descriptions" of participants, and chart their comments which are rather informative and promising even without bringing the numeric determinants into the analysis. The current stage of our meta-analysis shows that the beneficial effect of the cooperation has been seen in participants´ everyday work motivation. They have also reported their balance is bigger thanks to the possibility of discussing things through at our international meetings.

3. CONCLUSION

In the article above we have taken a look at our semi-online, blended learning principle international teacher cooperation projects. Supported by Nordplus, our Nordic and Baltic partner universities have joined us in working out joint learning materials for students, creating on on-line methodology and teaching related data-base for teachers (work in progress), doing research and giving out a collection of articles on teaching our subject, as well as teaching and methodology in general.

In addition to the immediate products of joint cooperation between Iceland, Estonia, Sweden and Latvia (in earlier projects also Lithuania and Finland), the discussions and analyses of our profession, institutional and culture based communication, and the teaching principles at large – form an important part of this semi-digital international cooperation and teacher development model.

We have found the principles of cooperation to be useful at both these levels:

1. The practical one – leading to joint production of learning materials, digital data pools, booklets and on-line resources.

2. The cooperation and further development one – from joint international learning seminars to joint international meetings, which allow participants to ponder upon and develop the perspectives of their personal and professional roles, as well as experience both work and academic culture as well as the society and culture in our common Nordic-Baltic area.

REFERENCES

Brophy, Jere. 2004. *Motivating Students to Learn.* Second edition. Lawrence Erlbaum Associates, Publishers. Mahwah, New Jersey, London.

Dörnyei, Zoltan. 2001. *Motivational Strategies in the Language Classroom.* Cambridge University Press.

Goleman, David. 2013. *Focus: The Hidden Driver of Excellence Harper*; Illustrated edition Conference homepage: https://www.elearning-conf.org/call-for-papers/

IATEFL journal *Voices https://www.iatefl.org/resources/iatefl-voices-issue-280*

Mullamaa, Kristina. 2009. Ethics in Teaching. Should We Do More? In: *Humanising Language Teaching Magazine*. Year 11; Issue 3; June 2009. Canterbury, England. http://www.hltmag.co.uk/.; http://www.hltmag.co.uk/jun09/sart01.htm 22-35. 13.

Mullamaa, Kristina. 2011. The challenges and opportunities of on-line language teaching. *Humanising Language Teaching*, 13(1), 12 - 24. England. 14.

Mullamaa, Kristina. 2015. Motivating aspects of modern e-learning – from individualisation to gamification. *Dubai. HMSU University. Innovation Arabia 8. Innovation Arabia* http://www.innovationarabia.ae/innovation-arabia-8/conference-proceedings. Conference proceedings.

Mullamaa, Kristina. 2016. Teachers as Supporters in the Modern Learning Process. In: *Emerging Horizons*. Eesti Ülikoolide Kirjastus. 15.

Mullamaa, Kristina. 2017. Motivational Aspects of Learner and Teacher Cooperation. Inted *2017 Proceedings. INTED2017 11th annual International Technology, Education and Development Conference*. 6th, 7th and 8th of March, 2017. Valencia, Spain. INTED2017 Proceedings. Digital library

Reeve, Johnmarshall. 2009. *Understanding Motivation and Emotion.* Fifth edition. John Wiley & Sons, INC. USA.

Schutz, Paul and Pekrun, Reinhard. 2007. *Emotion in Education*. Academic Press, Elsevier. Amsterdam, Boston, Heidelberg, London, New York, Oxford, Paris, San Diego, San Fransisco, Singapore, Sydney, Tokyo.

Schutz, Paul, Cross, Dionne et al. 2007. *Teacher Identities, Beliefs and Goals Related to Emotions in the Classroom*. In Schutz, Paul and Pekrun, Reinhard. 2007. In: Emotion in Education. Academic Press, Elsevier. Amsterdam, Boston, Heidelberg, London, New York, Oxford, Paris, San Diego, San Fransisco, Singapore, Sydney, Tokyo. 223-243.

ONLINE MS IN BUSINESS ANALYTICS BEST PRACTICES AND BEYOND

Rose Sebastianelli and Nabil Tamimi
Operations & Analytics Department, University of Scranton, Scranton, PA 18510, USA

ABSTRACT

The demand for talent in analytics has significantly outpaced supply, prompting institutions of higher learning to develop related curricula to address this unmet market need. One degree offering that has proliferated is the online Master of Science (MS) degree in business analytics. Unlike the more general MBA with a concentration in analytics, this specialized master's degree usually includes advanced coursework in areas related to the three subject area domains (descriptive, predictive, and prescriptive analytics). These types of courses, such as statistical modeling, optimization, and computer programming, involve technical and quantitative skills that can pose unique challenges in an asynchronous online learning environment. Moreover, with the explosive growth in job opportunities, prospective students may be drawn to the MS in business analytics without understanding fully the advanced nature of the curriculum or the prerequisite knowledge needed to succeed. With the goal of better positioning prospective students for success in the MS in business analytics, we approach the issue in two different ways. First, we benchmark the top ranked online MS business analytics programs in the US to uncover best practices with respect to curriculum as well as the online application and admissions screening process. Second, we consider strategies for increasing student engagement in asynchronous online learning environments suitable for use in advanced online business analytics courses.

KEYWORDS

Business Analytics, Benchmarking, Admissions, Student Engagement, Asynchronous Learning

1. INTRODUCTION

As organizations in every industry face the challenge of leveraging big data for competitive advantage, the demand for talent in analytics has significantly outpaced supply. Employers seek professionals that possess a mix of data analysis, information technology, and business skills. Several similar, yet distinct, academic programs of study have emerged to address industry needs. These include business intelligence, data analytics, data science, and business analytics. While the audience for data-oriented degrees is large and expanding, it is not homogeneous. Even within business analytics, programs differ in level (undergraduate versus graduate), focus (general versus specialized), and mode of delivery (traditional versus online).

The Institute for Operations Research and Management Science (INFORMS) defines analytics as the scientific process of transforming data into insights for the purpose of making better decisions (www.informs.org) and categorizes it into three domains (descriptive, predictive, and prescriptive). The Master of Science (MS) degree in business analytics is a specialized program with an emphasis on advanced techniques for analyzing structured and unstructured data to gain insights for better decision-making.

It has been suggested that the graduate level is ideal for business analytics education (due to the prerequisite of specific knowledge) and that business schools provide the needed integration of information systems, statistics, and management science with business domain knowledge (Hsinchun, et al., 2012). Parks, et al. (2018) analyzed MS programs in business analytics using the classification scheme proposed by Chiang et al. (2012). This scheme involves (1) IT knowledge and skills used for descriptive analytics, (2) analytical skills that integrate the disciplines of statistics and computer science used for predictive analytics, and (3) business knowledge and communication skills used to support prescriptive analytics. Of the 62 programs evaluated, only 22 were online. Among the most popular analytical skills were data mining, statistical analysis, optimization, and simulation. Although the *average* percentage of total credit hours devoted to each category was about the same, the study found substantial disparity across programs with respect to focus and topic coverage due, according to the authors, to a lack of alignment with industry needs.

Efforts have increased to identify (and close) the gaps between industry requirements and academic curricula. Paul and MacDonald (2020) stress that courses must address all core business analytic areas: Descriptive Analytics (i.e., what happened: data visualization techniques), Predictive Analytics (i.e., what might happen: forecasting and statistical modeling techniques), and Prescriptive Analytics (i.e., what actions should be taken: optimization and simulation techniques) (see Figure 1). Furthermore, they note the importance of ensuring student success in business analytics graduate programs by admitting those that possess the necessary prerequisite knowledge (e.g., statistics) and personality traits (e.g., recommending interviews as part of the application process).

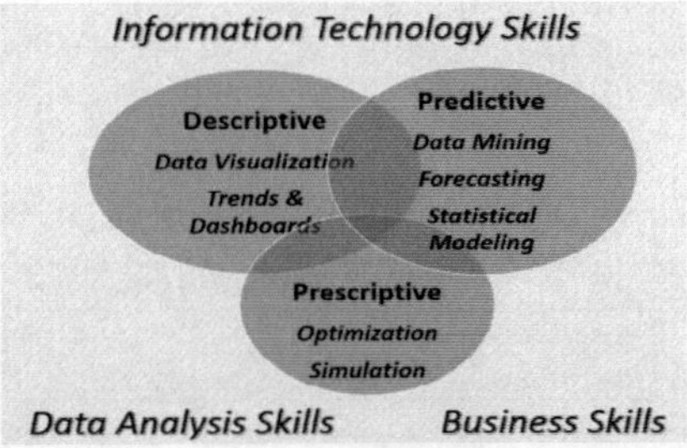

Figure 1. Three domains of business analytics within the framework of professional skills

With the goal of better positioning prospective students for successful completion of the online MS in business analytics, we undertake a benchmarking study of the top ranked programs in the US. Unlike previous studies, we focus solely on the *online* MS in business analytics delivered asynchronously. These programs have been increasing in number, not surprisingly, given the unmet and growing demand for analytics professionals as well as the advantages this mode of delivery offers (i.e., convenience and flexibility). However, the technical and quantitative nature of this specialized graduate program may pose challenges to students in an asynchronous online learning environment. Furthermore, the explosive rise in job opportunities may attract prospective students to the online MS in business analytics who do not understand fully the advanced nature of the curriculum, or the prerequisite knowledge needed to succeed. Consequently, our benchmarking study is not only directed toward examining best practices with respect to curriculum, but also with respect to the online application process. In our view, this has become increasingly important in light of the pandemic's impact on standardized testing (e.g., GMAT). With many institutions waiving the standardized test as a requirement, it is critical that the online application process both conveys and collects relevant information to attract and effectively screen suitable applicants for admission.

Engagement plays a key role in student learning, and it is evident that online students in remote asynchronous environments engage with content differently than students in traditional face to face settings (Meyer, 2014; Mayer, 2019). Business analytics is a STEM-designated program (Science, Technology, Engineering, Math). In order to enhance student engagement, we consider how strategies shown to be effective in other online STEM-designated programs (Al Mamun and Lawrie, 2023) may be adapted for advanced online business analytics coursework.

2. BEST PRACTICES

The programs included in the benchmarking study are those that appear on the most recent list of "best online master's in business analytics degree programs" as ranked by *Intelligent (https://www.intelligent.com/best-online-masters-in-business-analytics-degree-programs).* All institutions are non-profit (public or private) and many programs on the list are accredited by AACSB (Association to Advance Collegiate Schools of Business). The website for each online MS program in business analytics is visited and data are gathered on benchmarking criteria broadly classified as relating to (1) information available on the website and (2) the website itself (features and functionality).

Benchmarking criteria related to information available on the website are either program specific (e.g., completion time) or convey expectations about what is required to succeed in the program (e.g., preferred undergraduate major). Benchmarking criteria related to the website itself include features (e.g., ability to translate into multiple languages) and functionality (e.g., ease of navigation).

2.1 Benchmarking

The top 30 programs on the list are benchmarked for this study. The top 10 programs on the list are ascribed a "tagline" by *Intelligent.* These programs and their taglines are (#1) Boston University *(Intelligent Pick)*, (#2) Texas A&M University (*Best Public Institution*), (#3) University of Maryland (*Fastest Completion Time*), (#4) Indiana University (*Best for Adult Learners*), (#5) Arizona State University (*Most Flexible Start Terms*), (#6) Syracuse University (*Best Private Institution*), (#7) Oklahoma State University (*Best Electives*), (#8) Maryville University (*Most Affordable*), (#9) William & Mary (*Most Innovative Curriculum*), and (#10) Stevens Institute of Technology (*Best Job Placement*). Table 1 shows the remaining 20 programs that round out the top 30.

Table 1. Online MS Degree Programs on Intelligent List

#11 Iowa State	#16 MIT	#21 Ohio University	#26 U Mass Lowell
#12 Penn State World	#17 Drake University	#22 West Virginia Univ.	#27 U Tennessee Chattanooga
#13 American University	#18 U of Nebraska	#23 Quinnipiac	#28 Lewis University
#14 Purdue University	#19 Pepperdine	#24 Southern Methodist	#29 U Texas Dallas
#15 Worcester Polytechnic	#20 Carnegie Mellon	#25 NYU	#30 U N. Carolina Wilmington

Table 2 shows the percentage of programs that display specific types of program information on the website (for the 30 programs as well as just the top 10). Program specific information includes cost, credits, completion time, job opportunities, and curriculum. The top 30 programs require between 21 and 54 credits (average of 32.73), with all top 10 requiring at least 30. Costs range from $14,910 to $86,300 (the average being $43,116). In the top 10, Maryville tagged *Most Affordable*, costs more than Oklahoma State. Only 70% of the top 30 programs provide an estimate of completion time, which ranges from 9 months to 3 years. The University of Maryland, tagged *Fastest Completion Time*, lists 16 months. Over half of the programs include information about job prospects after graduation (job titles and salary) and all list specific courses on the website. Our focus is on six courses we believe best meet current industry needs and/or are considered innovative. Given that most employers of analytics professionals cite programming skills as desirable, it is not surprising that courses in Python and R are prevalent in the top 10 programs. Although we expected to find a course in data visualization as well as a capstone course across all programs, these are not offered in 23-30% of the top 30 (or in 40% of the top 10). Very few (none in the top 10) offer a course in data ethics. Finally, William & Mary, tagged as *Most Innovative*, does not stand out as substantially different from the others, at least with respect to these six courses.

In order to gauge the extent to which program websites convey to prospective students what is necessary to succeed in the online MS in business analytics, we examine information provided on admissions criteria (see Table 3). All programs list admissions criteria on their websites. All top 10 include the preferred undergraduate degree, which is important given the specific prerequisite knowledge needed for advanced coursework in business analytics. This drops to 83% among the top 30, although 93% do list the minimum acceptable undergraduate GPA. The majority no longer require applicants to submit scores on either the GMAT or GRE.

Table 2. Program Specific Information Available on Website

Specific Information	**Top 30 (% Yes)**	**Top 10 (% Yes)**
Job Titles Listed?	53 %	60 %
Job Salaries Listed?	50 %	60 %
Specific Courses Listed?	100 %	100 %
Data Visualization Course?	77 %	60 %
Python Course?	70 %	80 %
R Programming Course?	77 %	80 %
Capstone Project Course?	70 %	60 %
Artificial Intelligence Course?	73 %	60 %
Data Ethics Course?	17 %	0 %

Table 3. Admissions Information Available on Website

Specific Information	Top 30 (% Yes)	Top 10 (% Yes)
Admissions Requirements Listed?	100 %	100 %
GRE / GMAT Required?	33 %	40 %
Preferred UG Degree Required?	83 %	100 %
UG GPA Requirement?	93 %	90 %

Finally, we go beyond benchmarking program specific information to evaluating the website itself. Table 4 shows benchmarking results for website features and functionality. Not surprisingly, all websites provide the ability to apply online, yet 30% do not display a privacy policy. While the vast majority are easy to navigate and are engaging (e.g., graphics, multimedia, interactive content), a smaller percentage provides FAQs or online help/chat. Given the high level of interest in these programs among international students, it is surprising that so few provide the ability to translate website content into multiple languages.

2.2 Engagement

As a STEM-designated program, online MS programs in business analytics may benefit from adapting strategies used in other STEM-designated programs in order to increase student engagement with content in the asynchronous online environment. Evidence suggests that commonly used approaches that foster student-to-student interaction, such as discussion forums, are less effective for learning quantitative/technical subjects online than faculty guided student interaction (Sebastianelli and Tamimi, 2011).

An approach that has been particularly prevalent in STEM related courses, and one that offers viable instructional design ideas for online self-directed individual engagement with business analytics content, is inquiry-based learning (IBL) (Abd-El-Kahlick et al., 2004). IBL is exploratory by nature and has been effectively implemented in asynchronous online environments via POEE (predict, observe, explain, evaluate) pedagogical design (Al Mamun et al., 2020). This is an ideal framework within which faculty teaching online business analytics courses can create guided interactive content (via interactive multimedia) that provides high quality student engagement with all three subject area domains.

Table 4. Website Features and Functionality

Specific Feature / Functionality	Top 30 (% Yes)	Top 10 (% Yes)
FAQ?	67 %	100 %
Ability to Translate?	3 %	10 %
Online Help or Chat?	53 %	80 %
Privacy Policy?	70 %	70 %
Online Application?	100 %	100 %
Easy to Navigate?	77 %	90 %
Engaging?	70 %	80 %

3. CONCLUSION

Given the proliferation of online Master of Science (MS) programs in business analytics, it is important to understand what constitutes best practices, not only in terms of designing curriculum to meet (or anticipate) industry needs, but also in terms of how to increase the probability for student success in an asynchronous online learning environment. Toward this end, we benchmark the top ranked online MS programs in the US. We also recommend borrowing strategies from other STEM-designated disciplines to enhance student engagement with online business analytics content.

REFERENCES

Abd-El-Khaik, F. et al., (2004). Inquiry in Science Education: International Perspectives. *In Science Education,* Vol. 88, No. 3, pp. 397-419.

Al Mamum, M.A. et al., (2020). Instructional Design of Scaffolded Online Learning Modules for Self-Directed and Inquiry-Based Learning Environments. *In Computers and Education,* Vol. 144, p. 103695.

Al Mamun, M.A. and Lawrie, G., (2023). Student-Content Interactions: Exploring Behavioural Engagement with Self-Regulated Inquiry-Based Online Learning Modules. *In Smart Learning Environments,* Vol. 10, No. 1, pp. 1-31.

Chiang, R.H. et al., (2012). Business Intelligence and Analytics Education and Program Development: A Unique Opportunity for the Information Systems Discipline. *In ACM TMIS,* Vol. 3, No. 3, pp. 12.

Hsinshun, C. et al., (2012). Business Intelligence and Analytics: From Big Data to Big Impact. *In MIS Quarterly*, Vol. 36, No. 1, pp. 1293-1327.

Mayer, R.E., (2019). Thirty Years of Research on Online Learning. *In Applied Cognitive Psychology,* Vol. 33, pp. 152-159.

Meyer, K.A., (2014). Student Engagement in Online Learning: What Works and Why. *In ASHE Higher Education Report*, Vol. 40, No. 6, pp. 1-114.

Parks, R.F. et al., (2018). Harnessing Business Analytics: Analyzing Data Analytics Programs in U.S. Business Schools. *In Information Systems Education Journal*, Vol. 16, No. 3, pp 15-25.

Paul, J.A. and MacDonald, L. (2020). Analytics Curriculum for Undergraduate and Graduate Students. *In Decision Sciences Journal of Innovative Education,* Vol. 18, No. 1, pp. 22 – 58.

Sebastianelli, R. and Tamimi, N. (2011). Business Statistics and Management Science Online: Teaching Strategies and Assessment of Student Learning. *In Journal of Education for Business,* Vol. 86, No. 6, pp. 317-325.

THE TUTOR'S PERCEPTION ON THE ACTIVITY OF A STUDENT IN GROUP ACTIVITY IN REMOTE MODE

Bertil P. Marques[1], Marílio Cardoso[2] and Rosa M. Reis[1]
[1]*GILT/ISEP/IPP, Portugal*
[2]*SIIS/ISEP/IPP, Portugal*

ABSTRACT

This paper seeks to respond to one of the difficulties founded by a tutor in a situation of monitoring a group activity at a distance activity (remote mode), namely the perception of the individual activity of the students in the group activity. Based on a literature study on remote collective activity scenarios and group dynamics, the authors propose ways to facilitate the tutor's perception of individual activity. These means are based on a perception of the individual and group productions of the students, collective discussions, social behaviors as well as the sociometric analysis of the group. The authors confront these proposals for the real needs of the tutors of the group activities through interviews conducted with eight experts.

KEYWORDS

Open and Distance Learning, Group Learning, Tutoring, Student Monitoring

1. INTRODUCTION

When monitoring a collective activity online, the tutor is faced with difficulties related to the distance situation. In particular, the tutor needs to find the individual progress of the students involved in a distributed project in such a way as to support everyone in the collective task (Berger and Rieben, 2000). This work focuses on a twofold dimension of monitoring collective activity:

1) monitoring of the collective activity itself, that is, in order to achieve the objectives envisaged;
2) individual monitoring of each student involved in the collective activity.

Following each student individually assume everything, perceiving each student's activity beyond the group's activity. The task of perception is central because it is from the perception of the activities that the tutor can help the group and guide the students individually. The work is related to the issue of instrumentation of the tutor's activities to perceive the individual activity of the students, that is, to conceive and develop tools to understand the individual activity. For this, responses were developed to improve the perception of individual activity by the tutor from a study of the literature. The authors also provide information about what the tutor can rely on to understand the students' activity. These proposals were then confronted with the needs of the tutors by conducting interviews.

2. PROPOSALS FOR RAISING AWARENESS OF A STUDENT'S ACTIVITY IN A GROUP

In this section, we present the study of the literature that led to make proposals to facilitate the perception of a student in group activity. That proposals focused on ways to highlight:

1) the contribution of the student in the group activity through the study of various scenarios of the collective activity;
2) the involvement of a student in the functioning of a group, studying the work related to the analysis of social networks and group dynamics.

2.1 The Student's Contribution to the Group Activity

The study highlights that collective activity scenarios are often based on an organization in individual and collective contexts (Gaudioso *et al.* 2003). The individual phase allows the student to reflect on the problem and form an opinion that he can defend in the collective phase (Stahl, 2000), a reflection that enriches discussions and collective productions. The collective step consists of reinvesting what was produced by each student during the individual step. The discussion has a central focus since it allows the group to agree on the content of the production or on the solutions to its problem (Baker and Lund, 1997).

Through the previous study (Gaudioso *et al.* 2003), we see that a student is led to produce individually, that the group is brought in to discuss before producing collectively, and that these discussions can focus on the initial individual productions. Therefore, to perceive the activity of a student in the context of collective activity, we propose to focus on individual productions, discussions, and collective productions. The tutor can perceive the collective work through the consulting of the collective productions and the discussions. He can perceive a student's individual work by consulting their individual productions, their contributions to collective productions, and their participation in discussions. We believe that it is possible to offer the tutor ways to bring these different productions closer together, in order to take into account and see the strong dialectic that exists between individual and group monitoring.

2.2 Involvement of a Student in a Collective Activity

More and more works are focused on the analysis of social networks to better understand the process of collaboration (Reffay and Chanier, 2003). The analysis of social networks is a way to identify the roles played by members of the group (Beck et al., 2003). It seems to us, therefore, that one of the means of praising the activity of a student in a group is to perceive his place in this group in the social network. Currently, the analysis of group dynamics is often quantitative: information about the density of interactions and the centralization of interactions. This quantitative information is usually represented under graphs to provide a synthetic view to the tutor. The authors understand that while this type of quantitative information is a good point for the tutor, to understand the collaborative process is not enough. Other works concern the qualitative study of group dynamics. The first trend that seems interesting, concerns the study of social behaviors, and in particular, work in ethnology. After a series of studies in the field, with students, (Plety, 1996) four profiles of social behaviors were highlighted according to the volume of intervention, the type of intervention, communicative gestures, and reactions of other students.

The four social behaviors proposed are the Facilitator, the Auditor, the Collector and the Independent. Based on the work of George and Leroux (2002) a tool was developed to perceive the social behavior of students in distance situations in the context of group discussions. The particularity of the discussion tool is that it is based on a structuring of parallel discussion and the importance of linguistics. This structure makes it possible to determine the automatic pattern of the profiles of the linguistic acts used by the students during their interventions in a discussion. These calculations result in the generation of graphs that present the profiles of the students in this same discussion. A second current that draws attention is the sociometric analysis of a group. This analysis consists of highlighting the interpersonal relationships between the members of the group (Parbelas, 1992). To this purpose, each member must complete a questionnaire. The answers to the questionnaire provide information about how each one perceives the group (who he or she would like to work with, who he or she thinks is the leader of the group, etc.) and how he or she thinks he is perceived by the group. The sociometric analysis highlights the "latent structure" of the group at different times of collective activity: the status of the different members (Favorite, Isolated, Rejected, Solitary, etc.), the structure of the group into micro groups. Sociometric analysis can be represented in the form of graphs (centered on a particular student or not).

The authors could work on qualitative analyses because it is an aspect of group dynamics that may be of interest to the tutor. These analyses propose and offer the tutor points of view which provide him with qualitative information on the group dynamics, and the place of a student in the group, in order to reinforce his perception of the situation of the learner. The sociometric analysis is complementary to that of social behaviours since it offers a picture of students' interpersonal relationships from the perspective of those same students.

2.3 Summary of Proposals to Facilitate the Perception of the Activity

Table 1 represents the synthesis of the proposals to perceive individual activity in the group. This table highlights the need to perceive group activity in order to perceive a student's activity (and vice versa). The two types of perception are complementary to ensure the individual follow-up of the students involved in a collective activity.

Table 1. Proposals to facilitate the perception of a student's activity

Perception of the student in a group	Perception of its activity in an individual spot	Through his individual productions
	Perception of his activity in a collective task	Through his contributions in the collective contributions
		Through his contributions in the discussions
	Perception of his place in the group dynamics	Through his social behavior
		Through his sociometric status

3. COMPARISON OF PROPOSALS WITH THE NEEDS OF TUTORS

The authors brought together eight people with different experiences of collective distance learning mentoring (apprentice acts mixed or distance, apprentice fields act and different group sizes). It now presents the synthesis of interviews according to four types of needs pointed out by the tutors.

• Need to follow the student individually in the group activity.

The tutors said that they need individual information about students when they are in a situation of monitoring the collective activity, and this from the beginning of the collective activity: this allows them to identify the least active students, with more difficulties and the least motivated, to be able to intervene enough from the beginning of the activity with them. Tutors need tools to compare the information provided with their own perception of students. Currently, there is not a wide variety of tools to understand the behavior of an individual in a group, being the information of the quantitative type (tools for the consultation of links that indicate the dates and durations of interaction of a student). Tutors reported that this first level of perception of a student is essential, but not sufficient; A second level of perception of the individual in the group is required, through qualitative information (type of intervention in the exchanges, identification of the role of a student in the group).

• Need to understand group dynamics.

The tutors expressed that the functioning of the group depends on the individual motivation of the students: it is therefore important for the tutor to be able to locate exactly one student in the group dynamics, to approximate their behavior in the group and foster their involvement in productions and discussions. Tutors said they get an idea of the group dynamics from student exchanges but need formalized tools to identify the purpose of their perception, stating that currently distance learning platforms do not provide such tools. In addition, tutors need to notice the evolution of group dynamics during the collective activity.

• Need to understand sociometric analysis.

All tutors interviewed believe that using sociometric information about the group involves risks: disruption of the course of the collective activity, entropy introduced if the students do not know each other well enough to answer such questions, risk for the student to consider his position in the group as an objective.

However, most believe that such information can be useful for the tutor and students, as long as all the risks involved are taken into account. For the tutor, it would be a way of confronting his representation of the group to that of the students, as well as a way of acting with the students who are perceived as dysfunctional in the group. For a student, knowing the sociometric representations of the group would allow him to adopt a reflective attitude about the place he occupies in this group.

• **Need to understand social behaviors.**

On the question of social behaviour profiles, tutors explain that it is difficult to make a clear idea of students' social behaviours and that it may be interesting to provide tutors a rating, even if it is reductive. There, however, the tutors appreciate the idea of having more objective information than is deduced and confront the students with their own perception.

4. EVALUATION

Feedback from tutors reinforced the authors' goal of providing them with tools to perceive a student's activity in a group activity, and thus confront the students' own point of view. In particular, they confirmed the hypothesis of their need to perceive the dynamics of the group and be able to identify the student role in the group, which is in line with the authors' choices regarding the design of tools for visualizing group dynamics from the point of view of their originality and their usefulness to the tutor. Regarding sociometrics, the authors are aware of the risks that the use of remote entails, but it is believed that it is interesting to test this type of tool with experienced tutors. More generally, the interviews reinforced the authors' approach, which is to provide the tutor with information about the contribution and involvement of a student in a group, information that cannot be obtained directly by consulting the students' activity, but which can be useful to better understand the functioning of the student and the group.

5. CONCLUSION AND FUTURE WORK

These proposals to facilitate the perception of the activity will allow the authors to develop a model of perception of the individual activity, which they integrate into a broader model of individual monitoring of the students involved in a collective distance activity. This model serves as the foundation for the development of an IT environment for individual monitoring of students involved in a collective distance activity. The development of these tools is currently a work in progress (now in the development phase of the tools) hoping to move to the testing phase from the beginning of the next academic year with the students of the 2nd year of graduation (Academic Year 2023/2024, L2).

REFERENCES

Baker M., Lund F., 1997 - "*Promoting reflexive interactions in a computer-supported collaborative learning environment*", Journal of Computer Assisted Learning, Vol. 13, n°3, 1997, p. 175-193.

Beck R.J., Fitzgerald W.J., Paukszart B., 2003 - "*Individual Behaviors and social structure in the development of communication networks of self-organizing online discussion groups*", CSCL, Bergen, Norway, p. 313-322.

Berger J-F., Rieben P., 2000 - "*Environnements interactifs d'apprentissage sur Internet – Stratégies de conception et expérimentations*", TICE 2000, Troyes, France, 18-20 octobre, p. 185-194.

Gaudioso E., Santos O.C., Rodriguez A., Boticario J.G., 2003 - "*A proposal for Modeling a Collaborative Task in a Web-Based Collaborative Learning Environment*", International Conference on User Modeling, University of Pittsburgh, Johnstown, USA, p. 70-79.

George S., Leroux P., 2002 - "*An approach to automatic analysis of learners' social behavior during computer-mediated synchronous conversations*", ITS'02, Biarritz, France, p. 630-640.

Parlebas P., (1992) - *Sociométrie, réseaux et communication*, PUF, Paris, 1992.

Pléty R., 1996 - *L'apprentissage coopérant*, Lyon, 1996.

Reffay C., Chanier T., 2003 - "*Mesurer la cohésion d'un groupe d'apprentissage en formation à distance.*", EIAH'2003, Strasbourg, France, p. 367-378.

Stahl G., (2000) - "*A model of collaborative knowledge-building.*", International Conference of the Learning Sciences (ICLS 2000), Mahwah, NJ: Erlbaum, p. 70-77.

THE ROLE OF NEW TECHNOLOGY IN EDUCATION OF PROGRAMMING NOVICES

Mario Konecki[1], Mladen Konecki[1] and Ivana Ogrizek Biškupić[2]
[1]*Faculty of Organization and Informatics, University of Zagreb*
Pavlinska 2, 42000 Varaždin, Croatia
[2]*Algebra University College*
Gradišćanska 24, 10000 Zagreb, Croatia

ABSTRACT

Constant efforts to develop more and better information systems have become a critical factor of successful functioning of virtually all major systems. The world has become highly dependent on well-designed and well-developed information infrastructure. Education systems are constantly struggling to provide sufficient number of information system professionals, especially programmers. However, there is a constant lack of this kind of professionals on the market, and one of possible reasons for this kind of state is the fact that education of programming novices has proven to be extremely challenging and complex task for their educators. In this paper a discussion about the challenges and reasons for these challenges in education of programming novices is given. Possible aiding approach and technological solutions that could impact student motivation and work habits are presented and elaborated.

KEYWORDS

Programming novices, Education, Challenges, Motivation, New technology

1. INTRODUCTION

It is very common for educators to state that it is very hard and challenging to teach programming novices how to create functional programming solutions. Teachers have tried various approaches to this day but the problem of teaching novices how to program has persisted.

Students tend to perceive programming as very hard and abstract content, and frequently they are simply not motivated enough to deal with all challenges. This kind of situation results in the general opinion that programming is hard (Baldwin Kuljis, 2001; Bergin and Reilly, 2005; Gomes and Mendes, 2007; Hanks et al., 2004; Jenkins, 2002; Peng, 2010; Robins et al., 2003) but what is significantly more important is that this kind of situation results is quite high failure and dropout rates (Nikula et al., 2011; Yadin, 2011).

This problem itself is of high importance because lower education success rate results in lower number of highly skilled programming experts that will be able to continue with the rapid development of information systems that are of vital importance for virtually all major real-life systems today.

In order to really understand this problem, it is of great importance to understand all the factors that influence this kind of situation. On one side there are teachers with their own teaching preferences, technological and pedagogical knowledge and skills, characters and teaching approaches.

On the other side there are students who also have their individual learning preferences, characteristics, prior knowledge and skills regarding technology, programming and problem-solving. It is important to not only understand teachers but also students and the interaction that the education process implies. Another aspect that needs to be considered is the nature of the learning subject itself.

Programming is a specific learning matter that is abstract and complex. Because of this it requires a prolonged way of learning, and a lot of practical experience. This learning strategy is in many cases something that students are not used to because of different learning strategy that can be used on most of their other courses.

This means that it is even more time-consuming and energy-consuming for students to adapt to a learning style required to master programming as a skill. In this aspect motivation plays a major role because only motivated students will be willing to invest additional effort to learn this complex skill. Other research has also shown that motivation is of high importance for successful programming education (Kinnunen and Malmi, 2006).

2. CHALLENGES AND POSSIBLE APPROACHES

Many different challenges and questions can be stated when talking about programming education on both teachers' and students' side.

On the teachers' side some of the challenges and questions are:

- What programming language is best suited for programming novices?
- Is the popularity and the level of possible professional application of the programming language of greater importance than the novice-oriented style of programming language?
- Are students' prior skills and knowledge prerequisite for successful programming education, what skills and knowledge, and to what extent?
- Is "one size fits all" approach to teaching programming a suitable approach to education of programming novices?
- What is the best pedagogical approach to teaching programming novices?
- Is programming course content simply too much for some students, and should some elements be reduced?
- How to use technology to help students understand abstract and complex programming concepts?
- How to use technology to motivate students to deal with programming?
- How much practical work should be incorporated into the programming course?
- How to support development of problem-solving skills?
- etc.

As it can be seen, teachers have more than complex situation regarding decisions and effort they need to invest in order to create the best possible educational ecosystem for programming novices.

The situation is equally challenging on the students' side. Some of the challenges and questions are:

- How much programming practice is enough?
- What learning strategy will work best?
- How to visualize and understand abstract programming concepts?
- How to gain competence in problem-solving skills?
- How to manage learning time considering other courses?
- How to overcome fear and anxiety regarding programming?
- How to stay motivated and focused?
- etc.

As can be seen, the students face as much challenges as their teachers. Research has shown that fear of programming is one of key aspects that can form initial attitude of programming novices toward their programming courses (Rogerson and Scott, 2010).

One of the problems is that research has shown that many students who enroll programming courses simply do not have sufficient interest in programming activities. Research has shown the following (Bergin and Reilly, 2005):

- Only 22% of students enroll in programming courses because of interest.
- 40% of students enroll in programing courses because they think it is necessary to do so in order to have a prospective career options.
- As much as 35% of students enroll in programming courses just because they have no choice since these are mandatory courses.

As can be seen, the fact that so many students enroll programming courses not because of genuine interest but because of other reasons is a huge additional challenge for their programming educators. Teachers can use extrinsic motivation measures, but research has shown that intrinsic motivation is of even greater importance (Nikula et al., 2011) and this is something that is very hard and complex to influence.

Teachers have tried to use many programming languages in their search for the most suitable one for novices, and there are many programming languages that have been designed with the programming novices in mind, such as (Smith et al., 2000): Logo, Basic, Pascal, Smalltalk, and HyperTalk.

However, some authors claim that the programming language that would be completely suitable for programming novices does not exist and will never exist (Smith et al., 2000).

Research has shown that one of the problems is the way students are used to learn since most of their former education was based on memorization of facts, not adopting a new skill such as programming (Gomes and Mendes, 2007). This has created a gap between the way programming novices think and the way of thinking needed to understand algorithms.

Some authors claim that this gap is as wide as Grand Canyon (Norman and Draper, 1986). Many students try to learn programming not in a prolonged repetitive manner but in a short period of time (Lister et al., 2004). This results in gaining knowledge about the programming language structure but does little to increase skill level in the domain of algorithmic thinking and problem-solving skills.

When talking about possible solutions for challenges in education of programming novices then many different approaches can be identified (Konecki, 2014):

- Introduce additional programming course prior to introductory programming course that would promote algorithmic way of thinking
- Increase motivation of students for learning programming
- Explain to students that programming is a skill, not merely knowledge
- Introduce elements of constructivism into teaching process
- Introduce learning by example
- Introduce animation and other visualization techniques combined with interaction
- Introduce interactive visual simulations
- Include support for multiple learning styles

As can be seen, many different approaches can be taken but to this day no final solution has been identified. Whether this situation is persisting because of not enough measures being taken or whether the right measure has not yet been invented is something that is yet to be researched and concluded about.

3. NEW TECHNOLOGY IN EDUCATION

One of possible approaches to aiding programming novices in their programming education is usage of new technology to make abstract programming concepts clearer and more comprehensive to students, and what is of great importance is to try increasing students' motivation to learn programming. This is done by giving students tools that are of educative nature but also serve as examples of what can be achieved by development activities. There are many different technologies that can be used in education, and the same is true for the education of programming novices.

Some of already well-known and used technological solutions include: video materials, chats, forums, quizzes, self-assessment tools, presentations, various reading learning content, educative visual programming environments, such as Alice, Kodu, Scratch, and Greenfoot (Jahongir, 2022), visual programming languages (Tsai, 2019).

Interactive animations are also well-known but are less used than formerly mentioned content types.

Some of less known and less-commonly used technological solutions that represent latest forms of learning content are: virtual reality (Christou, 2010), artificial intelligence (Roll and Wylie, 2016), educative videogames (Lee et al., 2004), and gamified educational content (Caponetto et al., 2014).

Potential advantages of mentioned new technologies are: better visualization and representation options of various programming concepts, higher level of immersion and engagement in educative content, and multi-sensory experience of various educational elements.

4. METHODOLOGY AND RESEARCH RESULTS

In order to conclude about possible impact of new technology a research that included 87 information technology students who participated in the research has been conducted. All students have been given three new technology-based educational systems that included the following technologies:

- Virtual reality
- Artificial intelligence
- 3D modeling

Virtual reality has been incorporated in the form of custom virtual reality learning environment (shown in Figure 1) in which students were able to experience different programming concepts in the form of implemented videogame, and they were also able to test their knowledge through developed virtual reality self-assessment module.

Artificial intelligence has been incorporated in the form of virtual assistant (Konecki et al., 2015; Konecki and Kadoić, 2015) (shown in Figure 1) that enabled students to ask different questions related to programming. Virtual assistant has provided answers, along with additional learning content related to the question.

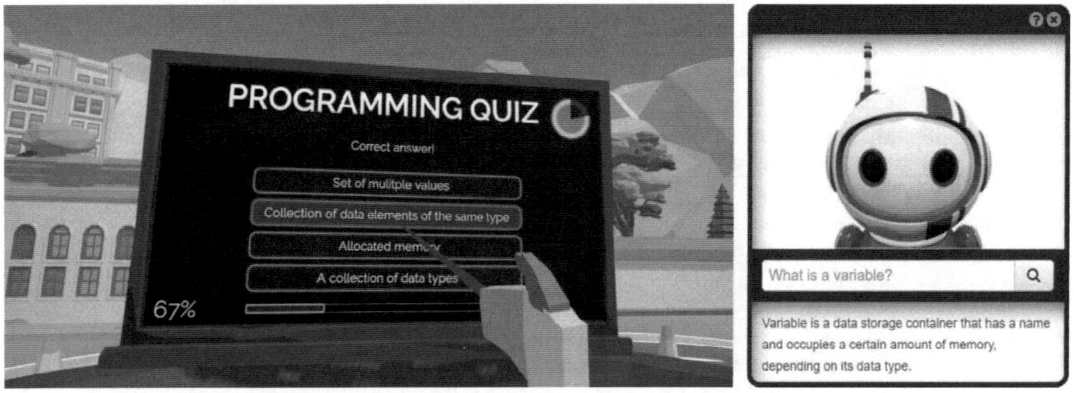

Figure 1. Virtual reality learning environment and virtual assistant

3D modeling has been incorporated in the form of virtual 3D realistic model of the faculty (shown in Figure 2). By using this interactive 3D model students were able to walk through the faculty, and to learn about its structure. Developed interactive 3D model served as a demonstration of what can be achieved by using programming knowledge and skills, but it also provided students with a fun and engaging way to learn about the faculty.

Figure 2. 3D realistic model of the faculty

After mentioned experiences a research in the form of questionnaire has been conducted to assess whether this new technology experience has affected students' motivation and whether they feel more ready to do programming exercises in a more prolonged manner compared to before. In order to do so, research has been conducted in two parts in order to conclude about the difference in their perception and motivation. Likert scale was used ranging from "Strongly disagree" (1) to "Strongly agree" (5). Research results are shown in Table 1 and Table 2.

Table 1. Research results (Students' attitude towards programming education)

R-1.1	Statement	$\bar{x}$	SD
1.	I am motivated to learn programming	3,11	1,05
2.	I am willing to learn programming in a prolonged way	3,01	1,14
3.	I am willing to do all programming exercises that are available to me as a part of my programming course each week	2,95	1,12
4.	I would like to know programming better	3,25	1,00

Results of the first part of the research have shown that students are moderately motivated to learn programing and to put additional effort needed to learn programming as a skill. Slightly more affirmative was the statement about wanting to know programming better.

Table 2. Research results (Students' attitude towards programming education)

R-1.2	Statement	$\bar{x}$	SD
1.	I am motivated to learn programming	3,41	1,02
2.	I am willing to learn programming in a prolonged way	3,32	1,12
3.	I am willing to do all programming exercises that are available to me as a part of my programming course each week	3,30	0,98
4.	I would like to know programming better	3,62	0,82

After the second part of the research was conducted, a paired samples t-test was performed to compare the strength of agreement with the four statements before and after students' experience with new technology in programming education.

There was a significant difference (p=0.015) in being motivated to learn programming before (M = 3.09, SD = 1.06) and after students' experience with new technology in programming education (M = 3.39, SD = 1.04).

The respondents on average were more willing to learn programming in a prolonged way (p=0.026) after engaging in experience with new technology in programming education (M = 3.29, SD = 1.12) than before (M = 2.99, SD = 1.14).

Also, the respondents on average were more willing to do all programming exercises that are available as a part of the programming course each week (p=0.009) after experience with new technology in programming education (M = 2.93, SD = 1.12) than before (M = 3.27, SD = 0.99).

The respondents on average agreed more (p=0.001) with the statement „I would like to know programming better" after experience with new technology in programming education (M = 3.23, SD = 1.02) than before (M = 3.59, SD = 0.87).

5. CONCLUSION

Programming is one of vitally important professions, especially in the context of the modern world and society. However, both programing novices and their teachers experience various difficulties and challenges in their programing courses and education.

In this paper, different challenges that both programing novices and teachers experience in programming education have been presented and discussed, along with the reasons for these challenges. Possible approaches and solutions have also been presented and elaborated. Research results about the impact of new technology on the motivation of programing novices and their working habits have been presented and analyzed.

New technology has proven to be a promising mean of aiding programing novices in their programming education, especially in raising their motivation to learn programming as a skill. More research is needed to conclude about whether these effects will be long lasting and whether the course results will follow the perception and impression of programming novices. This part will be the focus of the future research efforts.

REFERENCES

Baldwin, L. P. and Kuljis, J., 2001. Learning programming using program visualization techniques. *Proceedings of the 34th Annual Hawaii International Conference on System Sciences*, Washington, DC, USA, pp. 1051-1058.

Bergin, S. and Reilly, R., 2005. The influence of motivation and comfort-level on learning to program. *Proceedings of the 17th Annual Workshop on the Psychology of Programming Interest Group*, Brighton, UK, pp. 293-304.

Caponetto, I. et al., 2014. Gamification and education: A literature review. *European Conference on Games Based Learning, Academic Conferences International Limited*, Vol. 1, p. 50.

Christou, C., 2010. Virtual reality in education. *Affective, interactive and cognitive methods for e-learning design: creating an optimal education experience*, pp. 228-243.

Gomes A. and Mendes, A. J., 2007. An environment to improve programming education. *Proceedings of the 2007 international conference on Computer systems and technologies*, New York, USA, pp. 1-6.

Hanks, B. et al., 2004. Program quality with pair programming in CS1. *Proceedings of the 9th annual SIGCSE conference on Innovation and technology in computer science education*, United Kingdom, pp. 176-180.

Jahongir, Z., 2022. The Role of Visual Learning in Improving Students' High-Order Thinking Skills. *Barqarorlik va Yetakchi Tadqiqotlar onlayn ilmiy jurnali*, 2(2), pp. 252-256.

Jenkins, T., 2002. On the difficulty of learning to program. *Proceedings of the 3rd Annual Conference of the LTSN Centre for Information and Computer Sciences*, United Kingdom, pp. 53-58.

Kinnunen, P. and Malmi, L., 2006. Why students drop out CS1 course?. *Proceedings of the 2006 International Workshop on Computing Education Research*, pp. 97-108.

Konecki, M., 2014. Problems in programming education and means of their improvement. *DAAAM international scientific book*, pp. 459-470.

Konecki, M. et al., 2015. MIA: A multi-purpose intelligent assistant. *2015 38th International Convention on Information and Communication Technology, Electronics and Microelectronics (MIPRO)*, pp. 1109-1113.

Konecki, M. and Kadoić, N., 2015. Intelligent assistant for helping students to learn programming. *2015 38th International Convention on Information and Communication Technology, Electronics and Microelectronics (MIPRO)*, pp. 924-928.

Lee, J. et al., 2004. More than just fun and games: Assessing the value of educational video games in the classroom. *CHI'04 extended abstracts on Human factors in computing systems*, pp. 1375-1378.

Lister, R. et al., 2004. A multi-national study of reading and tracing skills in novice programmers. *ACM SIGCSE Bulletin*, 36(4), pp. 119–150.

Nikula, U. et al., 2011. A motivation guided holistic rehabilitation of the first programming course. *ACM Transactions on Computing Education (TOCE)*, 11(4), 24.

Norman D. A. and Draper, S. W., 1986. *User-Centered System Design: New Perspectives on Human Computer Interaction*, Erlbaum, Hillsdale, NJ.

Peng, W., 2010. Practice and experience in the application of problem-based learning in computer programming course. *Proceedings of the International Conference on Educational and Information Technology (ICEIT)*, pp. 170-172.

Robins, A. et al., 2003. Learning and Teaching Programming: A Review and Discussion. *Journal of Computer Science Education*, 13(2), pp. 137-172.

Rogerson C. and Scott, E., 2010. The fear factor: How it affects students learning to program in a tertiary environment. *Journal of Information Technology Education: Research*, 9(1), pp. 147-171.

Roll, I. and Wylie, R., 2016. Evolution and revolution in artificial intelligence in education. *International Journal of Artificial Intelligence in Education*, 26, pp. 582-599.

Smith, A. et al., 2000. Programming by example: novice programming comes of age. *Communications of the ACM*, 43(3), pp. 75-81.

Tsai, C. Y., 2019. Improving students' understanding of basic programming concepts through visual programming language: The role of self-efficacy. *Computers in Human Behavior*, 95, pp. 224-232.

Yadin, A., 2011. Reducing the dropout rate in an introductory programming course. *ACM Inroads*, 2(4), pp. 71-76.

THE INCLUSIVE EDUCATIONAL POTENTIAL OF VIRTUAL REALITY FROM AN ETHNO-RACIAL APPROACH: DEVELOPING DIGITAL SKILLS FOR THE CONTEMPORARY WORLD

Wanderlucy Czeszak[1] and Romero Tori[2]
[1]*Université TÉLUQ, Canada*
[2]*Universidade de São Paulo, Brazil*

ABSTRACT

The theme of this study is about the potential of Virtual Reality in the process of ethno-racial inclusion in Education, as a technological resource to assist in training the individual to use fundamental digital technologies in the contemporary world. This is a basic, exploratory, field and web-based research, with a mixed-methods approach and phenomenological, ethnographic and survey methodologies. Involving university teachers and students, its overall objective is to analyse to what extent Virtual Reality involves pedagogical practices that can add value in the development of inclusive digital skills.

KEYWORDS

Virtual reality, Ethnic and racial inclusion, Education, Pedagogical practices; Digital skills

1. PROBLEMS AND JUSTIFICATION

Developing digital competencies today is something urgent in all age groups and social strata, regardless of any differences between individuals, so that everyone can exercise their role in society and ensure their rights as citizens (Casero Béjar & Sánchez, 2022; Diz-Otero et al, 2022).

Digital resources have become increasingly accessible, either as a result of the popularization of technological artifacts (Tori & Hounsell, 2020) to some extent due to public policies, aimed at expanding access to the internet and training for the development of digital skills for the use of technologies (Sharata et al. 2022; de Cara et al, 2022).

Thus, if we can consider some positive aspect in the consequences arising from the tragic period of the Covid-19 pandemic, it is valid to point out that the need for confinement accelerated a process of virtualization of our daily lives in general (Belda-Medina, 2021). However, the remote learning adopted during confinement further increased the technological gap described by Castells (2005) in developing countries, as in the case of Brazil. Thus, despite the challenges, especially economic ones (Frolova et al., 2021), virtual environments have offered their users opportunities to interact differently from the face-to-face world. Individuals who feel more intimidated because they do not meet social standards, such as those belonging to some racial-ethnic groups, can relate on an equal footing, expressing themselves with more resourcefulness (Stoiciu, 2021). This possibility of reinventing themselves, breaking down barriers, can broaden participation and the construction of knowledge through access to academic and professional training, promoting self-confidence and self-realisation. This study project recognises the problems experienced by ethno-racial groups in Brazil, in a historical scope, which are the indigenous and the afro-descendants (de Oliveira et al. 2021), focusing, however, on the Afro-descendants, due to their more numerically significant representation of the population at the margins of Brazilian society, mistreated by low income and prejudice (Domingos & Nogueira, 2020; de Albuquerque Aquino, 2013). Moreover, the author of this study project is part of this group that, as a teacher and researcher, brings her own professional trajectory as an aggregator repertoire of reflections.

A brief first literature review conducted for the development of this study project (2013-2023) seams to reveal that studies involving issues related to the role of digital skills development in Education as a strategy for ethnic-racial inclusion are scarce. Google Scholar searches using the following formula - digital skills AND education AND racial inclusion AND virtual reality AND technologies AND Metaverse AND second life (in Portuguese); digital skills AND education AND racial inclusion AND virtual reality AND technologies AND Metaverse AND immersive AND second life (in English); compétences numériques ET éducation ET inclusion ET réalité virtuelle ET technologies numériques Et immersif Et second life (in French) - found only three items in Portuguese (two from Portugal and one from Cape Verde), 21 in English and none in French.

About inclusive ethno-racial pedagogical practices in Brazil involving concepts such as virtual reality, immersive resources and metaverse, nothing was found. Therefore, we believe that this project can shed light, providing students with more diverse, autonomous, and immersive educational experiences.

2. OBJECTIVES

Flexibility of time and space are particularly important factors in studies regarding the feasibility and practicality provided by the use of virtual environments in education. However, it is worth to opportunize reflections on other factors that can make viable the Virtual Reality in Education as the ethno-racial inclusion, to the extent that the use of resources such as Augmented Reality, the metaverse and the creation of avatars can promote the re-signification of its users for themselves and for others with whom they interact. Thus, we propose as a hypothesis of this project that Virtual Reality environments can contribute to ethno-racial inclusion, because they can provide their users different types of interaction than those that occur in face-to-face, re-signifying the relationships. In other words, physical aspects can be reinvented in the virtual world, and the way of expressing and relating, either verbally or through actions, can also be reinvented, generating opportunities to highlight behaviours, points of view and intellect, reducing influences related to prejudice and ethno-racial discrimination. The re-signification of ethno-racial differences from Virtual Reality may imply the deconstruction of the hierarchy of values that brings transformations in relationships in hybrid and face-to-face environments.

Thus, this study aims to reflect on the inclusive educational potential of Virtual Reality from a racial-ethnic approach, analyse the interaction between students and their peers and between students and teachers in virtual learning environments, identify how digital skills can be developed by individuals of African descent through the adoption of Virtual Reality resources, based on pedagogical strategies such as a) representativeness: creation of virtual environments and avatars that reflect the diversity of the students' ethnic-racial origins; b) building empathy: using virtual reality to create immersive experiences that allow students to see the world from the perspective of someone from a different ethnic-racial background; c) tutoring: virtual guidance of individuals from different ethnic-racial backgrounds; d) self-knowledge: promotion of reflection on their own identities and cultural origins; e) collaboration: collaborative virtual projects that bring together students from different ethnic-racial backgrounds.

3. METHODOLOGY

This research is based on Garrison, Anderson and Archer's (2000) CoI (Community of Inquiry Model), which focuses on the learning context and interactions that construct learning and specify social presence (learners identify their social and emotional presence during learning), cognitive presence (learners can establish the meaning of the learning experience) and teaching presence (instructors design, organise and facilitate the learning content).

A literature review will be conducted for the conceptualization and the relationship between the terms present in this study: Virtual Reality, Augmented Reality, Avatar, Immersive Process, Metaverse, Second Life, Digital Skills and ethno-racial inclusion. We intend, therefore, through the development of this research, to build knowledge, seeking to contribute with reflections on relationships still little explored between ethnic-racial approaches and Virtual Reality. Thus, the monitoring of the pedagogical activities of the supervisor of this project, as well as the discussions and interactions between students and their peers and students and teacher, will be fundamental to the reflections and the realization of this study.

Through the content analysis of Bardin (2016), in a qualitative and quantitative perspective, fieldwork will be conducted on the internet, through the analysis of the interactions occurring in virtual learning environments of courses taught by the supervisor of this project.

4. THEORETICAL FOUNDATION

New forms of sociability of cyberspace in virtual communities offer opportunities for remodelling the identity itself (Stoiciu, 2021) and re-signification of the subjects, especially in Virtual Reality environments, with or without the creation of avatars. This is a process that can serve students belonging to ethnic-racial minority groups conditions to reinvent themselves, and participate in discussions, free from the stigma of prejudice and discrimination arising from the standards imposed by society.

According to the research of Lorenzo et al (2012), in Spain, on the effectiveness of immersive multiuser environments for collaborative assessment tasks, the association with the community created in a virtual social space can neutralize isolation and may have a positive impact on learning outcomes, while the development of synchronous communication skills can promote an effective collaboration potential. For Tori (2022), the perception provided using the metaverse of being with other participants in the same space, even if virtual, makes us feel close and inserted in the same environment, generating important gains to the extent that it produces the sense of belonging and reduction of transactional distance. Koechli & Glynn (2014) show the effectiveness of an online role-playing educational website used in some classes at Ryerson University, Canada, demonstrating that Virtual Reality can serve as a tool to support learning.

The study by Rogerson-Revell et al (2012) in the UK on the uses of voice board and virtual world (Second Life) concludes that these resources can bring positive aspects such as reduced isolation, increased personalisation, improved support and feedback, as well as increased engagement, adding variety to the learning materials. Tori (2022) stresses the importance of the experience with Second Life, as an immersive virtual environment that is a precursor to other lighter and easier to navigate environments that have greatly enhanced this initial experience.

In a distance education aid program for First Nations (native) students at the Université du Québec analyzed by Rocheleau & Pouliot (2022), opportunities to collaborate with peers increased their sense of belonging to the group. The authors observed that team evaluations reduced the anxiety of the participating female students, allowing them to pool their strengths to meet the demands of the course. Being able to collaborate with people from different communities and with varied personal and professional experiences might allow the students to understand diverse realities.

The literature review in Dixon-Saxon & Buckley's (2020) study on how to support higher education student success in distance counselling in selection, development and retention suggests that several factors can enable student retention in online counsellor training: the admissions policy (broad access), the development of student and faculty digital competence; the creation of learning communities; faculty presence and engagement (quick feedback/comments, relevant and targeted opportunities, direct student follow-up, etc.). The authors propose the creation of a learning community to combat the feeling of isolation.

For Knittel et al (2022), the development of digital competences seems to be one of the main points for a more significant and successful support to the student, although in the Brazilian context, as well as in other developing countries, it is first necessary to invest in access to internet and quality equipment. Moreover, both in developing countries and around the world, well-planned teacher training can result in the adoption of teaching strategies that promote interactive and inclusive activities.

5. EXPECTED RESULTS

Through the development of this project, our aim is to contemplate inclusive pedagogical practices that offer ample opportunities for enhancing digital skills among students. Our goal is to foster an awareness of the importance of self-regulation and prepare them to become well-rounded citizens equipped to tackle the challenges of the contemporary world. In this era of growing prevalence of digital technological resources, virtual environments, virtual reality, and the concept of the metaverse, we strive to empower students with the necessary tools for success. Thus, we hope to comprehend the new forms of sociability that might be developed in cyberspace, and how the use of digital technological resources can provide students with situations in which they can express themselves and debate as equals with their peers, regardless of their ethnic-racial characteristics. Moreover, it is appropriate to reflect on how ethno-racial aspects, in a society as plural and diverse as the Brazilian one, can add value, contributing to awareness, reflection and inclusion and assisting in the deconstruction of prejudiced and discriminatory positions.

REFERENCES

Bardin, L. (2007). *L'analyse de contenu.* Presses Universitaires de France, Paris.

Belda-Medina, J. (2021). Enhancing multimodal interaction and communicative competence through task-based language teaching (Tblt) in synchronous computer-mediated communication (scmc). *Education Sciences*, *11*(11). Scopus. https://doi.org/10.3390/educsci11110723

Casero Béjar, M. O., & Sánchez Vera, M. M. (2022). Change from face-to-face to virtual mode during confinement due to covid-19: Perceptions of university students. *RIED-Revista Iberoamericana de Educacion a Distancia*, *25*(1), 243–260. Scopus. https://doi.org/10.5944/ried.25.1.30623

Castells, M. (2005). *A sociedade em rede.* Paz e Terra, São Paulo.

De Albuquerque Aquino, M. (2013). *A inclusão afrodescendente na era da informação | RDBCI: Revista Digital de Biblioteconomia e Ciência da Informação.* https://periodicos.sbu.unicamp.br/ojs/index.php/rdbci/article/view/1638

De Cara, G. R. B., do Carmo, G. T., & Bastos, D. F. (2022). Education and Innovation: Challenges for Teacher Continuing Education in Light of Post-pandemic Demands. *Smart Innovation, Systems and Technologies*, *256*, 803–811. Scopus. https://doi.org/10.1007/978-981-16-5063-5_65

De Oliveira, E. C., Bertrand, P., Roel Lesur, M. E., Palomo, P., Demarzo, M., Cebolla, A., Baños, R., & Tori, R. (2016). Virtual Body Swap: A New Feasible Tool to Be Explored in Health and Education. *2016 XVIII Symposium on Virtual and Augmented Reality (SVR)*, 81–89. https://doi.org/10.1109/SVR.2016.23

Dixon-Saxon, S., & Buckley, M. R. (2020). Student Selection, Development, and Retention: A Commentary on Supporting Student Success in Distance Counselor Education. *Professional Counselor*, *10*(1), 57–77. ERIC.

Diz-Otero, M., Portela-Pino, I., Domínguez-Lloria, S., & Pino-Juste, M. (2022). Digital competence in secondary education teachers during the COVID-19-derived pandemic: Comparative analysis. *Education and Training*. Scopus. https://doi.org/10.1108/ET-01-2022-0001

Domingos, L. T., & Nogueira, R. M. G. P. (2020). *Desafios da Presença de Conhecimentos e Saberes Africanos e Afrodescendentes*. Editora Appris.

Frolova, E. V., Rogach, O. V., Tyurikov, A. G., & Razov, P. V. (2021). Online Student Education in a Pandemic: New Challenges and Risks. *European Journal of Contemporary Education*, *10*(1), 43–52.

Garrison, D. R., Anderson, T., & Archer, W. (2001). Critical thinking, cognitive presence, and computer conferencing in distance education. *American Journal of Distance Education*, *15*(1), 7–23. https://doi.org/10.1080/08923640109527071

Knittel, T. F., Mattar, J., Czeszak, W. A. A. C., & Oliveira, N. A. A. de. (2022). A School Selfie: Digital competences and interaction with students during emergency remote teaching. *International Journal for Innovation Education and Research*, *10*(8), Article 8. https://doi.org/10.31686/ijier.vol10.iss8.3754

Koechli, L., & Glynn, M. (2014). Diving into Lake Devo: Modes of Representation and Means of Interaction and Reflection in Online Role-Play. *International Review of Research in Open and Distributed Learning*, *15*(4), 43–69. https://doi.org/10.19173/irrodl.v15i4.1528

Lorenzo, C.-M., Ángel Sicilia, M., & Sánchez, S. (2012). Studying the effectiveness of multi-user immersive environments for collaborative evaluation tasks. *Computers & Education*, *59*(4), 1361–1376.

Rocheleau, J., & Pouliot, E. (2021). Un concept infopédagogique innovant pour soutenir la motivation des étudiantes autochtones dans leurs études universitaires. *Éducation et francophonie*, *49*(1), 170–190. Érudit. https://doi.org/10.7202/1077007ar

Rogerson-Revell, P., Nie, M., & Armellini, A. (2012). An Evaluation of the Use of Voice Boards, E-Book Readers and Virtual Worlds in a Postgraduate Distance Learning Applied Linguistics and TESOL Programme. *Open Learning*, *27*(2), 103–119.

Sharata, N. H., Berezovska, T. V., & Kravchenko, T. P. (2022). Efficiency of the Method of Student Thematic Online Debates in the Development of Professional Competences in Future Specialists. *WSEAS Transactions on Environment and Development*, *18*, 191–203. Scopus. https://doi.org/10.37394/232015.2022.18.21

Stoiciu, G. (2021). Les Défis Épistémologiques De L'anthropologie Du Monde Virtuel: the epistemological challenges of the anthropology of virtual world. *Revista Romana de Sociologie*, *5/6*, 329–342.

Tori, R. (2022). *Metaversos e Avatares: O que são? Por Onde Andavam? Por Que Voltaram?* Bett Brasil 2022 - Portuguese. https://brasil.bettshow.com/bett-blog/metaversos-e-avatares-o-que-sao-por-onde-andavam-por-que-voltaram

Tori, R., & Hounsell, M. da S. (2020). *Introdução a Realidade Virtual e Aumentada | Sociedade Brasileira de Computação*. https://sol.sbc.org.br/livros/index.php/sbc/catalog/book/66/291/540-1

RELATIONSHIP BETWEEN ONLINE AND OFFLINE ATTENDANCE AND ACADEMIC PERFORMANCE

Victor K. Y. Chan
Macao Polytechnic University
Rua de Luís Gonzaga Gomes, Macao, China

ABSTRACT

This paper empirically investigated the relationship between students' online and offline attendance and their academic performance. A data sample of students' online and offline lecture attendance rates and their academic performance was collected from the author's university in the academic year 2022/2023 in Chinese-culture-oriented Macao, covering four senior undergraduate courses of disciplines ranging from software engineering to management. Students' academic performance in a course was assessed by their final examination scores or their coursework scores in the course, depending on whether the course concerned included any final examination. It was found that students' both online and offline attendance rates were positively associated with their academic performance, and students' both online and offline attendance rates predicted or, more specifically, improved their academic performance after controlling for the effect of each other. These findings offer guidance to universities, educators and students on the importance of lecture attendance.

KEYWORDS

Online attendance, Offline attendance, Academic performance, Examination, Coursework

1. INTRODUCTION

In offline education research, the impact of students' lecture (or class) attendance on their academic performance was found positive in most contexts (Millis, Dyson and Cannon, 2009; Fadelelmoula, 2018). However, in quite some studies, such positive impact was not statistically significant (Fadelelmoula, 2018). In some other recent studies, such positive impact was even found no longer existing (Kauffman *et al.*, 2018). In fact, some refined studies revealed that this impact was mediated by students' in-class participation level but not direct (Kim *et al.*, 2020).

Nonetheless, in recent years when quite some traditional offline educational paradigms are migrated to their online counterparts, controversies were found in some settings. Related studies on the interplay between online attendance and academic performance are hitherto so substantially rarer than those on corresponding offline attendance that no general results have been concluded. For example, Nieuwoudt (2020) established online attendance and academic performance varying positively whilst Diab-Bahman *et al.* (2022) discovered that such variation was negative. The former was specific to an enabling (non-mainstream) programme in an Australian university. In contrast, the latter was confined to junior and thus generic business management courses (for first- and second-year undergraduates) and an Arabic country/region. Therefore, there are still room for further work to extend such investigations to courses and programmes in other disciplines at more senior levels and in other cultural contexts. For instance, more senior courses or programmes other than pure business management ones in East Asian universities are candidates for such further work.

Amidst the COVID pandemic in the last few years, universities almost worldwide probably reluctantly resorted to online lecturing as Hobson's choice almost regardless of the courses' natures. Moreover, during the pandemic, it was not uncommon to see the same course's lectures being delivered partially online and partially offline, depending on the pandemic's development on different dates of the various lectures. This set the stage for collection of empirical data on both online and offline attendance for same courses, enabling comparison between these two paradigms of attendance in affecting academic performance. In particular, this study focused on a blend of software engineering and management courses administered to students of predominantly the Chinese culture and sought to uncover such relationship between students' online and offline attendance and their academic performance. It also compared online and offline attendance in their impacts on academic performance.

2. THE PRESENT STUDY

2.1 Materials

For the study, an empirical data sample was collected from the Faculty of Business at the author's university in the city of Macao[1] (Special Administrative Region), China during the first semester of the academic year 2022/2023, the semester's lecturing lasting from 19 August 2022 to 24 November 2022. The sample embraced the courses entitled "System Analysis and Design" and "E-commerce Project Management" and two courses entitled "Operations Management", all at the senior undergraduate year 3 level. The numbers of students in the four courses were respectively 20, 20, 19 and 19. Please note that both Macao local and mainland Chinese students in the courses are substantially Chinese-culture-oriented (Liu *et al.*, 2021).

It was a local legal mandate that attendance be taken for each student in each lecture at the author's university. Such records were stored on the university's specialized attendance information system, from which the corresponding data was retrieved for the study. In that extraordinary semester during the COVID pandemic, an appreciable proportion of the lectures were delivered online and the rest offline in traditional, physical venues, the exact proportion varying across courses in accordance with the exact timing of each course's lectures against the backdrop of the pandemic's development in society over time and thus the government's decrees on pandemic control that varied from time to time. The online lectures were conducted by the professor-in-charge hosting Zoom (*Zoom Online Video Conferencing | Zoom*, no date) sessions to which students were invited to join over the Internet through an "Assignment" on the Canvas learning management system (*Canvas LMS | Learning Management System | Instructure*, no date) of the university.

2.2 Methods

The attendance and academic performance assessment scores of the students in the above four courses were pooled together to constitute the data sample for investigation. For attendance, an online attendance rate (r_n) and an offline attendance rate (r_f) were calculated for each student in each course as follows:

$$r_n = \frac{\text{Total number of online lectures attended by the student for the course}}{\text{Total number of online lectures for the course}}$$

and

$$r_f = \frac{\text{Total number of offline lectures attended by the student for the course}}{\text{Total number of offline lectures for the course}}.$$

A Pearson correlation coefficient $\rho_{n,p}$ was computed between the online attendance rates r_n and the academic performance (i.e. the final examination scores or the overall coursework scores, as applicable) across all cases in the above data sample. Another correlation coefficient $\rho_{f,p}$ was calculated between the offline attendance rates r_f and the academic performance across all cases in the above data sample. The higher such a correlation coefficient, the stronger and positive the association between the online (or offline) attendance rate (r_n or r_f) and the academic performance for a student is. The corresponding z-statistics to test (two-tailed) the null hypotheses H_0: $\rho_{n,p} = 0$ and H_0: $\rho_{f,p} = 0$ were also calculated in order to verify with statistical significance whether the online (or offline) attendance rate (r_n or r_f) and the academic performance were actually statistically correlated based on all cases in the above data sample.

[1]"Macao" is sometimes also known as "Macau".

To further compare the online and offline attendance rates (r_n and r_f) as predictors of the academic performance of a student, regression as presented by the following equation was conducted:

$$p = \beta_0 + \beta_n r_n + \beta_f r_f + \epsilon \quad (1)$$

where p = the academic performance of the student in the corresponding case, i.e., the final examination score or the overall coursework score, as applicable, of the student;

ϵ = the residual in the corresponding case of the student; and

β_0, β_n and β_f were the constant regression coefficient to estimate.

Please note that in a very rigorous and formal sense, variables p, r_n, r_f and ϵ in equation (1) should be subscripted by, for example, *i*, which indicates each case *i* in the data sample. For the sake of simplicity, this subscript is omitted and understood.

The coefficient of determination (R^2) was calculated in order to estimate the proportion of the variation of p being explained the variations of r_n and r_f. The higher R^2, the better r_n and r_f predicted p. The goodness-of-fit F-statistic to test the null hypothesis H_0: $\beta_n = \beta_f = 0$ was also calculated so as to verify whether either r_n or r_f or both predicted p. The t-statistics to test the individual null hypotheses H_0: $\beta_n = 0$ and H_0: $\beta_f = 0$ was also computed so as to verify whether r_n and r_f individually predicted p.

3. RESULTS

Table 1 enumerates the Pearson correlation coefficients $\rho_{n,p}$, $\rho_{f,p}$ and their 95% confidence intervals (CIs) together with the corresponding z-statistics detailed in section 2.2 above and their p-values (two-tailed).

Table 1. $\rho_{n,p}$, $\rho_{f,p}$ and their 95% confidence intervals (CIs) together with the corresponding z-statistics and their p-values (two-tailed).

Pearson Correlation coefficient	95% CI	z (p-value)
$\rho_{n,p} = 0.483$	0.2876 to 0.6398	> 3.62 (0.000011**)
$\rho_{f,p} = 0.449$	0.2472 to 0.6134	> 3.62 (0.000054**)

** $p < 0.01$

Both the Pearson correlation coefficients $\rho_{n,p}$ and $\rho_{f,p}$ for this study's data sample as well as their 95% CIs were positive, and both of them in the population differed from zero ($p < 0.01$). It could be concluded with statistical significance that in the population, both the online and offline attendance rates of a student were statistically and positively correlated or associated with his/her academic performance. In other words, whether online or offline, a higher attendance rate was associated with better academic performance. Moreover, given that $\rho_{n,p} > \rho_{f,p}$, as far as the data sample was concerned, the online attendance rate had a stronger association with academic performance than the offline attendance rate did. In other words, a higher online attendance rate had a stronger effect on better academic performance than a higher offline attendance rate did. However, as the 95% CIs for $\rho_{n,p}$ and $\rho_{f,p}$ overlapped, there was no evidence to conclude that in the population, the online attendance rate had a stronger association with academic performance than the offline attendance rate did.

For the regression corresponding to equation (1), In view of R^2 being equal to 0.361, the variations of r_n and r_f explained 36.1% of the variation of p. The F-statistic was statistically significant ($p < 0.01$), so the online or offline attendance rates of a student or both predicted his/her academic performance. Also, both the t-statistics for β_n and β_f were statistically significant ($p < 0.05$) after controlling for each other. Therefore, in a sense, both the online attendance rate and the offline attendance rate of a student predicted his/her academic performance. Also, the CIs for both β_n and β_f did not span the zero point, so it could be confirmed with statistical significance that both a higher online attendance rate and a higher offline attendance rate tended to boost academic performance after controlling for the effect of each other.

Table 2 depicts the estimates of β_0, β_n, β_f, their 95% CIs, their two-tailed *t*-statistics (alongside their *p*-values and degrees of freedom), the R^2 and the *F*-statistic (alongside its *p*-value and degrees of freedom).

In view of R^2 being equal to 0.361, the variations of r_n and r_f explained 36.1% of the variation of p. The *F*-statistic was statistically significant ($p < 0.01$), so the online or offline attendance rates of a student or both predicted his/her academic performance. Also, both the *t*-statistics for β_n and β_f were statistically significant ($p < 0.05$) after controlling for each other. Therefore, in a sense, both the online attendance rate and the offline attendance rate of a student predicted his/her academic performance. Also, the CIs for both β_n and β_f did not span the zero point, so it could be confirmed with statistical significance that both a higher online attendance rate and a higher offline attendance rate tended to boost academic performance after controlling for the effect of each other.

Table 2. Estimates of β_0, β_n, β_f, their 95% confidence intervals (CIs), their *t*-statistics, the R^2 and the *F*-statistic for the regression of equation (1)

		t (*p*-value), df
β_0, CI	-9.572, -37.918 to 18.775	-0.673 (0.503), 72
β_n, CI	50.012, 26.519 to 73.505	4.244 (0.000**), 72
β_f, CI	53.462, 25.416 to 81.507	3.800 (0.000**), 72
R^2	0.361	
F (*p*-value), df	20.358 (0.000**), (2, 72)	

** $p < 0.01$
df = degrees of freedom

4. CONCLUSION

Against the background of the Chinese culture of the students concerned and the omnibus disciplines of the senior undergraduate courses included, both online and offline attendance were found in this study to be positively associated with academic performance, this discovery being in line with Nieuwoudt (2020) and various education theories, for instance, the agency theory (Westerman *et al.*, 2011).

In fact, both the online attendance and the offline attendance of a student could predict or, more specifically, improved his/her academic performance after controlling for the effect of each other.

This study is not without its critics. First, the portfolio of course disciplines, though multi-faceted, could still be extended. Second, the statistical results could have been more enlightening if the number of students in the study was increased.

Having said this, this study affirms that attendance, be it online or offline, is to the students' benefit in terms of their academic performance. These findings offer guidance to universities, educators and students on the importance of lecture attendance.

REFERENCES

Canvas LMS | Learning Management System | Instructure (no date). Available at: https://www.instructure.com/en-au/canvas (Accessed: 17 March 2023).

Diab-Bahman, R. *et al.* (2022) 'The effect of attendance on student performance: implications of using virtual learning on overall performance', *Journal of Applied Research in Higher Education*, 14(3), pp. 1175–1192. Available at: https://doi.org/10.1108/JARHE-04-2021-0135.

Fadelelmoula, T. (2018) 'The impact of class attendance on student performance', *International Research Journal of Medicine and Medical Sciences*, 6(2), pp. 47–49. Available at: https://doi.org/10.30918/IRJMMS.62.18.021.

Kauffman, C.A. *et al.* (2018) 'Relationship between classroom attendance and examination performance in a second-year medical pathophysiology class', *Advances in Physiology Education*, 42(4), pp. 593–598. Available at: https://doi.org/10.1152/advan.00123.2018.

Kim, A.S.N. *et al.* (2020) 'Understanding the impact of attendance and participation on academic achievement.', *Scholarship of Teaching and Learning in Psychology*, 6(4), pp. 272–284. Available at: https://doi.org/10.1037/stl0000151.

Liu, M.T. *et al.* (2021) 'Taking a break is for accomplishing a longer journey: hospitality industry in Macao under the COVID-19 pandemic', *International Journal of Contemporary Hospitality Management*, 33(4), pp. 1249–1275. Available at: https://doi.org/10.1108/IJCHM-07-2020-0678.

Millis, R.M., Dyson, S. and Cannon, D. (2009) 'Association of classroom participation and examination performance in a first-year medical school course', *Advances in Physiology Education*, 33(3), pp. 139–143. Available at: https://doi.org/10.1152/advan.00028.2009.

Nieuwoudt, J.E. (2020) 'Investigating synchronous and asynchronous class attendance as predictors of academic success in online education', *Australasian Journal of Educational Technology*, 36(3), pp. 15–25. Available at: https://doi.org/10.14742/AJET.5137.

Westerman, J.W. *et al.* (2011) 'The Relationship Between Undergraduate Attendance and Performance Revisited: Alignment of Student and Instructor Goals', *Decision Sciences Journal of Innovative Education*, 9(1), pp. 49–67. Available at: https://doi.org/10.1111/j.1540-4609.2010.00294.x.

Zoom Online Video Conferencing | Zoom (no date). Available at: https://explore.zoom.us/en/lp/video-conferencing/?creative=649288063628&keyword=webconferenceplatforms&matchtype=e&network=s&device=c&zcid=6095&utm_id=19735288070_649288063628&gclid=Cj0KCQjwn9CgBhDjARIsAD15h0B9FdK_h6JAkHpGIK60d6ylhxW1MraG0S380gFnavqaAEo (Accessed: 17 March 2023).

INVESTIGATION OF INTERACTIVE VIDEO-WATCHING PATTERNS AND VIEWS BY CONTENT TYPE

Cansu Gokrem[1] and Emine Sendurur[2]
[1]*Turkish Ministry of National Education*
Istanbul, Turkey
[2]*Ondokuz Mayis University*
Samsun, Turkey

ABSTRACT

The interactive features embedded into videos have the potential to improve engagement. This study aims to investigate this engagement from different content types: academic vs. non-academic. First, the interaction patterns were examined through logs and screen recordings. Second, they were interviewed to have insights into views about the contents. Forty high school students were involved in the study. The research design of the study is an explanatory sequential mixed design. Participants were exposed to two types of interactive video content. The interactive features and design principles did not vary across videos. The findings indicated that students have smoother video-watching experiences in non-academic content and are more successful in scores collected during the interactions. In general, students expressed their views positively, but some found academic content too challenging.

KEYWORDS

Video analytics, Interactive videos, Multimedia learning

1. INTRODUCTION

People can access different forms of information, and videos are famous for delivering data ranging from daily routines to scientific studies. This is an excellent opportunity to be informed about the current trends. Still, excessive irrelevant information might overwhelm people due to unfocused/irrelevant/poorly-organized content (Mohammed et al., 2018). In other words, the user might not constantly be exposed to what s/he is looking for; instead, their cognitive load increases. Therefore, the content's type and quality might affect users' interaction with digital materials like videos.

The videos offer audio-visual content; thus, the elements' design might require special attention. For example, the absence of textual and narrative aspects of the same content might cause increased cognitive load (Clark et al., 2011). In traditional videos, the flow is linear, but the interactive ones can provide non-linear and response-based flow in addition to customized content (Delen et al., 2014). Clicking, answering questions, selecting paths, speeding, skipping, etc., are some features of interactive videos. These features can serve as the basis of video analytics to keep records of learners. Recording user behaviors via interactive features is one of the common ways of keeping users' logs in the literature (Tacchino et al., 2021). In education, video analytics were used to investigate the instruction process (Yeun et al., 2020); and the role of interactive features (Kolås, 2015; Sancheti et al., 2022).

The studies focusing on interactive videos generally tend to investigate the effects of various types of interactions on specific variables, and studies focusing on behavioral patterns were limited in number (Beloufa et al., 2019; Martins et al., 2019). This study aims to contribute to this line of literature, explicitly emphasizing perceptions of learners varied by content type. One concern of this study is based on the quality of the videos. That's why the cognitive load guidelines were used to design the interactive videos.

According to the cognitive load theory, being a novice or expert in the content can affect cognitive load (Clark et al., 2011). Based on this assumption, this study investigates if similar patterns exist when the content type differs (academic vs. every day). The interactive features embedded into videos have the potential to transform passive watching into active engagement (Kolås, 2015), but it might not guarantee the same effects across subject domains. That's why the domains in this study split into two: academic vs. non-academic. To understand the differences in their views and interaction patterns, the following research questions were investigated:

1. How do students' interaction patterns differ across two different content types?
2. How do students' views vary across two different content types regarding difficulty?

2. METHOD

This study is designed as a mixed study, and the research pattern is an explanatory sequential hybrid design (Creswell, 2012). The participants' logs, including session duration, scores, number of pauses, and so on, were collected in the quantitative part of the study. In the following stage, the interviews were conducted. The study took place at a vocational high school. The target population was 11th graders who had not registered for any computer science course. The participants (N=40) were included in the study by convenience sampling method. They were all exposed to two interactive videos: academic vs. non-academic. The educational content was selected among computer science course topics, namely "informatics ethics." The non-academic content was selected among popular computer science topics, namely "metaverse." The duration of the videos was very close to each other (app. five min.), and the design principles followed were the same.

The embedded interactive features (eight questions of the same format) were identical for both videos. The feedback for the answers was designed in the same manner. Adobe After Effects were used to produce videos, and the H5P Moodle plugin was used to embed interactions. Then, they were converted into SCORM packages and uploaded to the SCORM Cloud platform to keep records of logs. The logs and screen recordings were collected during video-watching sessions. The quantitative data were analyzed descriptively to compare the interaction patterns across videos. The interviews were conducted at the end of each session. The qualitative data were analyzed using by content analysis technique.

3. FINDINGS

3.1 Interaction Patterns

The analysis of logs and screen recordings was examined to understand the patterns of interaction. Since the data is quite massive and the researchers continue to analyze the screen recordings, a few interaction features were included for this short paper. The *duration* of the videos was approximately 5 minutes, but it became extended as there were interactive questions. The overall time depends on the user because the user can give the correct answer, move on, give the wrong answer, and either try again, view the answer, or prefer not to answer. Compared to academic and non-academic videos, the session duration was recorded longer for educational content ($M_{academic}$=12 min. 43 sec., $M_{non\text{-}academic}$=9 min. 41 sec.). The *scores* collected during the session were higher for non-academic content than the academic one ($M_{academic}$=60.94, $M_{non\text{-}academic}$=75.31). The average *correct* answers were recorded higher in non-academic content out of eight questions ($M_{academic}$=4.9, $M_{non\text{-}academic}$=6.03). The pattern was vice versa for the *incorrect* answers ($M_{academic}$=3.1, $M_{non\text{-}academic}$=1.83). Although there were very few *empty* answers, the average was slightly higher in non-academic content ($M_{academic}$=.08, $M_{non\text{-}academic}$=.20). The interaction with buttons of "*try again*" and "*show me*" were much more frequent in academic content (n_{try}=118, n_{show}=63) than that of non-academic (n_{try}=62, n_{show}=27). The comparison is presented in Table 1.

Table 1. Interaction Patterns Across Content Types

Pattern Feature	Academic	Non-Academic
Duration	12 min. 43 sec.	9 min. 41 sec.
Score	60.94	75.31
Correct Answers	4.9	6.03
Wrong Answers	3.1	1.83
Empty Answers	0.08	0.20
Try Again	118	62
Show Me	63	27

3.2 Views of Students

When students watched both contents, they were asked which one was the harder to understand. Most students found the non-academic content hard ($n_{academic}$=30, $n_{non\text{-}academic}$=10). The following excerpt exemplify the reasons students found the academic content much more complex than the other content:

"Comparing the videos, for me, the ethics video was much more complicated. Metaverse not that much. I mean it is recent…I found interesting information." (K4)

The students explain the reason why they found non-academic content difficult as follows:

"…because there were too many concepts and I got confused…I could not have distinguished the concepts" (K12)

To have insights from students about the features that make the interactive videos difficult or easy, they were asked to explain their thoughts about the content. The *positive* aspects were more frequently mentioned for non-academic content ($n_{academic}$=23, $n_{non\text{-}academic}$=38) as follows:

"The issue of privacy, the privacy of personal life, the privacy of lots of things. I am happy that I have learned new things. I am also happy that it explained the things in a short time." (K8)

The *negative* aspects were more frequently mentioned for academic content ($n_{academic}$=14, $n_{non\text{-}academic}$=1). The following excerpt is an example of negative views:

"I got nervous in ethics video because I thought that I would not achieve…I felt bad as I failed at questions." (K34)

Some students preferred *neutral* words ($n_{academic}$=2, $n_{non\text{-}academic}$=2) as follows:

"Nothing was boring. There was nothing fun either. I watched videos to evaluate. I mean, they were not boring, at least. I mean, they were explaining something. The graphical content was good in the background. I mean, it was beyond looking at the monitor." (K12)

The details of the themes and subthemes are presented in Table 2.

Table 2. Views of Students about the Content

Thoughts	Academic (n)	Non-Academic (n)
Positive		
Entertaining	17	26
Interesting	2	3
Informative	4	6
Convenient	0	3
Negative		
Boring	5	1
Too challenging	8	0
Weird	1	0
Neutral		
Ordinary	2	2

4. CONCLUSION

The interaction patterns of participants differed in terms of certain features. It might be concluded that users showed smooth patterns of interaction during engaging with the non-academic content. They spent less time but got higher scores from the interactive questions. In addition, the percentage of correct answers in non-academic content was more than that of the academic one despite the high frequency of trying again for academic questions. In the literature, studies confirm the interactive features' contribution to achievement (Ploetzner, 2022). This study extends the literature that the level of such contributions might depend on the content exposed in the interactive videos. The same participants were exposed to two different contents, and their scores differed. Further statistical analysis can explain the differences in terms of significance. The effectiveness of interactive features might depend on perceptions (Palaigeorgiou & Papadopoulou, 2019). The overall views of participants were quite favorable for both content, but the non-academic one was favored more. Moreover, it was not perceived as complex as the academic content. Excessive unfamiliar words and context were mentioned as the distractors for both videos. As Clark et al. (2011) suggested, this might increase cognitive load, so the participants reported the videos as complex or too challenging. As interactive features were required to answer questions, some participants felt overwhelmed, which was expected to cause excessive cognitive load (Barut-Tugtekin & Dursun, 2022). On the other hand, the majority reported positive views, which might align with the previous findings of Antonis et al. (2022). This study says the preliminary analysis of the gathered data; thus, the results cannot be generalized.

ACKNOWLEDGEMENT

This project is funded by OMU BAP PYO (project number: EGF.1904.22.007).

REFERENCES

Antonis, K., Lampsas, P., Katsenos, I., Papadakis, S., & Stamouli, S. M. (2022). Flipped classroom with team-based learning in emergency higher education: methodology and results. *Education and Information Technologies,* 1-17.

Barut Tugtekin, E., & Dursun, O. O. (2022). Effect of animated and interactive video variations on learners' motivation in distance Education. *Education and Information Technologies, 27*(3), 3247-3276.

Beloufa, S., Cauchard, F., Vedrenne, J., Vailleau, B., Kemeny, A., Mérienne, F., & Boucheix, J. M. (2019). Learning eco-driving behaviour in a driving simulator: Contribution of instructional videos and interactive guidance system. *Transportation research part F: traffic psychology and behaviour*, *61*, 201-216.

Clark, R. C., Nguyen, F., & Sweller, J. (2011). *Efficiency in learning: Evidence-based guidelines to manage cognitive load*. John Wiley & Sons.

Creswell, J. W. (2012). *Educational research: Planning, conducting, and evaluating quantitative and qualitative research*. Pearson Education, Inc.

Delen, E., Liew, J., & Willson, V. (2014). Effects of interactivity and instructional scaffolding on learning: Self-regulation in online video-based environments. *Computers & Education*, *78*, 312-320.

Kolås, L. (2015, June). Application of interactive videos in education. In *2015 International Conference on Information Technology Based Higher Education and Training (ITHET)* (pp. 1-6). IEEE.

Martins, D. S., Cunha, B. C., Yaguinuma, C. A., Zaine, I., & Pimentel, M. D. G. C. (2019). Effects of interactive video annotations on students' browsing behavior and perceived workload. *ACM SIGAPP Applied Computing Review*, *19*(2), 44-57.

Mohammed, G. S., Wakil, K., & Nawroly, S. S. (2018). The effectiveness of microlearning to improve students' learning ability. *International Journal of Educational Research Review, 3*(3), 32-38.

Palaigeorgiou, G., & Papadopoulou, A. (2019). Promoting self-paced learning in the elementary classroom with interactive video, an online course platform and tablets. *Education and Information Technologies, 24*(1), 805-823.

Ploetzner, R. (2022). The effectiveness of enhanced interaction features in educational videos: a meta-analysis. *Interactive Learning Environments*, 1-16.

Sancheti, P., Sabnis, N., Kadam, K., Rode, Y., & Vidap, P. (2022). Interactive Learning Application for Indian Sign Language. In *Intelligent Sustainable Systems* (pp. 623-631). Springer, Singapore.

Tacchino, C., Impagliazzo, M., Maggi, E., Bertamino, M., Blanchi, I., Campone, F., ... & Casadio, M. (2021). Spontaneous movements in the newborns: A tool of quantitative video analysis of preterm babies. *Computer Methods and Programs in Biomedicine*, *199*, 105838.

Yeun, E. J., Chon, M. Y., & An, J. H. (2020). Perceptions of video-facilitated debriefing in simulation education among nursing students: Findings from a Q-methodology study. Journal of professional nursing, 36(2), 62-69

VR GAME FOR IMPROVING SPATIAL UNDERSTANDING

Abdullah Doksanbir, Fabian Dillenhöfer, Jan Lennart Kraske, Martin Möller and Bernd Künne
TU Dortmund University
Leonhard-Euler-Str. 5, D - 44227 Dortmund, Germany

ABSTRACT

Some engineering students at the TU Dortmund University have trouble with spatial understanding regarding technical drawings in engineering. This is an important skill that is needed for higher education studies and future professions. To support aspiring engineers in higher education to acquire those skillsets, a VR learning tool is being developed. A pretest was conducted with a small group of students to get their first impression. They then filled out a survey which is analyzed in this paper. Overall, this paper is a progress report about the development of this spatial-perception-assistant software. The key finding is that this VR environment is an effective tool to improve the spatial visualization ability. Since the state of the software is still in its early stages, the VR software is still too technical for some students. Therefore, more playful game elements should be implemented to make the game more appealing to students.

KEYWORDS

VR, Educational Game, Pretest Evaluation, Engineering, Technical drawing, Spatial-perception-assistant

1. INTRODUCTION

Engineering students at the TU Dortmund University are trained in a variety of areas from the basics of mathematics to more specific areas such as turbines. The Department of Machine Elements teaches students many disciplines early on, including but not limited to what standard parts exist, how to choose the right parts for different applications, and how to design machines with their acquired knowledge. Engineering students face challenges when trying to match 2D projections with real machine elements, often due to a lack of spatial awareness and limited opportunities to physically interact with machine parts. The Department of Machine Elements continuously seeks out new ways to combat this issue and improve education.

In the field of engineering education more and more VR and AR technologies are being used (Cheng et al., 2018; Scaravetti & François, 2021). At the same time, serious games are attracting more and more interest in the education sector (De Gloria et al., 2014). Serious games are interactive tools designed to educate while providing entertainment. Players engage in these games to enhance their knowledge and practice their skills by overcoming various challenges presented within the gaming environment (Zhonggen, 2019). In response to the increasing digitization of teaching and the potential of VR technology, the Department of Machine Elements has decided to enhance a workshop in which students assemble a ball-sorting machine in small groups by offering it as a serious game in a VR environment named ViSAs (Virtual Sorter Assembly). This decision is based on the findings of Soliman et al. (2021), who demonstrated that VR technology improves students' understanding of engineering topics and offers positive cognitive and pedagogical benefits (Pantelidis, 2009). Additionally, VR incorporates elements of gamification, making the learning experience more engaging and motivating. This approach not only enhances motivation but also accommodates different learning styles, catering to individuals with varied preferences and strengths in their learning process (Laura Freina & Michela Ott, 2015; Rojas Sanchez et al., 2022). By leveraging VR technology, the department aims to provide an immersive and interactive learning experience, which is a preference and requirement of the current generation of engineering freshman (Abulrub et al., 2011; Dinis et al., 2018). The motivation for the creation of a serious game is that a serious game increases the effectiveness of learning (Wouters et al., 2013). In the areas of declarative knowledge, knowledge retention and procedural knowledge, serious games are found to be more advantageous than traditional teaching methods (Riopel et al., 2019).

In the updated workshop, students will be able to assemble the machine as a serious game in a 3D environment. They can interact with the virtual components, examine their functionality, and observe how they fit together, all while gaining a better understanding of how the technical drawing translates into a real machine. Abdelhameed (2013) showed these VR benefits in the context of architectural design. To evaluate the effectiveness of the VR tutorial, students will be interviewed to assess the extent to which the tool has improved their spatial understanding and understanding of engineering drawings. This feedback will help the Department of Machine Elements gauge the impact of VR technology on the learning outcomes of mechanical engineering students and determine its potential as a valuable educational tool in the field. This innovative approach not only enhances students' spatial visualization skills but also prepares them for future engineering challenges in an increasingly digitized world (Salah et al., 2019).

At the beginning of the paper, the VR game is introduced. The students were able to test the current development state of the support tool and filled out an anonymous survey afterwards. According to this, an evaluation is discussed in chapter two, where the students' satisfaction and the rating of the support tool is analyzed. Last but not least, a short conclusion on how the support tool or respectively the VR environment are capable of raising motivation through the innovative software.

2. VR GAME - VISAS

To improve the independent skill training for the assessment of technical drawings, a virtual reality game is being developed as part of the *CrossLab* initiative. The purpose of this game is to incentivize students through the use of VR technology as an innovative tool for higher education. Constructive Alignment, which is a didactic concept developed by John Biggs, is used as a basis for the development of the game. The primary target group are undergraduate students, namely students studying mechanical engineering, logistics and industrial engineering.

The game will consist of three scenes. In the first scene, students will be asked to find the correct machine element in an assortment of different parts through the use of a technical drawing. This scene will help students recognize parts by means of technical drawings. It will also act as a tutorial that will teach students how the controls inside the game work. In a second scene, the functions of different machine elements, like ball bearings, will be asked so that the students understand the purpose of these parts. In the last scene the assignment will be to assemble a ball-sorting machine. All the knowledge that has been gained in the previous scenes will be necessary to finish the last assignment. In the following tests, only the last scene was used since the first two scenes are still early in development.

For the development of the game, the game engine Unity version 2021.3.9f1 in combination with Virtual Reality Toolkit (VRTK) is used. The prototype of the game that was used in the following tests can be seen in Fig. 1. The main goal of the game is to assemble the given sorting-machine with the help of a technical drawing and a parts list. Two VR controllers, which are visualized as hands inside the VR environment, are used to grab, move and combine objects so that the assembly can be completed. The main parts of the assembly are situated on a table where most of the assembly work is done. On the left side of the player, different standard parts that are required for the assembly can be taken out of labeled storage boxes. A white board with the technical drawing of the assembly and the parts list is positioned on the right side of the player (see Fig. 2). The controls of the game are displayed behind the table. An assembled ball-sorting machine is also placed on the table as a comparison, although this assembly will be removed in the final version.

Different functions have been implemented to support the player inside the game. First of all, grabbed objects will be highlighted on the parts list, as seen in Fig. 2. These parts still have to be associated with the technical drawing by the player. Secondly, markers and an eraser can be used on the white board to draw and erase comments on the technical drawing and the parts list. Lastly, parts that are near the intended position, given that the assembly order is correct, will be highlighted to signalize the assembled position on the release of the object. The purpose of these supports is to guide the students to the correct solution without interrupting the learning process. While the main functionality of the third scene is already implemented (the assembly of the ball-sorting machine), it still lacks game elements. One such element that is planned is a time trial where students will be able to time their attempts at assembly. These times will be saved on a score board to further motivate the usage of the software.

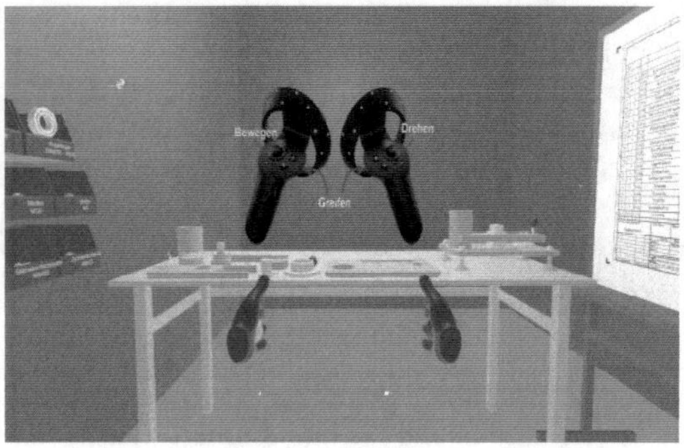

Figure 1. Overview of the VR environment

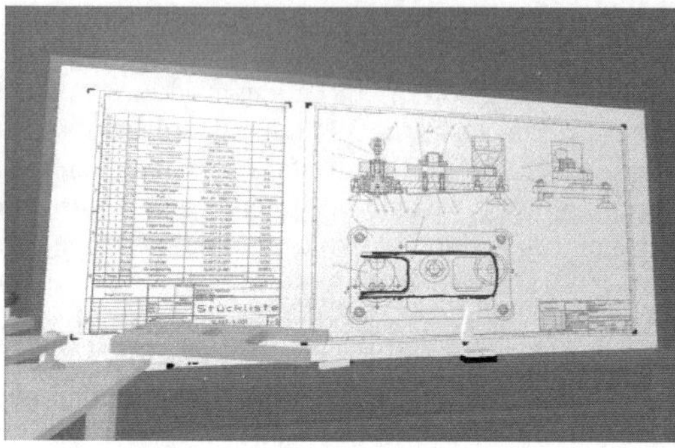

Figure 2. Whiteboard with a technical drawing and a parts list

3. PRETEST AND EVALUATION

A pretest was conducted with a small group of students (13) for excellent academic teaching purposes. The results should help the developers to focus on important target goals of the target group. These survey results (see Table 1) are now being described in the following paragraphs. Around 80 % of the students study mechanical engineering or other programs such as logistics, industrial engineering and computer science. All students who participated in the survey completed the module of technical drawings and should have the necessary knowledge for this topic (see Q1). When it comes to rating their own spatial understanding, the results are scattered, because they have different prior knowledge and experiences (see Q2). This should be considered to adapt the difficulty of the progress and assignments inside the program. Due to the different skills, the answers to the question as to whether there are any problems in imagining the 2D technical drawing in three dimensions differ.

Table 1. Survey results (m: arithmetic mean, σ : standard deviation, very good: 1 – very poor: 5)

Q1: How do you rate your spatial visualization awareness?

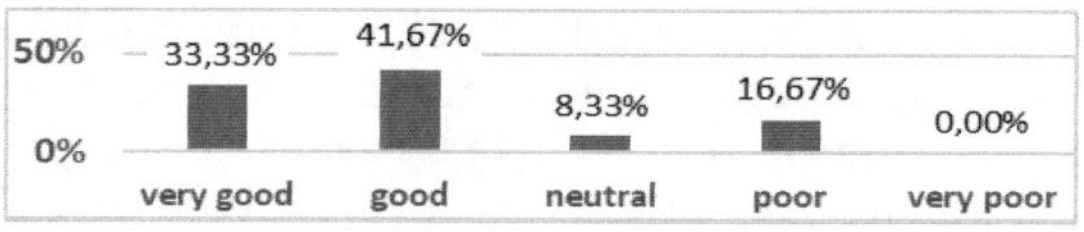

$m = 2.08$
$\sigma = 1.08$

Q2: Did you have any problems imagining the technical drawing in three dimensions?

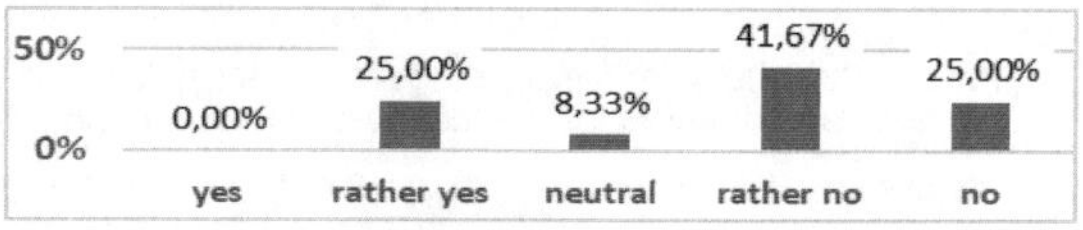

$m = 3.67$
$\sigma = 1.15$

Q3: Have you improved your three-dimensional imagination in general as a result of using the program?
How well did the three-dimensional visualization of the two-dimensional technical drawing work out?

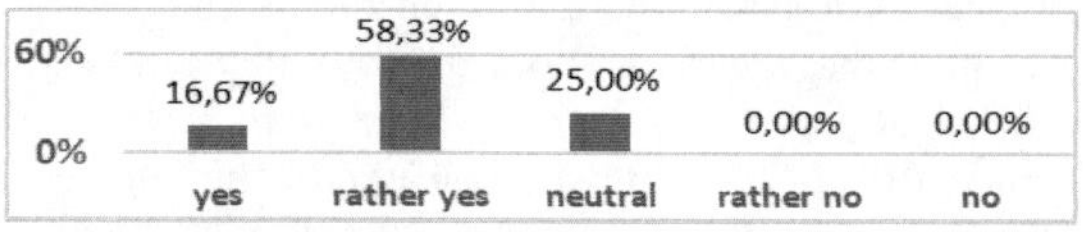

$m = 2.08$
$\sigma = 0.67$

Q4: How well did the three-dimensional visualization of the two-dimensional technical drawing work out?

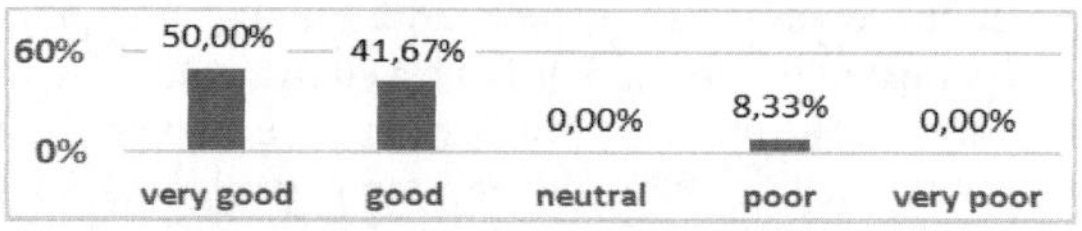

$m = 1.67$
$\sigma = 0.89$

Q5: Is the game fun to play?

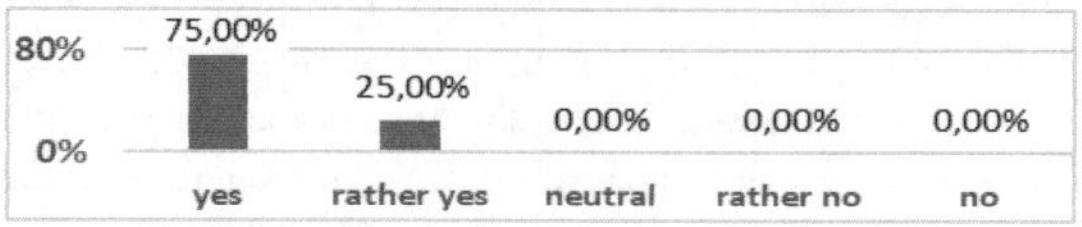

$m = 1.25$
$\sigma = 0.45$

Q6: How do you rate the implementation to achieve the learning objectives (Improved ability to read and interpret technical drawings) vs. playful implementation within the software? The game design is...

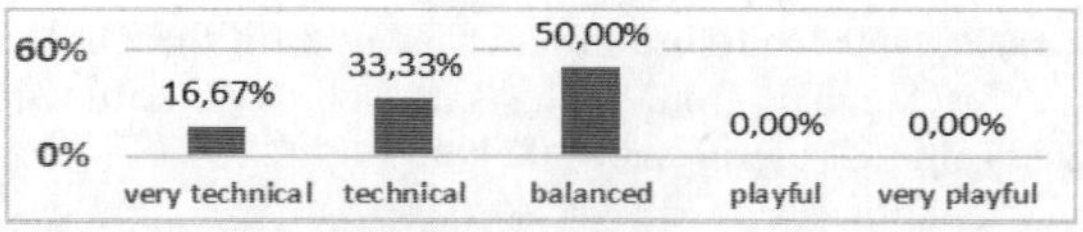

m = 2.33
σ = 0.78

Q7: Does the program help you apply the skills you learned in the lectures to other problems?

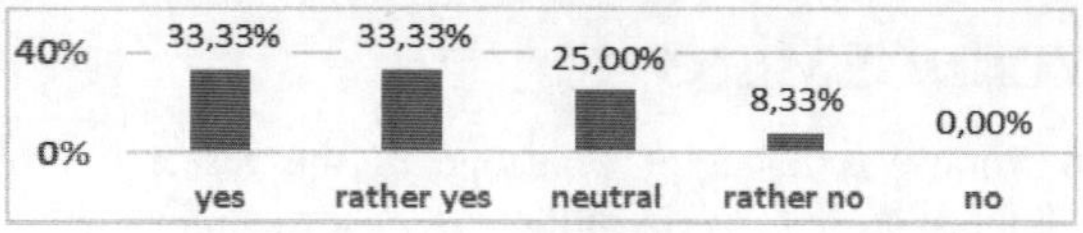

m = 2.08
σ = 1

Q8: Do you own a VR headset?

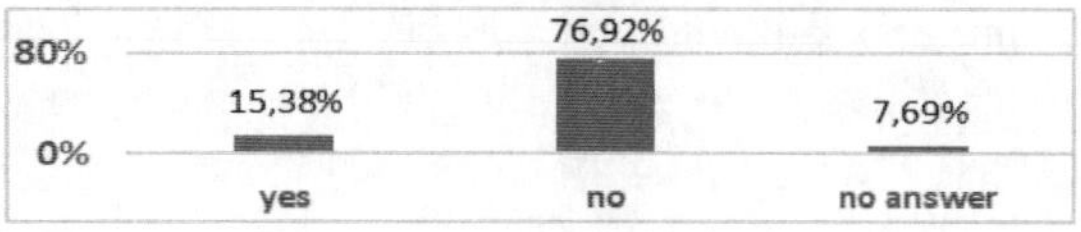

Furthermore, the spatial understanding is an important skill that engineering students need for future studies and professions. All students have more or less improved their spatial visualization ability by using the software (see Q3) and they rate the three-dimensional visualization quality as good or very good. Only a few students (around 10 %) rated the visualization inside the game as poor (see Q4). Maybe the latter will be convinced in the next stages of the development of the program when the focus lies on the design.

Moreover, the majority of students agreed that the game is fun to play (see Q5). It also helps achieving several learning objectives. However, the current development status is stated as technical, where one improvement could be the implementation of a strategic gameplay to make the software more playful (see Q6).

On the other hand the software helps also to apply interdisciplinary skills (see Q7). The opinions of the students are split in this case, so it could be useful to make those interdisciplinary objectives more transparent that the students realize and understand those aims. Secondly, some more interdisciplinary learning objectives from other sources could be added. To be able to use the software, access to the VR technology is a key factor. Currently, most of the students do not own a VR headset (around 80 %, see Q8). This is where institutions should step in and make this technology available by providing the free use in their learning spaces. The TU Dortmund University for example offers a VR lab for students.

These are the most occurrent answers in the open box, where the students gave the following feedback:

- Students wish more game-like elements such as a timer or a scoreboard for gameplay purpose,
- Students wish a more realistic visualization regarding part details like threads by adding textures.

4. CONCLUSION

The results of the surveys show that there is an interest in a VR solution for improving the spatial understanding of students. All students had fun using the VR software and the visual representation of the objects compared to the technical drawings seems to be well received, which indicates a clear advantage of such a solution. Nevertheless, there is still a lot of potential for enhancements. First of all, the first two scenes that act as a tutorial and prepare the player for the assembly assignment need to be completed. Secondly, a system to assist the player if they have issues should be implemented. Some students couldn't progress due to the difficulty of the assignment which meant that supervisors had to give out tips on how to progress. This shouldn't be necessary in the final version of the software since the independent skill training is one of the key points of this project. Furthermore, the pretest has shown that not every function that was available inside the VR environment has been used by the students. For example, most students didn't realize that objects that have been picked up by them were shown on the parts list. This can be improved with the use of audio cues that direct the player's attention to the white board once they pick up an object. Lastly, more game elements like high scores should be implemented to attract the students' attention and motivate the use of the software. After the implementation of these enhancements another evaluation with a larger group of students will be performed. In this test, students without the knowledge of technical drawings should also participate so that the software can be reviewed more accurately. Moreover, there is the problem of the availability of VR headsets. Not only should the Department of Machine Elements provide setups to use this software, but it should also be made available to use in other faculties of the TU Dortmund University, for example at a VR lab. After the feedback gained from the tests as well as other function have been implemented, the software will be made accessible for use in early 2024.

ACKNOWLEDGEMENT

The presented work was done in the scope of the research project "CrossLab–Flexibel kombinierbare Cross-Reality Labore in der Hochschullehre: zukunftsfähige Kompetenzentwicklung für ein Lernen und Arbeiten 4.0", funded by Stiftung Innovation in der Hochschullehre (funding code: FBM2020-VA-182-3-01130).

REFERENCES

Abdelhameed, W. A., 2013. *Virtual Reality Use in Architectural Design Studios: A Case of Studying Structure and Construction.* In Procedia Computer Science, Vol. 25, pp. 220-230, doi.org/10.1016/j.procs.2013.11.027.

Abulrub, A. G. et al., 2011. *Virtual Reality in Engineering Education: The Future of Creative Learning.* 2011 IEEE Global Engineering Education Conference (EDUCON). Amman, Jordan, p. 751-757, doi: 10.1109/EDUCON.2011.5773223.

Cheng, J. et al, 2018. *Construction of Interactive Teaching System for Course of Mechanical Drawing Based on Mobile Augmented Reality Technology.* In International Journal of Emerging Technologies in Learning (iJET), Vol. 13 No. 2, pp. 126–139, doi.org/10.3991/ijet.v13i02.7847.

De Gloria, A. et al., 2014. *Serious Games for education and training.* In International Journal of Serious Games, Vol. 1, No. 1, p. 1, doi: 10.17083/ijsg.v1i1.11.

Dillenhöfer, F., Künne, B., Willms, U. (2022, November 14). Neologised teaching concept and materials combine remote teaching and hands-on activities. 2022 IEEE German Education Conference (GeCon), Article 9942764. Berlin. https://doi.org/10.1109/gecon55699.2022.9942764.

Dinis, F. M. et al, 2018. *Disseminating Civil Engineering through Virtual Reality: An Immersive Interface.* In International Journal of Online and Biomedical Engineering, Vol. 14, No.5, p. 230, doi.org/10.3991/ijoe.v14i05.7788.

Ott, M. and Freina, L., 2015. *A literature review on immersive virtual reality in education: state of the art and perspectives.* 11th International Conference eLearning and Software for Education.

Pantelidis, V. S., 2009. *Reasons to Use Virtual Reality in Education and Training Courses and a Model to Determine When to Use Virtual Reality.* In Themes in Science and Technology Education, Vol. 2, No. 1-2, p. 67.

Riopel, M. et al., 2019. *Impact of serious games on science learning achievement compared with more conventional instruction*: an overview and a meta-analysis. In Studies in Science Education, Vol. 55, No. 2, pp. 169-214, doi: 10.1080/03057267.2019.1722420.

Rojas-Sánchez, M. A. et al., 2022. *Systematic literature review and bibliometric analysis on virtual reality and education.* In Education and Information Technologies, Vol. 28, No. 1, pp. 155–192, doi.org/10.1007/s10639-022-11167-5.

Salah, B. et al., 2019. Virtual Reality-Based Engineering Education to Enhance Manufacturing Sustainability in Industry 4.0. In *Sustainability*, Vol. 11, No. 5, article number: 1477. doi.org/10.3390/su11051477.

Scaravetti, D. & François, R., 2021. Implementation of Augmented Reality in a Mechanical Engineering Training Context. *Computers* 2021, Vol. *10*, No. 12, article number: 163, doi.org/10.3390/computers10120163.

Soliman, M.et al., 2021. The Application of Virtual Reality in Engineering Education. In *Applied Sciences*, Vol. 11, No. 6: article number: 2879, doi.org/10.3390/app11062879.

Wouters, P. et al., 2013. A Meta-Analysis of the Cognitive and Motivational Effects of Serious Games. In Journal of Educational Psychology. Vol. 105, pp. 249-265. doi: 10.1037/a0031311.

Zhonggen, Y., 2019. *A Meta-Analysis of Use of Serious Games in Education over a Decade.* In International Journal of Computer Games Technology, vol. 2019, Article ID 4797032, doi.org/10.1155/2019/4797032.

USING PARTICIPATORY ACTION RESEARCH TO IMPROVE STUDENT OUTCOMES IN FASHION SUSTAINABILITY EDUCATION

Alexandra Howell Abolo[1] and Elizabeth Shorrock[2]
[1]*Drexel University, Philadelphia, PA, USA*
[2]*West Virginia University, Morgantown, WV, USA*

ABSTRACT

Using participatory action research (PAR) the current, ongoing, study aims to engage undergraduate fashion students at two U.S. institutions of higher education in extending their understanding of sustainable fashion. Through data collection processes wherein the students act as both researcher and participant, they will self-report on their clothing purchase and disposal habits. Upon completing of the data collection period, students will engage in a reflective process through group discussion with the aim of coming to conclusions about the growth of their own knowledge through this experiential learning assignment. The research question guiding this study asks, does participatory action research (PAR) improve student understanding about sustainable practices in the fashion industry?

KEYWORDS

Education, Fashion, Participatory Action Research, Pedagogy, Sustainability

1. INTRODUCTION

Sustainability in fashion curricula is nearly ubiquitous. The scope of what sustainability factors are covered vary greatly in both depth and breadth. To that end, sustainability often becomes a buzz word in student lexicons, but carries little meaning beyond the classroom. This creates a troublesome dynamic where faculty are preparing students for an industry that is a major contributor to the current climate crisis, but students are under prepared to advocate for change, "textile production is one of the most polluting industries, producing 1.2 billion tonnes of CO2 equivalent (CO2e) per year, which is more emissions than international flights and maritime shipping combined" (*Energy, Climate Change & Environment: 2016 insights,* 2016; as cited in *A New Textiles Economy*, 2017, p. 20). Furthermore, American consumers continue to robustly participate in the fashion economy, apparel purchases increasing 60% between 2000 and 2014; the life of the garments in U.S. households lasted only 50% as long (Buzzo & Abreu, 2019). The gap for undergraduate fashion students is not in understanding the gravity of the current situation, but rather how to enact meaningful change in their field of study. Undergraduate students demonstrate cognitive dissonance about clothing usage, consumption, and disposal relative to the impact it has toward the climate crisis. Notably, the current generation of undergraduate students are the most aware and vocal about the damages of the climate crisis (Tyson, et al., 2021). The purpose of the current study is to utilize participatory action research (PAR) to engage undergraduate students enrolled in fashion design and merchandising courses in understanding their impact on apparel and textile usage, consumption, and disposal.

1.1 Research Question

Does participatory action research (PAR) improve student understanding about sustainable practices in the fashion industry?

2. BODY OF PAPER

2.1 Literature Review

Informed by Jestratijevic & Hillery (2023) this study expands the "clothing mountain." First, exposing students to their habits with clothing purchase and disposal, and further asking students to analyze what materials their clothing is made of, where their clothing is being made, and from who they gather new information about apparel and textiles. Ultimately, leading to the "will," asking student to reflect on their willingness to change.

To execute on this process, researchers employed participatory action research (PAR) because it engages participants as researchers in the process (Chevalier & Buckles, 2013). In this study undergraduate students enrolled in two fashion courses at separate institutions, act as both researchers and study participants. This research method emphasizes a focus on a need for change within a community and inquiry is driven by the members themselves. Utilizing PAR aims to reduce cognitive dissonance or general apathy toward clothing purchase and disposal among students studying fashion at the university level. Further, employing the Burns (2015) model of sustainable pedagogy provides a framework for a more meaningful data collection and analysis process. Chevalier & Buckles (2013) indicate the purpose of PAR centers on community change guided by members themselves; Burns equally emphasizes the goal of change through transformational learning, "The central goal of this model is to provide opportunities for transformational learning, in which learners are motivated and inspired to shift their values, and make sustainable and authentic changes in their own lives, and within their communities and places" (Burns, 2015, p. 263). Using these frameworks cohesively creates a landscape of student-centered-and-driven data collection and analysis.

2.2 Sample

Undergraduate students enrolled in a textile science course at a midsize, metropolitan university on the East Coast, institution A, and in an introduction to fashion merchandising course at a large land-grant institution in Appalachia, institution B.

2.3 Methods

Over a 5-week period, January 9 to February 12 2023 students (n=69) contributed information on a privately shared excel file about their clothing consumption and disposal habits, denoting the following information, a description of the garments purchased, the cost, the reason for their purchase, a description of the garments disposed of, initial cost of the garment at purchase, the method of disposal, and the reason for disposing of the garment. Students were asked to research their current wardrobes, investigating their three most worn clothing items, denoting the following information, a description of each item, the country each item was made in, and the fiber content of the garment or accessory. Students reflected on where they get the most information about apparel and textiles, denoting the following information, the name of the media side and the author, creator, or account handle. Finally, students were asked to respond to the qualitative reflection question, will I change my apparel purchase, consumption, and disposal habits after participating in this project; why or why not? At the conclusion of the data entry period, all student names were removed from the excel file to maintain confidentiality.

On February 14 and 16, 2023 each group of undergraduate students engaged in a guided discussion about the outcomes of the excel data (questions listed in table 1). The discussion format intended to enable students to hear from their peer group at institution A and B, respectively. On February 14, 2023, students at institution B recorded group responses to the questions using the virtual software Canva. Those recorded responses were shared with students at institution A. On February 16, 2023, students at institution A, listened to the recordings, and then completed their own reflective responses. All groups' responses were analyzed for nascent themes grouped by questions listed in table 1.

Excel data are analyzed for frequency, average, and purchase reason; the total number of garments purchased, the average total cost of each garment category, the range in amount (USD) spent within garment categories, and purchase reasoning were totaled and organized between participants responses as "want," or "need" (Table 2). This analysis process is ongoing for disposal of garments, fiber content and country of origin, and information gathering about apparel and textiles.

Table 1. Small Group Discussion Questions

Did tracking your buying habits change the way you make clothing and accessory purchases in the future? Why or why not?
What are your disposal methods for clothing and accessories, and why do you choose those methods?
Do you feel more informed about the fiber content and country of origin of your clothing and accessories; and will that impact future purchases? Why, why not?
Are most of your clothing and accessories purchases from a need or want? Explain.
Do you ever experience buyer's remorse? If so, what do you do with those clothing and accessories?
Considering the media sites you use to gain information about clothing and accessories, do you see any correlation with what you buy based on that information?
Do you ever feel like the media sites you gain fashion information from are compelling you to make a purchase? If yes, why?
What specifically about this project impacted your relationship with clothing the most?

Table 2. Student Tracking Journal – Excel Data

Category	Total Number	Total Cost	Average Cost	Range	Reason for Purchase
Bottoms	26	$749.00	$28.80	$149.50	20 Wants 6 Needs
Jeans	5				
Sweats/leggings	8				
Skirts	3				
Pants/cargo	6				
Shorts	4				
Tops	39	$216.00	$5.54	$69.00	32 Wants 7 needs
T-shirts	10				
Tank top	5				
Sweatshirt/hoodie	6				
Button up blouse/shirt	4				
Sweater	4				
Other /fashion tops	10				
Dress	8	$182.00	$22.75	$39.00	4 wants. 4 needs
Outerwear	20	$850.00	$42.50	$102	15 Wants. 5 needs
Accessory	7	$184.50	$26.36	$56.50	3 Wants 4 needs
Shoes	9	$990.00	$110.00	$251.00	8 wants 1 need
Sneakers	5				
Other	4				
Undergarments	1	$15.00	$15.00	$15.00	1 need
Other	3	$42.00	$14.00	$6.00	3 wants
Thrifted	20				
Total	116				

2.4 Findings and Discussion

Initial analysis indicates that students enrolled in these courses actively participate in the data collection process. A total of n=70 students are enrolled courses at Institution A and Institution B, and there was n=42 usable data entries. Students' responses were removed from purchase data if they entered "zero purchases" for the allotted period, this accounted for n=22 students. During informal discussion at Institution A with students about nothing purchased, many students referenced the recent holiday and financial pressure to not spend money in the new year, as reasons for not making purchases. Furthermore, during the small group discussions on February 14 and 16, students reiterated some of those reasons for not purchasing items.

2.5 Exploration of Consumption Habits: Purchases

From January 9 through February 12, students at institution A and B, purchased 116 total items, 22% or 26 items categorized as "bottoms" (jeans, sweats/leggings, skirts, pants/cargo, or shorts); 33% or 39 garments categorized as "tops," (t-shirts, tank top, sweatshirt/hoodie, button up blouse/shirt, sweater, other/fashion tops), 17% or 20 garments categorized as outerwear, 6% of 7 items categorized as accessories, 8 garments categorized or 6.89% as "dresses," and 3 items or 2.5% categorized as other. There were 9 items or 7.75% categorized as shoes, (sneakers or other). 20 of 116 garments or accessories purchased (17%) were thrifted. The total number of purchased items is relatively low for the student population enrolled. Based on analysis from the small group discussion it appears this number is intentionally low, students making conscious decisions to purchase less or infrequently. In total, students spent $3,228.50 (USD) on garments and accessories within the period of the study, students spend the most on shoes, $990 (USD), outwear, $850 (USD), and bottoms $749 (USD). The most spent on one item was shoes, $260 (USD) and the least was bottoms at $.50 (USD). The purchased items were also categorized by want or need, 86 wants and 28 needs; purchases categorized as needs were for weather, work, or functionality.

The purchase analysis reflects the small group discussion where students expressed an understanding about the negative impact of hyper consumption the clothing industry has on the global economy. Notably, students expressed various relationships to this reality, some expressing fervent beliefs "There is no ethical consumption under capitalism," (student from institution B), and others with more practical approaches, students at institution A explained not having the disposable income to continually purchase new clothes, and therefore had fewer garments, and purchased less. Students in both groups expressed desires to keep purchased items longer, and repair and mend items when damaged as a means of product lifecycle longevity. Through small group discussion students also explored the relationship between need and want driven clothing purchases. Students at both institutions expressed the feeling that at times wants are categorized as needs based on external influences such as social media, social environments, or appealing advertisements. Similarly, students recognized that they had experienced buyer's remorse, and that was tied directly to not wearing an item they purchased and feeling badly about it. However, students responded in equal measure that they continually try to purchase more consciously to avoid such outcomes. It is notable that students responded to the final reflective question with continued confidence that participating in this study was eye opening in some ways, but moreover reinforced what they already knew. Students at institution B, indicated new ways they planned to change some purchase habits such as not purchasing, purchasing secondhand, or abstaining from purchasing polyester as a means for improvement.

3. CONCLUSION

This ongoing study provides an expanded framework for helping students understand their clothing purchase and disposal choices that reflect their knowledge regarding sustainability in the fashion industry. While expanding on the clothing mountain, reported by Jestratijevic & Hillery (2023), the current study limited the scope of information collected and analyze to be manageable for student researchers in a single academic term. Initial analysis indicates that students who participated in this study as both researchers and participants have improved understanding about sustainable practices in the fashion industry. This is evidenced through ideas about how purchase habits can change and improve to have a more lasting impact on the industry overall. Further analysis on disposal, garment knowledge, and information gathering will expand on this result and guide authors toward future inquiry. An expansion of this study should extend multiple academic terms with a variety of undergraduate institutions. This extension would increase participant diversity in multifaceted ways and provide more data for richer analysis.

ACKNOWLEDGEMENT

A special thank you to the students who participated in this research study, your contributions made it possible.

REFERENCES

Burns, H. L. (2015). Transformative sustainability pedagogy: Learning from ecological systems and indigenous wisdom. *Journal of transformative education,* 13(3) pp. 259 – 276.

Chevalier, J.M. and Buckles, D.J. (2013) *Participatory Action Research: Theory and Methods for Engaged Inquiry*, UK: Routledge.

Ellen MacArthur Foundation (2017). *A New Textiles Economy: Redesigning Fashion's Future.*

International Energy Agency (2016). *Energy, Climate Change & Environment:2016 insights.*

Jestratijevic, & Hillery, J. L. (2023). Measuring the "Clothing Mountain": Action Research and Sustainability Pedagogy to Reframe (Un)Sustainable Clothing Consumption in the Classroom. *Clothing and Textiles Research Journal*, 41(1), pp. 10–25. https://doi.org/10.1177/0887302X221084375

Muthu, S.S. (eds). (2019). *Fast fashion, fashion brands & sustainable consumption. Singapore: Springer.*

Tyson, A., Kennedy, B., & Funk, C. (26 May 2021). Gen Z, millennials stand out for climate change activism, social media engagement with issue. Pew Research Center. https://www.pewresearch.org/science/2021/05/26/gen-z-millennials-stand-out-for-climate-change-activism-social-media-engagement-with-issue/

LIFE CYCLE ASSESSMENT OF THE EXTRACTION OF CONDENSED TANNINS FROM ACACIA BARK RESIDUES

Ricardo Simões [1], Carla L. Simões[1], Ricardo Ferreira[2] and Ana C. Rodrigues[2]
[1]*School of Technology, Polytechnic Institute of Cavado and Ave*
Campus do IPCA, Vila Frescaínha S. Martinho, 4750-810 Barcelos, Portugal
[2]*PROMETHEUS, Unidade de Investigação em Materiais, Energia e Ambiente para a Sustentabilidade, Escola Superior Agrária, Instituto Politécnico de Viana do Castelo, 4900-347 Viana do Castelo, Portugal*

ABSTRACT

The environmental impact of the production of condensed tannins from acacia bark was investigated from gate-to-gate perspective. Of the six process steps in the production system that were assessed, the spray drying and adsorption (purification) steps were identified as the main sources of the environmental impact of the whole system. This is due to intensive consumption of energy and consumption of makeup ethanol, respectively. The most important action to improve the environmental performance of the system is to reduce water consumption in the extraction, as it will reduce energy demand in the spray drying step. This LCA shows the most important steps to facilitate process development towards lower environmental impact.

KEYWORDS

Acacia Bark, Tannins, Extraction, Adsorption, Evaporation, Life Cycle Assessment

1. INTRODUCTION

Conventional water treatment using chemical coagulants always generates some kind of impact on the environment, therefore studies of alternative water treatments, especially with natural coagulants, are important. This is directly linked to the issue of sustainability and its development.

Natural coagulants are compounds that may or may not come from plants, which have the ability to add to contaminants and thus enable their removal. They are compounds easily found in nature and their use has a strong environmental, economic and social appeal, since they improve the quality of life of society and mitigate environmental damage or impact (Lima et al, 2020).

Among the most used natural coagulants is the group of tannins, which are organic compounds that can be extracted from plant barks such as Black Acacia, acting as coagulants and flocculants in water treatment (Fiorentini, 2005). Tannin is a phenolic substance with a complex molecular structure, classified as hydrolysable and condensed, it acts in colloidal systems, neutralizing charges and forming bridges between these particles, thus enabling the formation of flocs and their sedimentation (Coral et al, 2009). Tannins exhibit excellent properties in removing a wide range of metals. These metals can be removed by the tannins when they are adsorbed. In addition to metal adsorption, tannins can be used to coagulate and flocculate materials in wastewater. Tannins have many useful properties, such as biodegradability, interaction, low toxicity, low cost and easy application (Martins et al, 2014).

In this work, we are developing a new method to produce condensed tannins yield after post extraction treatment from bark residues to be applied in the process of water treatment. During the assessment of new production paths, it is important to understand the environmental impact of different choices making the best decisions possible and focusing development on potential improvements. An established method to evaluate a system's environmental performance is Life Cycle Assessment (LCA) (Bauman and Tillman, 2004; Finnveden et al., 2009; Pennington et al., 2004; Rebitzer et al., 2004). However, as far as we are aware, only one LCA study has been published on extraction of tannins from Norway spruce bark (Ding et al, 2017). The results of this study show that the extract drying process is the primary contributor to the environmental impact of tannin production. Another LCA study was performed, where it is analyzed the production of a flocculation agent based on actionized tannins from the bark of Norway spruce, including the production of spruce trees to the finished flocculation product ready to leave the factory gate (Carlqvist et al, 2020). This study shows that the main environmental impact stems from the reagents used in the cationization step. The purification step could also be a significant issue depending on the possibility of reusing the eluent (ethanol) and the lifetime of the resin used. Also, a LCA study was performed comparing in large-scale production scenario three different technologies: hot water extraction (HWE), ultrasound assisted extraction (UAE), and supercritical fluid extraction (SFE) for the extraction of phenolic compounds from spruce bark (Carlqvist et al, 2022). The LCA results show that the simpler HWE process has a lower environmental impact per amount of phenolic compound extracted than UAE and SFE unless the extraction yields of the latter are more than about 5 times that of HWE. The reason for this result is mainly the environmental burden caused by the consumption of ethanol in the UAE and SFE processes, which accounts for more than 70% of the total environmental burden in most impact categories. Other LCA studies have been published where extraction of tannins from softwood bark (such as maritime pine, mimosa) is also performed as a part of a system. However, the extracted tannins are used in the production of leather dyes, adhesives, resins, foams, etc (Arias et al., 2022; Conde et al., 2022; González-García et al., 2016). In this context it is noted a lack of studies in the application of the LCA methodology in the analysis of the environmental impact of the extraction of tannin-based natural coagulants for water treatment.

The present study aims to analyze the process of extraction and use of tannin-based natural coagulants for water treatment. This work provides a gate-to-gate LCA of the extraction and production of condensed tannins from acacia bark residues from the north Portugal region.

2. LIFE CYCLE ASSESSMENT

The main objective of the present work is to evaluate the potential environmental impact in the extraction and production of condensed tannins from acacia bark residues. Furthermore, it was defined as a secondary aim to identify the life cycle phase that contributes the most to its environmental impact. The LCA methodology, which includes all stages of a product's life (Bauman and Tillman, 2004) was used in accordance with the standards from the ISO 14040 and ISO 14044 (ISO, 2006a; ISO, 2006b).

2.1 Goal, Functional Unit and System Boundary

The objective of this LCA study is quantifying the environmental impacts of extraction and production of condensed tannins from acacia bark residues from the north Portugal region. The aim of the environmental assessment is to identify the most promising pathways and to identify optimization potential. The functional unit (FU) of this LCA is 1 kg of condensed tannins yield after post extraction treatment from bark residues.

Laboratory scale data was used to obtain data at pilot scale using SuperPro Design software simulation. It is assumed that extraction yield and extract composition on a pilot scale matches the laboratory results under the same process conditions. It was used a gate-to-gate approach, the acacia cultivation and acacia bark collection stages were excluded because it is assumed that it is a mandatory task of the forest management system and must be conducted in any circumstances. Also, the acacia bark is an invasive species therefore it grows without human help. Capital goods were also excluded.

The Acacia bark is assumed to be dried and shredded before being loaded to a tannin extraction unit. The extract in water is submitted to purification through adsorption and removal of solvent in an evaporation step. Finally, the extract is spray dried to remove remaining humidity (Figure 1).

2.2 Inventory Analysis

The Life Cycle Inventory and Life Cycle Impact Assessment were conducted with the SimaPro software. Unless otherwise specified, Ecoinvent was used as a main secondary data source, and missing data were collected from literature and laboratory data. An overview of mass and energy flows to produce the FU 1 kg of condensed tannins yield after post extraction treatment from bark residues is presented in Figure 1.

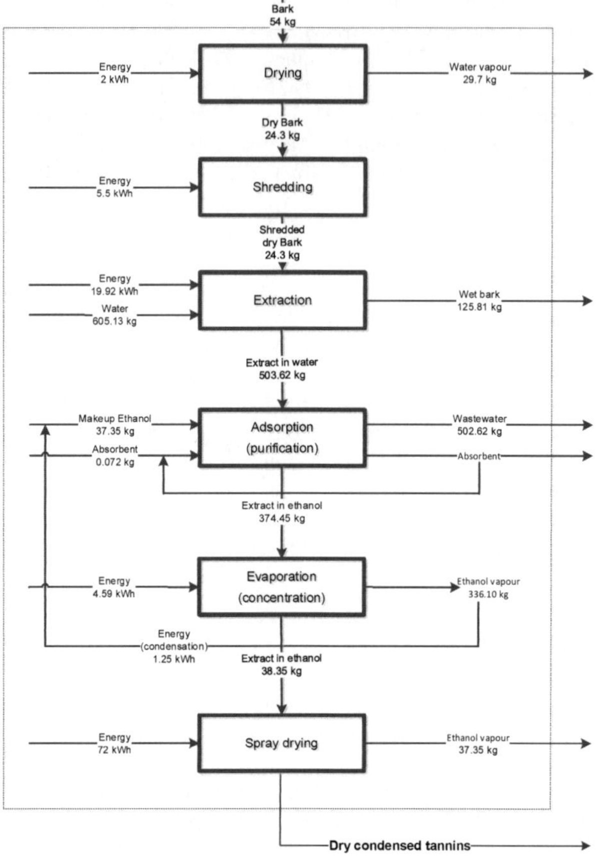

Figure 1. System boundary, showing the production system with mass and energy flows for 1 FU

2.3 Environmental Impact Assessment

Impact categories and characterization factors were obtained from the ReCiPe midpoint Hierarchist method. The ReCiPe model is well adapted to European conditions, which is consistent with the geographical scope of this LCA study. Of particular concern is the global warming potential (GWP), based on the current focus on environmental and sustainability goals. This category will therefore be evaluated first in detail and the others summarized in an overview.

An overview of the relative contribution of the potential impact on GWP, for each process step, is shown in Figure 2. The spray drying step has the largest potential impact on GWP in the present analysis and is the main hotspot in the overall process for many of the ReCiPe impact categories, as shown later in the summary of all 18 impact categories. This is due to intensive consumption of energy (the Portuguese country electricity mix is assumed). The second largest potential impact on GWP is the adsorption step, due to consumption of makeup ethanol and adsorption resin, even though bioethanol (based on lignocellulose) is assumed to be used. The remaining process steps (drying, shredding, extraction and evaporation) are also energy intensive.

In a broader perspective, the relative contribution to the impact categories of ReCiPe for the overall production of condensed tannins is shown in Figure 3. The pattern of the burden distribution is similar in many of the impact categories, dominated by the spray drying and adsorption processes. Except for Stratospheric ozone depletion, Marine eutrophication, Terrestrial ecotoxicity, Human non-carcinogenic toxicity and Land use, where the adsorption process is the main burden. These results are in line with previous published studies (Ding et al, 2017; Carlqvist et al, 2020), when taking in consideration the FU and system boundary. Hence a possible strategy to improve the environmental performance of the system is to try to reduce the water consumption in the extraction step since both the extraction and spray dying steps have high energy demands and by reducing the water content the energy demand will be reduced. Data for the adsorption step highlights the importance of re-using ethanol for elution and using a resin with long lifetime.

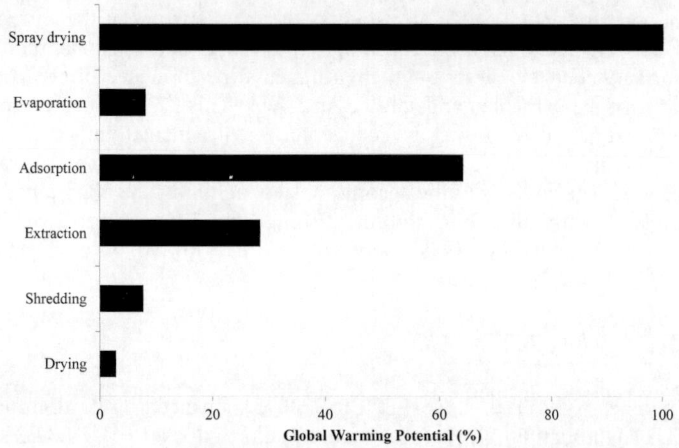

Figure 2. Relative contributions (in %) to the GWP for each process stage for the production of 1 FU

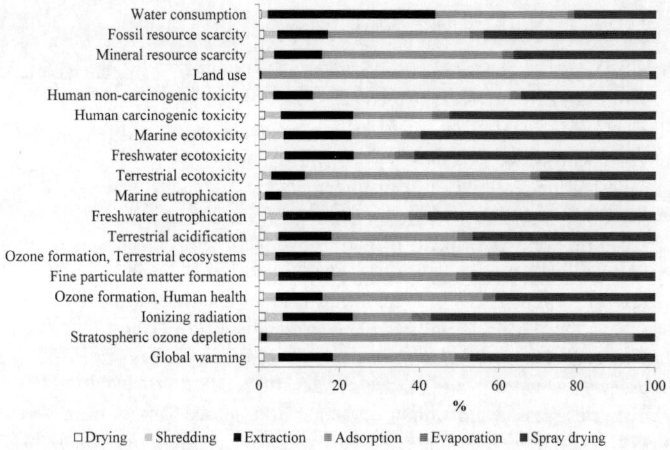

Figure 3. Relative contributions (in %) to the impact categories of ReCiPe per subsystem in the overall production of 1 FU

It should be mentioned that this work will be further developed in the future, in order to better understand the environmental impact of the new method proposed to produce condensed tannins yield after post extraction treatment from bark residues to be applied in the process of water treatment. Once the main hotspots of the production of condensed tannins from acacia bark have been identified, alternative formulation routes and a sensitivity analysis should be developed to determine the degree of improvement achievable in the environmental profiles. Possible aspects for future work include the possibility of eliminating the spray drying process step, to lower the energy consumption, since the main objective is to obtain condensed tannins to be used in the water treatment. Therefore, the drying process may be eliminated since the dry condensed tannins are hydrated again in subsequent steps. Also, the acacia cultivation and acacia bark collection stages were excluded from the study because it is assumed that this is a mandatory task of the forest management system and must be conducted anyway. However, on a larger scale, acacia will most likely have to be cultivated and not just collected from forests, in which case the acacia cultivation process could affect the environmental impact of the condensed tannins production and its potential impact on the system should be assessed.

3. CONCLUSION

In this study the environmental impact of the production of condensed tannins from acacia bark was investigated from gate-to-gate perspective. Of the six process steps in the assessed production system, spray drying and adsorption (purification) were identified as the main sources of the environmental impact of the whole system, respectively due to intensive consumption of energy and consumption of makeup ethanol.

The most important action to improve the environmental performance of the system is to improve the environmental profile of the extraction and spray dying steps. This LCA points towards a reduction in water consumption in the extraction as a way to reduce the environmental impact of the process. Both the extraction and spray dying steps have high energy demands and by reducing the water content the energy demand will be reduced. The analysis of the adsorption step highlights the importance of re-using the ethanol for elution and using a resin with a long lifetime. Through this LCA, it was possible to identify the most important steps that will facilitate process development towards a lower environmental impact.

ACKNOWLEDGEMENT

Project TECH (reference NORTE-01-0145-FEDER-000043), financed by the European Fund for Regional Development (FEDER) through Programa Operacional Regional do Norte (NORTE2020).

REFERENCES

Arias et al., 2022. Tannin-based bio-adhesives for the wood panel industry as sustainable alternatives to petrochemical resins, *Journal of Industrial Ecology*, Vol. 26, pp. 627-642.

Bauman H. and Tillman A.-M., 2004. *The Hitch Hiker's Guide to LCA - An orientation in LCA methodology and application*, Publisher: Studentlitteratur, Sweden.

Conde et al., 2022. Exploring the feasibility of substituting mimosa tannin for pine bark powder. A LCA perspective, *Cleaner Engineering and Technology*, Vol. 7, No. 100425.

Carlqvist K. et al, 2020. Life-cycle assessment of the production of cationized tannins from Norway spruce bark as flocculants in wastewater treatment. *Biofuels, Bioproducts and Biorefining*, Vol. 14, pp. 1270-1285.

Carlqvist K. et al, 2022. Life cycle assessment for identification of critical aspects in emerging technologies for the extraction of phenolic compounds from spruce bark. *Journal of Cleaner Production*, Vol. 333, No. 130093.

Coral L. A. et al, 2009. *Estudo da viabilidade de utilização do polímero natural (TANFLOC) em substituição ao sulfato de alumínio no tratamento de águas para consumo*. Key elements for a sustainable world. São Paulo - Brasil.

Ding T. et al, 2017. Life cycle assessment of tannin extraction from spruce bark. *iForest*, Vol. 10, pp. 807-814.

Finnveden et al., 2009. Recent developments in Life Cycle Assessment. *Journal of Environmental Management*, Vol. 91, pp. 1-21.

Fiorentini V., 2005. Uso do tanino no processo de tratamento de água como melhoria em sistema de gestão ambiental.

González-García et al., 2016. Environmental sustainability of bark valorisation into biofoam and syngas, *Journal of Cleaner Production*, Vol. 125, pp. 33-43.

ISO 14040: 2006, 2006a. Environmental Management – Life Cycle Assessment – Principles and Framework. ISO 14000 International Standards Compendium. Geneva, Switzerland: International Organization for Standardization (ISO).

ISO 14044: 2006, 2006b. Environmental Management – Life Cycle Assessment – Requirements and Guidelines. ISO 14000 International Standards Compendium. Geneva, Switzerland: International Organization for Standardization (ISO).

Lima P. R. et al, 2020. Os diferentes tipos de coagulantes naturais para tratamento de água. *Evidência*, Vol. 20, pp. 9-12.

Martins A. A. et al, 2014). Potencial de uso de compostos orgânicos como, coagulantes, floculantes e adsorventes no tratamento de água e efluentes. *Periódico Eletrônico Fórum Ambiental da Alta Paulista*, Vol. 10, No. 12.

Pennington et al., 2004. Life cycle assessment Part 2: Current impact assessment practice, *Environment International*, Vol. 30, pp. 701-720.

Rebitzer et al., 2004. Life cycle assessment Part 1: Framework, goal and scope definition, inventory analysis, and applications, *Environment International*, Vol. 30, pp. 721-739.

REFLECTION

ENGAGING WAYS OF LEARNING: ESPORTS AND VIDEO GAMES AS A LEARNING ENVIRONMENT FOR STUDENTS IN BRAZIL

Andréa Corrêa Lagareiro and Claudio Alexandre Kovacs Pires de Godoi
Pontifícia Universidade Católica de São Paulo – PUCSP, Brazil

ABSTRACT

Video games can represent a motivational environment for teaching and learning processes, considering the non-compulsory, healthy competition, and audiovisual resources they provide. There are many skills proven to be enhanced by virtual reality learning. Gaming and video-game-based approaches regarding learning processes were associated with improved motor, social, and interpersonal skills metrics, demonstrating potential as complementary approaches in education. Therefore, including video games as part of the academic curriculum can be a valuable resource for learning processes. In time, comprehensive and up-to-date research and systematic reviews are necessary to confirm and enrich this discussion. This paper aims to start a broader and deeper research looking to increasingly incorporate eSports as a learning approach in the scholar curriculum in the Brazilian public education system.
A literature survey of PubMed, Google Scholar, Elsevier, and SiBiUSP was conducted in January 2023, filtering publications by keywords such as eSports, gaming, education, curriculum, discovery-based learning. This included literature from different academic fields (education, psychology, and music, among others).

KEYWORDS

Esports, Gaming, Education, Curriculum, Discovery-Based Learning

1. INTRODUCTION

A world constantly evolving towards a more technological future demand new and sophisticated skills from students in the 21st century. There are also new pieces of knowledge that can help children develop skills such as problem-solving, interpersonal conflict resolution, mathematical and logical thinking, crisis management, emotional intelligence, cooperative behavior, resource management, and decision-making (Zhong et al., 2022, p. 3). By bringing different approaches to learning processes that are aligned with contemporaneity this paper aims to briefly present and better understand the possibilities for learners, as well as methods, their limitations, and the potential that rests in eSports.

As gaming elements are being carefully and progressively incorporated into education, an ecosystem serving the purpose of learning can be created in a space as interesting and safe as possible for kids.

2. BRAZILIAN EDUCATIONAL SCENERY: ADDRESSING THE GAP BETWEEN THEORY AND PRACTICE

As the world becomes increasingly more technological, it is mandatory to look into our education strategies (Toner, 2011, p.56). In Brazil, the current model of primary and secondary education is established by National Curricular Common Base (BNCC). The report was created by the Guidelines and Bases of National Education Act (most recently revised in 2018). It states, among other aspects, that the bases of Brazilian education should promote competency learning and prepare students for the market. (Brazil, 2018).

Even though the concept of competencies is paramount for the development of the future workforce (Dondi et al., 2021), some competencies seem to be under-prioritized by BNCC - not in the document *per se,* but in its application (Morais, 2021, Ferreti & Silva, 2017). The focus on standardized testing and a narrow curriculum limit students' creativity and interests. This model also fails to address students' diverse needs and backgrounds, contributing to a high dropout rate and a lack of equity in educational opportunities in a country impacted by social inequality (Ferreti & Silva, 2017, pp. 394-398).

It is paramount to discuss assisted discovery, gamification, and challenge-based learning approaches. The benefits of gaming applicability and the eSports presence in educational settings must be investigated.

3. GAMIFICATION APPROACHES IN NUMEROUS LEARNING PROCESSES

3.1 Playing, Gaming, and Learning

Gaming and playing provide leisure and/or casual settings to address learning issues. On account of sets of procedures, backdrop simulations, or even peer participation, gaming strategies in education have found a prolific field and seem to stimulate students to engage in their education. Drawing on narratives, scenery, characters, and specific goals, students find a space and time to materialize, implement, and transfer their knowledge in many areas, while connecting and expressing their feelings, likes, and dislikes. As a result, applied gaming strategies encompass different sets of skills to promote controllable and assisted setups that can be secure, captivating, and non-compulsory for students (Zhong et al., 2022).

3.2 Challenge-Based Gamification Strategies

Challenge-based gamification such as the application of points and levels systems, leaderboards, and clear goals, tasks, and quests, have been shown to improve learning outcomes. The gamification tools suggest that, when combined with or integrated into traditional approaches in education, gamification of learning improves students' performance. (Zixiu, Cahalane & Carbonie, 2022, p. 2-5).

As pointed out in research, groups who experienced challenge-based gamification performed better than those who only participated in traditional methods such as attending the class or reading the material (Legaki et al., 2020; Nah et al., 2014).

Merging assisted discovery-based learning, gaming, and challenge-based approaches and the effectiveness of said combination must be further investigated.

3.3 Esports: Benefits and Limitations

eSports (competitive electronic games) have become increasingly popular in recent years and have been associated with health benefits to their players, including educational contexts.

Research has shown that video games can help improve motor coordination, precision, and reaction time (Gupta, et al., 2021). The possibility to incorporate gaming resources into learning processes can be particularly useful for people who need to improve their skills, such as students, doctors, and drivers (González-González, et al., 2018; Mallow, et al., 2020).

Virtual reality tools can also aid in developing stress-reducing abilities and improving self-esteem and self-efficacy, forming healthy and result-oriented habits (Huang, et al., 2017).

Other researchers point out that eSports can promote socialization and teamwork conditions, being an effective tool for individuals who struggle with social interactions, live in remote areas, or can benefit from collective learning ecosystems (Cañabate, et al., 2021). These abilities are crucial for success in school and life, at every age. The promotion of teamwork and collaboration settings seems to be possible by the *principle of not-leaving-anyone-behind* and incorporating each one's best skills to put them into game strategies. For example, role-playing games (RPGs) use this very principle. They can integrate multiple skills and knowledge from various fields, fostering teamwork, problem-solving, critical thinking, and creativity (Soares, et al. 2015). These skills are essential for success in both the classroom and the workforce (Cañabate, et al. 2021).

Moreover, eSports can be an engaging way to teach academic content, such as math, languages, and history. These games can help students become interested in the content and learn more effectively (Caponetto et al.,2014). As such, gaming as an educational strategy represents a motivational way to learn, considering the activeness of the setting, the gamers' non-compulsory adherence to the proposal, and the self-determined goal. All of that indicates that learning often extends beyond just playing and passing time. (Anderson, et al., 2018).

Besides learning and refining delicate professional skills or even how to cope with emotions, the game Rocksmith is a good example of how eSports and other games can enhance learning.

3.4 Rocksmith: A Case to Analyze

Rocksmith is a game that utilizes a real guitar or bass as the controller, serving learning purposes by allowing players to learn and play along with a variety of popular songs while gradually building up their skills through dynamic difficulty adjustment (Havre, et al., 2019).

The game creates an environment based on 5 aspects to develop player skills, relying on gamification (Havre et al., 2019):

1. Learn-a-song mode: Rocksmith has a dynamic difficulty system that reads a player's skill level, adapts to it, and suggests lessons that help practice difficult parts at a slower rhythm.
2. Lesson mode: interactive video lessons are shown to the player. The mode provides feedback and is pointed to new challenges if they are overcome.
3. Jam-session mode: the mode simulates jamming with the virtual band, allowing improvisation skill practice.
4. Arcade mode: the player can practice skills and techniques such as playing the right chords to shoot alien spaceships or zombies.
5. Amplifier-mode: allows the player to use the game as a virtual amplifier with a vast selection of effects.

This approach aims to refine students' skills, increase adherence to the learning process, and foster a desire to learn through gaming experience. The model could include a competitive component, utilizing the game as a fair referee to teach students skills such as focus, resilience, and emotional regulation during a performance. Furthermore, it could also promote socialization, teamwork, and collaboration, capitalizing on the positive aspects of eSports.

4. CONCLUSION

This preliminary research proposes that the Brazilian educational curriculum needs updates and can benefit from incorporating assisted gamification strategies to enhance student learning and performance. The presence of assisted video games and eSports in the academic setting stimulates interest and curiosity among students, and also has the potential to improve student engagement and motivation during the learning process. Role-playing games (RPGs) can be particularly useful for integrating different subjects and fostering teamwork, problem-solving, critical thinking, and creativity, which are essential skills for students' academic and professional development. However, the implementation of these strategies must consider the diverse needs of students and the limitations of contexts, and be used in conjunction with proper methods as assisted approaches. Advanced technologies, such as eSports and virtual simulation, have the potential to enhance student engagement and provide new educational experiences but must be closely monitored to ensure children and adolescents are not exposed to harmful content. Therefore, additional research is needed to improve effective technology strategies in education, and the national curriculum must shift towards a student-centered approach that emphasizes critical thinking, problem-solving, and creativity, integrating technology and digital skills.

REFERENCES

Anderson, C.G., Tsaasan, A.M., Reitman, J.G., Lee, J.S., Wu, M., Steel, H., Turner, T., & Steinkuehler, C. (2018). Understanding Esports as a STEM Career Ready Curriculum in the Wild. (2018) 10th International Conference on Virtual Worlds and Games for Serious Applications (VS-Games), 1-6.

Cañabate, D., Bubnys, R., Nogué, L., Martínez-Mínguez, L., Nieva, C., & Colomer, J. (2021). Cooperative Learning to Reduce Inequalities: Instructional Approaches and Dimensions. Sustainability, 13(18), 10234. https://doi.org/10.3390/su131810234

Caponetto, I., Earp, J. & Ott, M. (2014). Gamification and Education: A Literature Review. Proceedings of the 8th European Conference on Games-Based Learning - ECGBL 2014. 1. 50-57.

Dondi M. Klier, J, Panier, F. & Schubert, J. (2021). Defining the skills citizens will need in the future world of work. McKinsey & Company. Geneva, Switzerland.

Ferreti, C. J., & Silva, M. R. d. (2017). Reforma do ensino médio no contexto da medida provisória nº 746/2016: Estado, currículo e disputas por hegemonia. Educação & Sociedade, 38(139), 385–404. https://doi.org/10.1590/es0101-73302017176607

González-González, C., Río, N. G., & Navarro-Adelantado, V. (2018). Exploring the Benefits of Using Gamification and Videogames for Physical Exercise: a Review of State of Art. International Journal of Interactive Multimedia and Artificial Intelligence, 5(2), 46. https://doi.org/10.9781/ijimai.2018.03.005

Guo, Z., Cahalane, M. & Carbonie, A. (2022): Online gaming with a purpose: exploring positive personal development achieved through esports play, Behaviour & Information Technology, DOI: 10.1080/0144929X.2022.2154266

Gupta, A., Lawendy, B., Goldenberg, M. G., Grober, E., Lee, J. Y., & Perlis, N. (2021). Can video games enhance surgical skills acquisition for medical students? A systematic review. Surgery, 169(4), 821–829. https://doi.org/10.1016/j.surg.2020.11.034

Havre, S., Väkevä, L., Christophersen, C., & Haugland, E. (2019). Playing to learn or learning to play? Playing Rocksmith to learn electric guitar and bass in Nordic music teacher education. British Journal of Music Education, 36(1), 21-32. doi:10.1017/S026505171800027X

Huang, J., Yan, E., Cheung, G., Nagappan, N., & Zimmermann, T. (2017). Master Maker: Understanding Gaming Skill Through Practice and Habit from Gameplay Behavior. Topics in cognitive science, 9(2), 437–466. https://doi.org/10.1111/tops.12251

Mallow, C., Shafiq, M., Thiboutot, J., Yu, D. H., Batra, H., Lunz, D., Feller-Kopman, D. J., Yarmus, L. B., & Lee, H. J. (2020). Impact of Video Game Cross-Training on Learning Bronchoscopy. A Pilot Randomized Controlled Trial. ATS scholar, 1(2), 134–144. https://doi.org/10.34197/ats-scholar.2019-0015OC

Morais, G. A. de, Leme, V. B. R., Falcão, A. O., & Pereira-Guizzo, C. de S. (2021). Habilidades sociais e expectativas de futuro como preditores da autoeficácia para a escolha profissional. Psico, 52(2), e32374. https://doi.org/10.15448/1980-8623.2021.2.32374

Legaki, N., Xi, N., Hamari, J., Karpouzis, K. & Assimakopoulos V. (2020). The effect of challenge-based gamification on learning: An experiment in the context of statistics education, International Journal of Human-Computer Studies, Volume 144, 102496, ISSN 1071-5819, https://doi.org/10.1016/j.ijhcs.2020.102496.

Soares, A. N. et al. The Role-Playing Game (RPG) as a pedagogical strategy in the training of the nurse: an experience report on the creation of a game. Texto & Contexto - Enfermagem, v. 24, n. Texto contexto - enferm., 2015 24(2), pp. 600–608, abr. 2015.

Toner, P. (2011). Workforce skills and innovation: an overview of major themes in the literature. OECD Directorate for Science, Technology, and Industry (STI) Centre for Educational Research and Innovation (CERI). https://www.oecd.org/sti/inno/46970941.pdf

Zhong, Y., Guo K., Su J. & Chu, S. K. W. (2022). The impact of esports participation on the development of 21st-century skills in youth: A systematic review, Computers & Education, Volume 191, 104640, https://doi.org/10.1016/j.compedu.2022.104640

AN OVERVIEW OF SOCIAL NETWORKS IN EDUCATION: HIGHLIGHTING THE ROLE OF ONLINE FEEDBACK

Elham Akbari
Assistant Professor at the Faculty of Humanities
Tarbiat Modares University, Iran

ABSTRACT

This review study aims to evaluate the integration of social networks in education and its impact on teaching and learning. While traditional learning management systems lack the advantages of social networks, such as peer interaction, social networks have the potential to positively influence the learning process, with feedback being a key component. The study suggests that social networks can increase feedback efficiency and improve university students' learning outcomes. However, the effectiveness of social networks may be influenced by individual learning styles and personalities. The study highlights the potential benefits of integrating social networks in education.

KEYWORDS

Education, Feedback, Social Networks, CMS/LMS, and Learning Language

1. INTRODUCTION

Today, online social networks are among the most important communication tools, connecting millions of people with shared interests and ideas all over the world (Zaidieh, 2012; Akbari, 2016). Results of various studies indicated that over three billiard users are engaged in social networks and the rate of social networks usage by young and university students is over 1 hour daily (Rogers, 2009, Akbari, Naderi, Yazdi, Simons& Pilot, Akbari,2019).

Today, there are more than 350 online social networks (Wikipedia, 2020). The most famous online social networks in the world include Facebook, YouTube, WhatsApp, Instagram and WeChat. In Iran, Instagram stands in the first place and is the most favorite, followed by WhatsApp and Twitter, respectively

1.1 Importance of Online Social Networks in Education

Due to the rapid growth in the social networks research over the past decade, the empirical and conceptual advances in education have been transformed. New educational possibilities can be created due to the flexible environment and various capabilities of these networks. According to Selwyn (2007), online social networking applications contain good educational technologies. However, there are some challenges. New educational standards need to be compiled and the existing studies need to be more comprehensive for implementing an optimal use of these networks for educational purposes

This study aims at highlighting the use of social networks in higher education. Furthermore, by reviewing the related studies, we try to investigate social networks in terms of (a) the learning processes (support, educational processes, communication and collaboration enhancement, academic performance) on the part of students and educators; (b) the users' personality type and style of learning; (c) online learning platforms (LMS—learning management system); and (d) their application in higher education. The investigation shows enhancement in all of the above-mentioned items. Therefore, a wider use of online social networks (OSNs) in higher education is quite likely in the future. Nevertheless, online social networks (OSN) are not employed by teachers and higher education institutions in their activities. Smart Education Research encompasses a wide range of topics, including advanced methods of learner profiling, dynamic and self-paced learning environments created with the help of technology, and the use of educational applications through mobile devices (Visvizi, Lytras, & Daniela, 2018; Lytras, Visvizi, & Naeve, 2015). Another important aspect of Smart Education Research is the development of methodologies for social learning and community-based learning in Higher Education, which is influenced by social networks. Learning analytics is a significant area of research in Smart Education, as it allows for the identification of hidden patterns in big data related to learners and educational institutions, including patterns related to motivation (Laaser & Concha, 2018). Student-centered learning analytics research also has focus on enhancing motivation and engagement. Massive Open Online Courses (MOOCs) have emerged as a popular option for open learning, and the provision of learning software has created new challenges for policy makers, learning administrators, and faculty.

Due to the fact that the youth are eagerly interested in social media, researchers investigate their application in education (Beemt, Thurlings, and Willems, 2020). Educational institutes integrate social media in the classroom to enable critical thinking, collaboration, and self-learning (Boateng & Amankwaa, 2016). A literature review revealed that social development is the most essential element in social networks (Akbari, Pilot & Simons, 2015; Balakrishnan& Gan, 2016). This fact increased the number of social networks, especially those related to education (Demirbilek, 2015). Moreover, various research findings (Akbari and Simon, 2018) indicated that social networks foster collaborative learning. Therefore, learners will learn beyond the classroom (Macià & García, 2016). According to Selwyn (2007), online social networking applications share many qualities of good educational technologies. Yet, realizing these features will require compiling new educational standards compatible with these networks. However, although relevant, existing studies need to be more comprehensive for implementing an optimal use of these networks for educational purposes. There are crucial advantages of social media in promoting student performance in education. They include:

- Communication and collaboration. Connection is possible via smartphone, tablet, or laptop with anyone at any time (Putnik, et al., 2016).
- Parental involvement. Through social media, parents can stay involved in their children's learning (Akbari, 2022).
- Distance learning opportunities. Online social media such as Skype or Webinar Jam can make students involved in distance learning which is an important element in recent education.
- Feedback. Interaction via online social networks results in feedback (Akbari,2021).

According to Al-Said, Al Said, and Hattab (2020), the use of social networks for instructional purposes will have a positive effect on student achievements; yet, higher education institutions resort to traditional platforms such as Course and Learning Management Systems (CMS/LMS) that do not include the pros of social networking, such as the ability to keep in touch with peers. Yet, during the Covid19 era, it is found that they are used more frequently and are more effective than LMSs. (Akbari,2021).

In general, there are four different areas that indicate the positive impact of social networks: (a) the learning processes (support, enhancement of educational processes, communication and collaboration, educational performance) from the students' and instructors' perspectives; (b) the user's personality and learning style (techniques or approaches of attaining knowledge, academic culture, efficiency); (c) the social networks including online learning platforms like learning management system (LMS); and (d) their application in institutions (administrational and educational uses of OSNs by faculty). The following paragraphs elaborate on these four areas in more detail.

1.2 Importance of Online Social Networks in Language Learning

Social networking sites not only profoundly impacted our social structure and intra-social interaction but also affected education in general and learning English language in particular. The easy access to information and people can, in particular be helpful in language education, e.g. online social networks allow foreign-language learners to get access to native speakers. In addition, these networks provide for authentic interactions. Researchers have recognized the various opportunities that social media provides for language learners and teachers. Many articles appear on the educational use of online social networks and researchers (Akbari, 2016, 2021; Muftah, 2022; Al Jahromi, 2020) argued that online social networks have a good potential for language learning. Moreover, I agree with Blattner and Fiori (2009) that Facebook can be used as a venue for authentic language interaction and that it can increase students' motivation. I concur with Kabilan, Norlida, & Mohd Jafre (2010) that Facebook can be used as an environment to facilitate the learning of English. However, research to develop and evaluate different ways of using online social networks is needed. The current research aims to investigate the potential of online social networks in language learning and teaching. That is, there are increased advantages of social networks with regard to educational outcomes, facilitating communication, the realization of a sense of identity, and belonging, and the direct interactions that happen in discussion groups, conversations, and videos. Within the extensive environment of worldwide interaction, not only are linguistic skills improved but also an international culture is created (Pikhart and Botezat, 2021).

Overall, mixed-method research designs are used in the studies investigated. Moreover, they all use university students as participants and report the positive influences of social media in language teaching and learning. Most of these studies are carried out in Far Eastern countries (mainly China), and the majority of English speakers are originally from Taiwan and China. The research is predominantly conducted in an EFL context. Yet, a number of studies are carried out in the context of Chinese and Spanish spoken as second languages.

To sum up, the studies reviewed in the present research indicate the significance of social media in language teaching and learning. This study will hopefully contribute to the integration of social media in education in general and in language instruction in particular through reviewing different studies conducted in recent years.

1.3 Using Social Networks in English Language Teaching

With the advent of Web 2s in 2005, potentially new developments in foreign language education were formed. Yet one of the most important types of web 2s is social networks. Move of teachers and students toward these networks caused their methods and experiences to develop and yet change (Planas, 2010). Applying these networks by language teachers indicates how learning happens in this new era of digital literacy and due to the resulting changes, the classic system of education has to coordinate with it (Lockard & Pegrum 2007). Some research efforts related to using these networks in English language learning follow:

1.4 Definition of Feedback

Feedback as a concept is composed of a series of information and data which are created through factors such as teachers, books, parents, friends and colleagues, experience, and the person himself (Hattie & Timperley, 2007) and this information are related to the people's level of understanding or their performance in relation to a particular subject.

Of course, the role of feedback in e-learning is more important than in traditional learning (Lynch, 2002; Palloff & Pratt, 2001). The online peer feedback increases the relevance of feedback for both givers and receivers of feedback (Akbari, 2021; Akbari, Simons, Pilot& Naderi, 2017). By participating in an online discussion, students acquire better understanding of their peers' viewpoints. Moreover, online interaction promotes motivation in the target language.

2. DISCUSSION AND CONCLUSION

The article reviews various studies that examine the impact of social networks on education and highlights the potential benefits of using these networks as a tool for enhancing learning outcomes. One key finding is that social networks can be effective for providing feedback to learners. The article also highlights the importance of using social networks to create a collaborative learning environment. Another key finding of the article is that social networks can help to bridge the gap between formal and informal learning.

In conclusion, the article suggests that social networks have the potential to revolutionize the education process by providing learners with access to a wide range of feedback sources and creating a collaborative and engaging learning environment.

REFERENCES

Akbari, E. (2021). Challenges and Effectiveness of Using the SHAD Social Network During COVID-19 According to Teachers, Parents and Students. Electronic Journal of e-Learning, 19(4), pp296-304.

Akbari, E. (2021). Examining the effectiveness of peer feedback in the context of learning English in a Web2. 0 environment. (peer feedback2. 0). Foreign Language Research Journal, 11(3), 556-570.

Akbari, E. (2022). Rethinking of The Zone of Proximal Development: The Role of Parental Engagement in Children's Language Learning Skills.

Akbari, E., & Simons, R. J. (2018). Efficacy of using social networks in learning and teaching based on self-determination theory: an interventional study. Interdisciplinary Journal of Virtual Learning in Medical Sciences, 9(4).

Akbari, E., Pilot, A., & Simons, P. R. J. (2015). Autonomy, competence, and relatedness in foreign language learning through Facebook. Computers in Human Behavior, 48, 126-134.

Akbari, E., Simons, P. R. J., Pilot, A., & Naderi, A. (2017). Peer feedback in learning a foreign language in Facebook. Global Journal of Human Social Science, 17(2), 31-44.

Al Jahromi, D. (2020), "A quantitative study of the perceived impact of social media networks on Bahraini users' English language learning", Teaching English with Technology, Vol. 20 No. 4, pp. 23-40

Al-Said, K. M., Al Said, N., & Hattab, E. (2020). Impact of Social Networks in Educational Media. Journal of information and communication convergence engineering, 18(4), 230-238.

Balakrishnan, V., & Gan, C. L. (2016). Students' learning styles and their effects on the use of social media technology for learning. Telematics and Informatics, 33(3), 808-821.

Beemt, A., Thurlings, M., & Willems, M. (2020). Towards an understanding of social media use in the classroom: a literature review. Technology, Pedagogy and Education, 29(1), 35-55.

Boateng, R., & Amankwaa, A. (2016). The impact of social media on student academic life in higher education. Global Journal of Human-Social Science, 16(4), 1-8.

Demirbilek, M. (2015). Social media and peer feedback: What do students really think about using Wiki and Facebook as platforms for peer feedback? Active Learning in Higher Education, 16(3), 211-224.

Kabilan, M. K., Ahmad, N., & Abidin, M. J. Z. (2010). Facebook: An online environment for learning of English in institutions of higher education? The Internet and higher education, 13(4), 179-187.

Laaser, W. L., & Concha, U. R. (2018). MOOCs, A Phenomenon with Many Faces: Success and Failures. International Journal of Smart Education and Urban Society (IJSEUS), 9(3), 27-39.

Livemocha (2008). Homepage, viewed 15 December 2008 http://www.livemocha.com.

Lockard J & Pegrum M (2007), Brave new classrooms: Democratic education and the Internet, Peter Lang.

Lynch, M. M. (2002). The Online Educator: A Guide to Creating the Virtual Classroom. NewYork: RoutledgeFalmer.

Lytras, M. D., Mathkour, H. I., Abdalla, H., Al-Halabi, W., Yanez-Marquez, C., & Siqueira, S. W. M. (2015). An emerging–Social and emerging computing enabled philosophical paradigm for collaborative learning systems: Toward high effective next generation learning systems for the knowledge society. Computers in Human Behavior, 51, 557-561.

Macià, M., & García, I. (2016). Informal online communities and networks as a source of teacher professional development: A review. Teaching and teacher education, 55, 291-307.

Muftah, M. (2022). Impact of social media on learning English language during the COVID-19 pandemic. PSU Research Review.

Pikhart, M. and Botezat, O. (2021), "The impact of the use of social media on second language acquisition", Procedia Computer Science, Vol. 192, pp. 1621-1628.

Planas, Sergi Roura, (2010), Social Networking Sites as Technology Enhanced Learning Communities towards autonomous language learning, presented at International Congers of Didactic.

Putnik, G., Costa, E., Alves, C., Castro, H., Varela, L., & Shah, V. (2016). Analysing the correlation between social network analysis measures and performance of students in social network-based engineering education. International Journal of Technology and Design Education, 26(3), 413-437.

Selwyn, N. (2007). "Screw blackboard…does it on Facebook!" An investigation of students' educational use of Facebook." Presented at the "Poke 1.0 – Facebook Social Research Symposium," University of London.

Visvizi, A., Lytras, M. D., & Daniela, L. (2018). (Re) defining smart education: Towards dynamic education and information systems for innovation networks. In Enhancing knowledge discovery and innovation in the digital era (pp. 1-12). IGI Global.

Zaidieh, A. J. Y. (2012). The use of social networking in education: Challenges and opportunities. World of Computer Science and Information Technology Journal (WCSIT), 2(1), 18-21.

PROPOSAL FOR A SERIOUS GAME ON THE THEME OF PERSONAL INFORMATION

Rin Kokubo and Jun Iio
Faculty of Global Informatics, Chuo University
1-18 Ichigaya-tamachi, Shinjuku-ku, Tokyo 162-8478, Japan

ABSTRACT

With the advent of the Internet, our society has evolved into a place where personal information can be easily shared and collected. To explore people's attitudes towards this reality, we developed an interactive online game to analyze these attitudes and promote digital literacy. We conducted a survey with 36 university students using this game. This paper provides an overview of the game, details the experiment, and discusses the findings.

KEYWORDS

Serious game, Personal information, Privacy, Gamification

1. INTRODUCTION

With the widespread use of information and communication networks, along with the accelerated expansion of information processing capabilities, the handling of vast amounts of personal information has become a reality. However, this has also amplified concerns over the potential leakage and misuse of privacy-related information. Advances in search technology have further complicated matters, as even fragmented pieces of information available publicly on the Internet can be pieced together with other data to increase the probability of identifying individuals. These issues underscore the crucial need for privacy protection in the digital era.

Therefore, we sought to investigate people's awareness and understanding of privacy issues. Instead of utilizing a traditional questionnaire, we developed an interactive game centered around personal data. By leveraging this engaging format, our goal was to stimulate individual involvement in privacy issues based on their personal experiences and enhance their overall digital literacy on the topic.

2. GAME STRUCTURE AND RULES

In the development of games centered around personal information, incorporating the viewpoints of those who might exploit such information is crucial. Games uniquely offer role-playing experiences that aren't typically encountered in day-to-day life. Thus, we designed a game format in which participants, acting as either the holders of personal information (defenders) or those attempting to exploit it (attackers), engage in competition.

The attacker is given three cards, each depicting a different scenario of information exploitation, from which they choose one. Conversely, the defender possesses five cards, each assigned a set score corresponding to the card chosen by the attacker. Without knowing the content of the attacker's selected card, the defender must decide on three cards to reveal. This arrangement mirrors real-world scenarios where a potential offender cannot fully anticipate the details of their prospective offense; in our game, the information is unveiled under conditions where the opponent doesn't know the specifics of the selected card.

The attacker's score is computed as the total points of the three cards passed on by the defender, while the defender's score is calculated as the sum of the points from the two remaining cards in their hand.

An overview of the game is depicted in Figure 1.

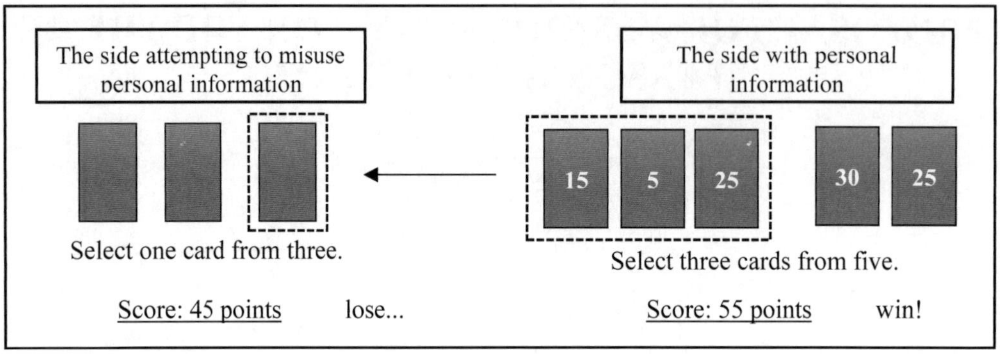

Figure 1. Game rule

3. OVERVIEW OF THE GAME SYSTEM

This section presents the game's system architecture and flowchart. Our game was developed as a web application, enabling online player interaction, and streamlined data collection.

Figure 2 provides a schematic representation of the system's architecture. The database meticulously captures all player decisions and the resulting game outcomes.

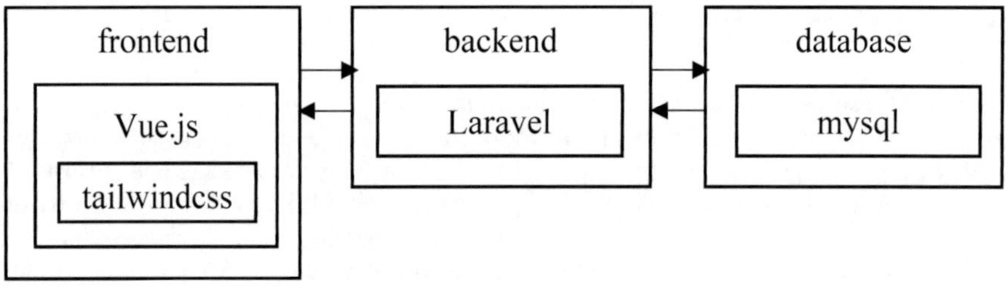

Figure 2. System configuration

The gameplay sequence is depicted in the flowchart provided (refer to Figure 3). Participants have the option to assume roles as either attackers or defenders.

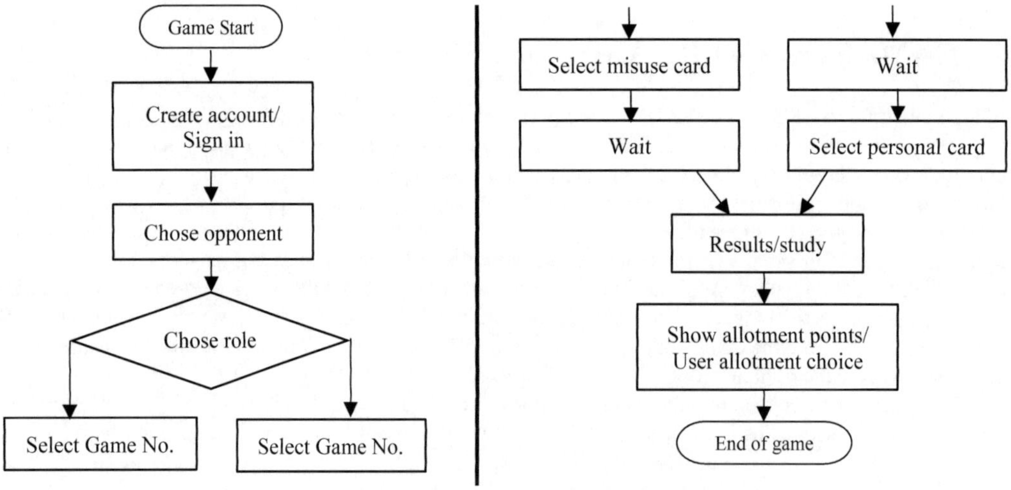

Figure 3. Game progression chart

4. OUTLINE OF EXPERIMENTS AND ANALYSIS OF RESULTS

This section delineates the conducted experiments and the ensuing analysis of the outcomes. Our experimental findings are further contextualized with comparisons to public opinion surveys concerning personal information, as provided by the Japanese government.

4.1 Outline of Experiments

We conducted two experiments: an initial trial in June 2022 with 12 students, and a subsequent one in December with 24 participants. Data was collected from 24 games in the first trial and 72 in the second. Prior to and following the experiments, participants completed a questionnaire on information literacy and game experience.

4.2 Analysis of Results

Table 1 summarizes the results of 93 valid matches from the first and second experiments. (Three games were excluded due to system issues rendering the results invalid.)

Table 1. Win rage by role

Role	Win	Lose	Draw	Win Rate
Attacker	67	21	5	0.72
Defender	21	67	5	0.23

The results indicate a higher winning rate for the attackers, as anticipated given their advantage of acquiring more cards towards the end.

Subsequently, we evaluated players' reluctance to share various personal information by analyzing the percentage of players unwilling to disclose each piece.

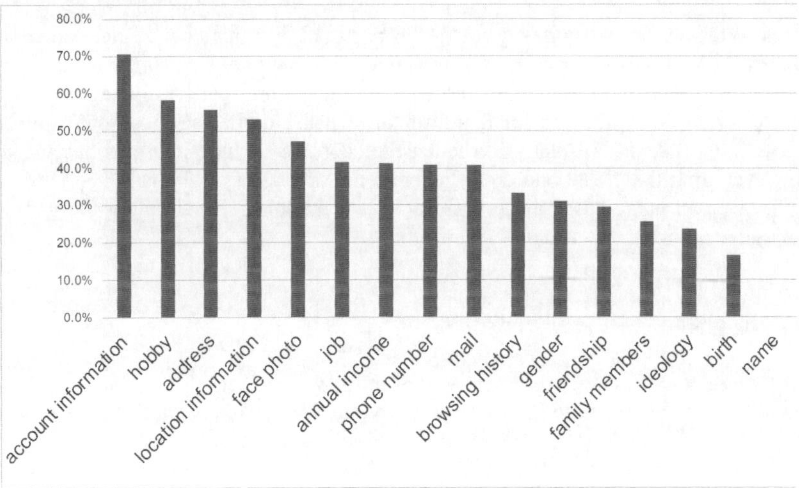

Figure 4. Percentage of players who did not want others to know

Figure 4 presents the types of personal information that players were most reluctant to share, ranked from most to least private: account information, hobbies, and addresses. On the other hand, participants were more comfortable with sharing information such as their names, dates of birth, and ideologies.

4.3 Comparison of Public Opinion Polls and The Results of The Experiment

In 2006, a public opinion survey conducted by the Japanese government (Cabinet Office of Japan, 2006) yielded results somewhat like our study. The survey found that 88.7% of respondents were most unwilling to disclose account and credit card numbers, making it the most guarded piece of personal information. Coming second was annual income, property status, and tax payments (74.2%), followed by photographs of faces (53.3%).

A noteworthy divergence between the government survey and our study concerns the data on hobbies. In our survey, hobbies emerged as the second most protected category of information among college student participants, which was an unexpected outcome. Although knowing an individual's hobbies can increase the chances of falling prey to crimes such as phishing scams and fraudulent websites, our data suggested that the high level of reluctance to share hobbies was not necessarily driven by crime-related fears. Rather, many individuals consider their hobbies to be a private matter.

5. RELATED WORK

Antonaci et al. (2017), Sailer et al. (2020), and Oroszi (2020) have all demonstrated the efficacy of gamification in education and security awareness, highlighting its role in enhancing IT literacy, student engagement, learning outcomes, and user security awareness.

Liz-Domínguez et al. (2020) and Smiderle et al. (2020) both explored the impact of gamification in educational contexts. While the former used gamification to collect data to improve education, the latter showed the effects of gamification varied according to students' personality traits. Differently, our study implements gamification to assess attitudes towards personal information, in addition to providing literacy education.

6. CONCLUSIONS

While there is no legislation categorizing hobbies as personal information, societal perceptions may not align with current regulations. With advancements in technology, one's hobbies can now be discerned from search histories, social media analytics, purchase records, and behavioral patterns. These developments warrant a reassessment of what constitutes personal information in the digital age. Although our study focused on Japanese university students, we plan to broaden the scope to include international and multi-generational participants, leveraging the global and cross-generational appeal of online gaming. This study also lays the foundation for a novel game-based awareness survey methodology, positioning the game as a competitive card-style activity.

REFERENCES

Antonaci, A., Klemke, R., Stracke, C. M., Specht, M., Spatafora, M., and Stefanova, K. (2017). Gamification to Empower Information Security Education. *In P. Tuomi, & A. Perttula (Eds.), Proceedings of the 1st International GamiFIN Conference (Vol. 1857, pp. 32-38). CEUR Workshop Proceedings (CEUR-WS.org)*

Cabinet Office of Japan (2006). Public Opinion Poll on Personal Information Protection. (in Japanese)

Liz-Domínguez. M., Caeiro-Rodríguez. M., Llamas-Nistal. M., and Mikic-Fonte. F. (2022). Exploring the Synergies between Gamification and Data Collection in Higher Education. *Learning Analytics Summer Institute Spain (LASI Spain)* 2022.

Oroszi. E. D., (2020). Using Gamification to Improve the Security Awareness of Users: The Security Awareness Escape Room. *ISACA Journal,* Volume 4, 2020.

Sailer, M. and Homner. L. (2020). The Gamification of Learning: a Meta-analysis. *Educational Psychology Review* 32, 77-112.

Smiderle, R., Rigo, S. J., Marques, L. B., Coelho, J. A. P. M., and Jaques., P. A. (2020). The impact of gamification on students' learning, engagement and behavior based on their personality traits. *Smart Learning Environments* volume 7, Article number: 3

THE ROLE OF DOMAIN FAMILIARITY TO LEARN INTRODUCTORY PROGRAMMING

Polat Sendurur and Emine Sendurur
Ondokuz Mayis University
Samsun, Turkey

ABSTRACT

Learning introductory programming concepts can be challenging for novice learners. Providing tasks related to the main field of study might serve as a facilitator during the learning process. This study aims to observe how a group of university students, who enrolled in a programming class for the first time, perform programming tasks by domain familiarity—110 university students involved in this mixed study. One group performed the domain familiar task while the other one performed the unfamiliar one. In the second activity, the groups were reversed. The quantitative analysis indicated that domain familiarity does not affect programming performance.

KEYWORDS

Programming Education, Domain Familiarity, Learning How to Code

1. INTRODUCTION

Changing paradigms of education have brought about new roles for learners. Today's learners are expected to manage their learning (Reigeluth, 2016), so such higher-order thinking skills as problem-solving might be required. 21st-century skill set combines many skills combining problem-solving and information science. Computational thinking, new media literacy, and virtual collaboration can be listed among these skills (Curzon et al., 2009; Mohaghegh & McCauley, 2016). To improve these skills, interdisciplinary approaches might be needed. Based on this motivation, the current study aims to explore the effects of domain familiarity on programming performance, referring to a group of computational thinking skills, including formulating problems, organizing and analyzing data, and abstraction (ISTE, 2011).

The practice is generally based on domain-specific approaches in traditional instructional settings. On the other hand, such disciplines as programming education might borrow various practices of different disciplines. For example, the creation and testing of an algorithm might apply in many domains ranging from biology to language classes. Programming pedagogy is evolving, so integrating known instructional approaches might contribute to its development. Moving beyond the boundaries of the taught domain is an approach triggering creativity, motivation, and learning (Baer, 1998; Bong, 2004; Ding et al., 2013; Shen et al., 2008). It is known that creativity is essential in the development of algorithmic thinking skills (Curzon et al., 2009; Grover, 2013).

Moreover, creativity is also related to domain familiarity. According to Mishra et al. (2012), breaking the boundaries of domains can contribute to creativity. On the other hand, the lack of domain knowledge might hinder creativity (Sendurur et al., 2018).

In summary, the literature about domain familiarity still needs to be more conclusive, and no studies investigate domain familiarity in the programming education context. This study assigned domain-familiar or domain-unfamiliar tasks to students without computer programming experience. Unlike previous studies, this study has a different perspective on utilizing domain familiarity. In other words, domain specificity was used as a content delivery method instead of the content itself, which might shed light on programming education pedagogy. This study aims to observe how a group of university students perform programming tasks by domain familiarity with the help of the following research question:

1. Do domain-familiar tasks affect novice undergraduate students' programming performances?

2. METHOD

A mix of qualitative and quantitative techniques was planned to utilize in the research. However, only quantitative results are presented as the work is in progress. 110 sophomore university students from the department of Math Education participated in the study. They attended the programming course, which is a must-course aiming to teach introductory programming topics. The participants do not have any programming experience. They were assigned randomly to one of the two groups. Each group completed two different activities; one was domain-specific, and the other was non-domain-specific. The following table summarizes the activities, their covered programming topics, and the group-activity matching. Participants completed four different activities, as shown in Table 1.

Table 1. Programming Activity-Group Matches

Group	Domain Familiarity	Activity Description	Covered Programming Topic
Group1	Familiar	Odd/even numbers	Variables, conditionals
Group2	Unfamiliar	Tools with/without electricity	Variables, conditionals
Group1	Unfamiliar	Vowels/consonants	Variables, conditionals, loops, random numbers
Group2	Familiar	Symmetric/non-symmetric geometric objects	Variables, conditionals, loops, random numbers

Multiple data sources were utilized multiple to observe the performance change across domains. Students were assigned to complete tasks within the Scratch environment, and their final products were evaluated according to task-specific rubrics. The students were also required to explain the code block's function and contribution to the program. Independent samples t-test conducted on groups' rubric scores and content analysis was applied to their explanations about code block's function and contribution. However, this short paper reports only the quantitative results as the content analysis is still in progress. The quantitative data were analyzed in SPSS 21.0. To understand the differences between groups, two independent samples t-tests were run on the scores gained through activities 1 and 2.

3. FINDINGS

The independent samples t-test conducted on the first activity scores showed that domain familiarity has no significant effect on programming performance, $t\,(127) = 1.12$, $p = .26$. The group structure was reversed in the second activity. Participants who completed the first activity in the domain familiar format participated in the second activity in the non-domain familiar form. The other group similarly took the second activity in domain familiar format. The independent samples t-test based on the scores of the second activity also did not indicate a significant difference between the two groups, $t\,(123) = -1.09$, $p = .28$. In other words, domain familiarity has not been found significant in participants' performances in the second activity.

4. CONCLUSION

Quantitative findings indicated that domain familiarity did not significantly affect programming performance. There are several possible reasons for such a conclusion. The fact that the programming course was at the introductory level may have caused this situation. The programming concepts within the scope were simple, which may have prevented observing the domain familiarity effect. Another point is that quantitative findings may need to be triangulated with more detailed investigations, such as qualitative findings. A thorough examination of the participants' codes may provide more information about domain familiarity. For this reason, in the following study processes, the code structure of the participant products will be subjected to content analysis, and the process will be examined in more detail. This study reports the preliminary analysis of the gathered data; thus, the findings cannot be generalized.

REFERENCES

Baer, J. (1998). The case for domain specificity of creativity. *Creativity research journal*, *11*(2), 173-177.

Bong, M. (2004). Academic motivation in self-efficacy, task value, achievement goal orientations, and attributional beliefs. The Journal of Educational Research, 97, 287–297.

Curzon, P., Black, J., Meagher, L. R., & McOwan, P. (2009). cs4fn. org: Enthusing students about Computer Science. *Proceedings of Informatics Education Europe IV*, 73-80.

Ding, H., Sun, H., & Chen, A. (2013). Impact of expectancy-value and situational interest motivation specificity on physical education outcomes. *Journal of Teaching in Physical Education*, *32*(3), 253-269.

Grover, S. (2013). OPINION: Learning to Code Isn't Enough, from https://www.edsurge.com/n/2013-05-28-opinion-learning- to-code-isn-t-enough

ISTE (2011). Opeartional Definition for Computational Thinking. Retrieved from https://www.iste.org/explore/computational-thinking-all.

Mishra, P., Henriksen, D., & The Deep-Play Research Group. (2012). Rethinking technology and creativity in the 21st century: on being in-disciplined. TechTrends, 56(6), 18–21.

Mohaghegh, D. M., & McCauley, M. (2016). Computational thinking: The skill set of the 21st century.

Reigeluth, C. M. (2016). Instructional Theory and Technology for the New Paradigm of Education. Revista De Educación a Distancia, (50). Retrieved July 20, 2019 from https://revistas.um.es/red/article/view/270781

Sendurur, E., Ersoy, E., & Çetin, I. (2018). The Design and Development of Creative Instructional Materials: The Role of Domain Familiarity for Creative Solutions. *International Journal of Technology and Design Education*, *28*(2), pp. 507-522.

Shen, B., McCaughtry, N., & Martin, J. (2008). The influence of domain specificity on motivation in physical education. *Research Quarterly for Exercise and Sport*, *79*(3), 333-343.

BACK TO THE FUTURE: ENTREPRENEURIAL EDUCATION IN A BRAVE NEW WORLD

Kathryn Cormican[1], Suzana Sampaio[1], Özge Andiç Çakır[2], Firat Sarsar[2] and Manon Van Leeuwen[3]
[1]Enterprise Research Centre and LERO—The Irish Software Research Centre, School of Engineering, University of Galway, Ireland, H91 TK33
[2]EGE University
35100 Izmir, Türkiye
[3]EOLAS SL,
Carretera de Sevilla KM 6 06080 Badajoz, Spain

ABSTRACT

It is becoming increasingly apparent that we need to reimagine our approach to entrepreneurial education to meet the needs of Industry 5.0. This reflection piece discusses some of the challenges with entrepreneurial education and advocates pivoting towards a user-centred, design-oriented approach that adopts a real-world perspective and is supported by digital tools. However, we acknowledge that educators must be equipped with relevant user-centred processes, methods and tools to do this. We conclude by presenting initial findings from the DeThink which seeks to design, develop and validate an entrepreneurial programme for educators that is supported by a digital toolkit. This open-source resource adopts a pragmatic perspective and focuses on equipping educators with the methods, templates and tools to easily absorb and apply good practices.

KEYWORDS

Entrepreneurial education, Challenges, Design thinking, Digital tools, DeThink project

1. INTRODUCTION

Entrepreneurial education is fraught with challenges. Students often don't understand the rationale or end goal, they lack self-efficacy and find it difficult to take risks. Educators struggle with student engagement and find it difficult to foster entrepreneurial mindsets. The pedagogy of entrepreneurial education is still in its infancy. Many entrepreneurship education programmes focus on learning about new venture creation rather than using tools to actively engage in these activities. Students are often required to come up with a business idea very quickly at the beginning of the semester and then perform planning activities and show economic viability at the end of the programme. This approach is no longer fit for purpose in today's digital age. Prior work suggests that insufficient attention is paid to solving problems, coming up with ideas and acting on them (Sarooghi et al., 2019, Hägg & Gabrielsson, 2020, Lynch et al., 2021). Entrepreneurial education must move away from approaches that focus on "about" and "for" and focus more on the "how" and "through" (Linton & Klinton, 2019, Hajer & Hatem, 2022) to prepare for a brave new world. We must step back and reimagine how we can deliver high-quality education in a digital world. We argue that we must pivot towards a process-based approach. Students or learners need to understand the problem before developing a solution; they need to know "why" and "how" the problem emerges and the impact and implications of these problems before they start thinking about a solution. This paper reflects on some of the challenges in entrepreneurial education and advocates a new experimental approach that fosters self-efficacy, empowerment and autonomy. It concludes by presenting an overview of the DeThink project, supported by the European Commission, that aims to equip educators with the self-efficacy and competencies required to develop entrepreneurial mindsets and empower innovative behaviour.

2. CHALLENGES IN ENTREPRENEURIAL EDUCATION

An analysis of the extant literature reveals that the obstacles to effective entrepreneurial education are multifaceted and diverse. The current state of knowledge suggests that entrepreneurial education is complex and plagued with obstacles. Several studies identify the impediments or barriers in this domain. While it is evident that there are several challenges to entrepreneurial education, three overarching constructs stand out as particularly significant and deserve further investigation. These are presented in Table 1.

Table 1. Challenges to entrepreneurial education

Challenges	Description
Students attitudes	Students often view entrepreneurship education as monotonous or irrelevant to their career objectives (Bacigalupo et al., 2016, Huang et al., 2021, Wolf et al., 2022)
	Students lack awareness and comprehension of the entrepreneurial process, perceive negative stereotypes, and extoll difficulties in starting and managing a business (Larsen, 2018, Jena 2020)
Pedagogical issues	There is no universal pedagogical approach regarding the consistency of content, training, and resources (Hägg & Gabrielsson, 2020).
	There is a narrow focus on the theoretical and conceptual understanding of entrepreneurship (Gabrielsson et al., 2020, Kozlinska et al., 2020)
	Current curricula lack an experimental process approach (Mukesh et al., 2020, Pepin et al.,2021)
Systems constraints	There is a lack of resources, training, and funding available for entrepreneurship education (Gedeon 2020, Peschl et al., 2021)
	Entrepreneurial education does not harness digital tools effectively (Lynch et al., 2021)

Based on this review, we conclude that we must move away from a content-heavy business perspective toward a dynamic process-based approach. While it is still important to help educators develop some traditional entrepreneurial competencies (e.g., identifying markets and generating business models), this is no longer sufficient. Entrepreneurial programmes must shift away from a focus on abstract theory building towards approaches that foster tangible skills such as empowered mindsets, methods and tools. Instructors must be endowed with the appropriate digital tools and techniques to employ effective classroom practices that foster entrepreneurial mindsets and skills. Students must have access to relevant models and purpose-built instruments to support their entrepreneurial development.

3. REIMAGINING ENTREPRENEURIAL EDUCATION

The development of entrepreneurs in the classroom necessitates the creation of an enterprising learning environment where entrepreneurial aptitudes and skills can flourish alongside business acumen and knowledge. Current pedagogies are frequently viewed as ineffectual in aligning students' skill expectations with their skill acquisition; consequently, a new approach that emphasises interactive and practical learning appears necessary. Entrepreneurship education or teaching entrepreneurship requires a multi-dimensional and cross-disciplinary approach with an emphasis on dynamic processes that can expose students to the complexity of entrepreneurial activities in such a way that their actions can be examined and understood relative to the context of their entrepreneurial development. An analysis of the literature reveals that entrepreneurship education requires self-efficacy, creative thinking, soft skills, collaboration skills and perseverance (Çeviker-Çınar et al., 2017, Linton, & Klinton, 2019, Alabduljade et al., 2020, Lynch et al., 2021)

We argue that entrepreneurship education and learning should mirror the entrepreneurial process by emphasising action, real-world experiences, and reflection and that students should learn through practice and iteration. This is supported by many scholars (e.g. Grivokostopoulou et al., 2019, Ahmed et al., 2020, Lynch et al., 2021). Our analysis reveals that it is imperative to create an environment conducive to creativity, problem-solving and risk-taking to engage students while ensuring to not leave anyone behind.

In recent years educators are turning to the concept of design thinking as a mechanism to enable this environment. A design thinking approach focuses on building empathy, discovering problems and experimenting with many ideas to solve problems and challenges. It advocates an iterative approach with key stages e.g., empathise, define, ideate, prototype and test. Essentially it is about showing people how to look with fresh eyes, identifying opportunities and problems, and testing and implementing ideas. A design thinking approach to entrepreneurial education allows teachers and students to follow a process and apply templates and tools. It enables them to understand the mindsets of entrepreneurs and develop a tangible skillset. It fosters a student-centred process where work is graded on the development of competencies rather than output. Experimentation is a central part of this perspective. It is proven to foster an entrepreneurial attitude by promoting engaging, effective learning experiences and allowing students to experiment and practice (Grivokostopoulou et al., 2019).

This approach has been widely used to design solutions in industry, yet it has received little attention in schools. A design thinking approach to entrepreneurial education allows students to follow a process, and apply templates and tools to come up with problems and ideas. It is organized in a systematic process that fosters ongoing learning for innovative problem-solving. According to Huq & Gilbert (2017), it can create a learning environment with humour, role-play, gamification and fewer barriers between students and teachers, empowering the students and thereby contributing to enhanced student satisfaction and learning outcomes. Table 2 presents some of the benefits of a design thinking approach to entrepreneurial education.

Table 2. Benefits of design thinking approach to entrepreneurial education

Advantages	References
A hands-on, participative, collaborative approach Process-oriented model for entrepreneurship education User-driven and interdisciplinary	Sarooghi et al., (2019)
Facilitates meaning or mental connections between aspects of "what" learners know and "how to" implement or use this knowledge Promotes interdisciplinary projects that can broaden individual understanding of different perspectives	Tselepis & Lavelle, (2020)
Transforms ideas into concrete entities in the form of prototypes, sketches, or flowcharts Supporting and simplifying collaboration Promotes deep thought and reflection	Biffi et al., (2017)
Increased student motivation Real-world learning	Lynch et al., (2021)

From the above analysis, it can be concluded that design thinking offers a welcome approach to entrepreneurial education, mainly due to its focus on human needs, experimental learning, and interdisciplinary approach to problem-solving.

4. TOWARDS A DIGITAL TOOLKIT

The DeThink project aims to address some of the challenges to entrepreneurial education and develop a suite of digital tools for educators. The partnership comprises organizations from different countries in Europe with complementary skills in the field of user-centred design, instructional technology, entrepreneurial education and design thinking. The goal is to equip educators with the ability to integrate a structured, easy-to-implement entrepreneurial programme supported by a digital toolkit into their courses. It adopts a pragmatic perspective and focuses on equipping teachers with the methods, templates and tools to easily absorb and apply good practices in the classroom. To do this, user requirements were captured, analysed and prioritized to ensure that the toolkit is relevant, right-sized and easy to digest. The approach is flexible and aims to provide educators with a suite of options that can be customized to a specific context. Key deliverables from this project include:

1. A Curriculum and Syllabus based on a detailed analysis of the voice of the user. User needs were identified, refined and prioritized to define the learning objectives, curriculum and content. The curriculum adheres to Universal Design for Learning (UDL) principles and includes learning and assessment tasks that support learner choice in how they are completed. This approach fosters an inclusive approach and helps to ensure that all students can demonstrate their abilities.
2. The Learning Content focuses on the process, methods, templates and digital tools, required to satisfy the learning objectives. Attention is paid to showing teachers how to engage and empower students to become active players in innovation and entrepreneurship. Content centres on an iterative process using defined methods and templates e.g.,
   Team building skills [e.g., team canvas]
   Define the challenge or problem [e.g., problem definition, user stories, persona, initial journey map]
   Brainstorm for ideas [e.g., creative problem-solving techniques]
   Prototype solutions [e.g., screen ideas, storyboard, sketches, Lego mockups]
   Develop the solution [e.g., revised journey map, lean canvas]
   Pitch the idea [e.g., selling the idea and the people]

   Much of the supporting material is visual and comprises pictures and symbols rather than text following the principles of Universal Design for Learning to enable easy access and foster inclusion.
3. The eLearning Toolkit houses the curriculum, learning content and a suite of digital tools. The current prototype employs a novel ontology to allow teachers to search and retrieve relevant material to meet their specific needs. It is modular to facilitate the addition and movement of content and is also compatible with existing elearning solutions It is currently being developed and will be tested by educators via a comprehensive pilot study.
4. A suite of Recommendations shows educators how to use, uptake and coherently implement the toolkit. These guidelines promote a shift from considering the educator as an "imparter of knowledge" towards a "coach and facilitator".

5. CONCLUSION

Design thinking has been suggested as a promising approach to teaching within entrepreneurship education. Entrepreneurial education is embracing a design thinking lens, yet there is limited understanding of the conceptual underpinnings and the implementation of such practices. There is a dearth of good-quality pedagogical resources and intuitive digital tools to support educators. This reflection piece discusses some of the challenges and benefits of this approach and presents work in progress that aims to equip educators with the methods, templates and tools so that good practices can be easily absorbed and applied in the classroom. It comprises a digital toolkit and a suite of options that can be customized to a specific context. It hopes to provide educators with the self-efficacy and competencies to encourage new mindsets and enable empowered students.

ACKNOWLEDGEMENT

Our work is funded by the European Commission and Leargas by DeThink: Design Thinking for Entrepreneurial Education 2021-2-IE01-KA220-SCH-000049697.

REFERENCES

Ahmed, T., Chandran, V.G.R., Klobas, J.E., Liñán, F., & Kokkalis, P. (2020). Entrepreneurship education programmes: How learning, inspiration and resources affect intentions for new venture creation in a developing economy. *The International Journal of Management Education*, *18*(1), 100327.

Alabduljader, N., Solomon, G. T., Kang, J.H., Choi, D.Y., & Al-Abduljader, S.T. (2020). Cognitive styles and entrepreneurial intentions: A cross-cultural comparison. *Journal of Small Business Management,* 1–31

Biffi, A., Bissola, R., & Imperatori, B. (2017). Chasing innovation: A pilot case study of a rhizomatic design thinking education program. *Education + Training, 59*(9), 957–977.

Çeviker-Çınar, G., Mura, G., & Demirbağ-Kaplan, M. (2017). Design thinking: A new road map in business education. *The Design Journal, 20* (sup1), S977-S987.

Gabrielsson, J., Hägg, G., Landström, H., & Politis, D. (2020). Connecting the past with the present: the development of research on pedagogy in entrepreneurial education. *Education+ Training, 62*(9), 1061-1086.

Gedeon, S. A. (2020). Theory-based design of an entrepreneurship microcredentialing and modularization system within a large university ecosystem. *Entrepreneurship Education and Pedagogy, 3*(2), 107-128.

Grivokostopoulou, F., Kovas, K., & Perikos, I. (2019). Examining the impact of a gamified entrepreneurship education framework in higher education. *Sustainability, 11*(20), 5623.

Hägg, G., & Gabrielsson, J. (2020). A systematic literature review of the evolution of pedagogy in entrepreneurial education research. *International Journal of Entrepreneurial Behavior & Research, 26*(5), 829-861.

Hajer C and Hatem D 2022 Combining Teaching for and through Entrepreneurship and Interaction between Developed Skills *on Education, Science, and Technology 2022*

Huang, Y., An, L., Wang, J., Chen, Y., Wang, S., & Wang, P. (2021). The role of entrepreneurship policy in college students' entrepreneurial intention: the intermediary role of entrepreneurial practice and entrepreneurial spirit. *Frontiers in Psychology, 12*, 585698.

Huq, A., & Gilbert, D. (2017). All the world's a stage: Transforming entrepreneurship education through design thinking. *Education+ Training.*

Jena, R.K. (2020). Measuring the impact of business management Student's attitude towards entrepreneurship education on entrepreneurial intention: A case study. *Computers in Human Behavior, 107*, 106275.

Kozlinska, I., Rebmann, A., & Mets, T. (2020). Entrepreneurial competencies and employment status of business graduates: the role of experiential entrepreneurship pedagogy. *Journal of Small Business & Entrepreneurship*, 1-38.

Linton, G., & Klinton, M. (2019). University entrepreneurship education: A design thinking approach to learning. *Journal of Innovation and Entrepreneurship, 8*(1), 1-11.

Lynch, M., Kamovich, U., Longva, K. K., & Steinert, M. (2021). Combining technology and entrepreneurial education through design thinking: Students' reflections on the learning process. *Technological Forecasting and Social Change, 164*, 119689.

Mukesh, H. V., Pillai, K. R., & Mamman, J. (2020). Action-embedded pedagogy in entrepreneurship education: An experimental enquiry. *Studies in Higher Education, 45*(8), 1679-1693.

Pepin, M., Audebrand, L. K., Tremblay, M., & Keita, N. B. (2021). Evolving students' conceptions about responsible entrepreneurship: a classroom experiment. *Journal of Small Business and Enterprise Development, 28*(4), 570-585.

Peschl, H., Deng, C., & Larson, N. (2021). Entrepreneurial thinking: A signature pedagogy for an uncertain 21st century. *The International Journal of Management Education, 19*(1), 100427.

Sarooghi, H., Sunny, S., Hornsby, J., & Fernhaber, S. (2019). Design thinking and entrepreneurship education: Where are we, and what are the possibilities? *Journal of Small Business Management, 57*, 78-93.

Tselepis, T. J., & Lavelle, C. A. (2020). Design thinking in entrepreneurship education: Understanding framing and placements of problems. *Acta Commercii, 20*(1), 8.

Wolf, P., Frederiksen, M.H., Wilhøft, A., Cormican, K., Kunz, C., Andiç-Çakır, Ö., Sarsar, F. Ulus, H.E., van Leeuwen, M. (2022) Higher education engineering educators perspective on challenges and upskilling needs in teaching online, *23rd International CINet Conference*, 11th – 13th September, Pisa, Italy.

AUTHOR INDEX